AF294869

S. M. Rytov   Yu. A. Kravtsov
V. I. Tatarskii

Principles of Statistical Radiophysics 2

Correlation Theory of Random Processes

With 54 Figures

Springer-Verlag Berlin Heidelberg New York
London Paris Tokyo

Professor Sergei M. Rytov
Corresp. Member of the USSR Academy of Sciences,
Department of General Physics and Astronomy, USSR Academy of Sciences,
SU-117901 Moscow, USSR

Professor Yurii A. Kravtsov
General Physics Institute, USSR Academy of Sciences,
SU-117942 Moscow, USSR

Professor Valeryan I. Tatarskii
Corresp. Member of the USSR Academy of Sciences, Inst. of Atmospheric Physics,
USSR Academy of Sciences, Pyzhevsky Per., SU-109017 Moscow, USSR

Translator

Alexander P. Repyev
Samarkandsky Boulevard 13-3-63, SU-109507 Moscow, USSR

Title of the original Russian edition: *Vvedenie v statisticheskuyu radiofiziku I*, sluchainuie protsessui
2. revised and enlarged edition
© Nauka, Moscow 1976

ISBN-13: 978-3-642-64804-5   e-ISBN-13: 978-3-642-61351-7
DOI: 10.1007/978-3-642-61351-7

Library of Congress Cataloging-in-Publication Data. Rytov, S.M., 1908-. Principles of statistical radiophysics.
I. Translation of: Vvedenie v statisticheskuiu radiofiziku. I. Bibliography: p. Includes index. Contents: v.
1. Elements of random process theory–v. 2. Correlation theory of random processes. 1. Radio waves–Mathematics. 2. Stochastic processes. I. Kravtsov, Yurii Aleksandrovich. II. Tatarskii, V. I. (Valerian Ilich). III. Title.
IV. Title: Principles of statistical radiophysics. 1.
QC661.R9213   1987   537.5'34   87-4644

© Springer-Verlag Berlin Heidelberg 1988

Softcover reprint of the hardcover 1st edition 1988

2157/3150-543210

Foreword

Principles of Statistical Radiophysics is concerned with the theory of random functions (processes and fields) treated in close association with a number of applications in physics. Primarily, the book deals with radiophysics in its broadest sense, i.e., viewed as a general theory of oscillations and waves of any physical nature[1]. This translation is based on the second (two-volume) Russian edition. It appears in four volumes:

1. Elements of Random Process Theory
2. Correlation Theory of Random Processes
3. Elements of Random Fields
4. Wave Propagation Through Random Media.

The four volumes are, naturally, to a large extent conceptually interconnected (being linked, for instance, by cross-references); yet for the advanced reader each of them might be of interest on its own. This motivated the division of the *Principles* into four separate volumes.

The text is designed for graduate and postgraduate students majoring in radiophysics, radio engineering, or other branches of physics and technology dealing with oscillations and waves (e.g., acoustics and optics). As a rule, early in their career these students face problems involving the use of random functions. The book provides a sound basis from which to understand and solve problems at this level. In addition, it paves the way for a more profound study of the mathematical theory, should it be necessary[2]. The reader is assumed to be familiar with probability theory.

In progressing from one volume to the next, the reader will see that the physical problems under consideration become more complex and the mathematical machinery more involved. This results quite naturally from the course-oriented origin of the book, based as it is on summarized lecture notes. Their extensive teaching experience has convinced the authors that such an approach is preferable to a more uniform presentation. Each chapter is followed by a

---

[1] It should be noted that certain questions in statistical radiophysics are not covered in this book. For example, it is neither concerned with quantum radiophysics and quantum electronics, nor with statistical phenomena related to the propagation of waves in nonlinear media, see [1].

[2] Treatments of the mathematical theory are given, for instance, in [2–9], and more specialized applications of the theory to radiophysics and engineering are to be found in [10–19].

number of problems worked out in detail. The problems not only serve as exercises, but in many cases contain additional theoretical material and further literature references.

Moscow, March 1986

S.M. Rytov
Yu.A. Kravtsov
V.I. Tatarskii

Preface

Volume 2 is devoted to the correlation theory of random processes (a generalization to random fields is to follow in Volume 3). Although it is a rather limited section of the general theory of random functions, this theory finds wide use in treatments of a number of fascinating physical problems. These include interference and temporal coherence of oscillations, and also polarization in the vector addition of oscillations.

It is well known that all of these problems rely heavily on harmonic spectral representations. The latter, therefore, receive primary emphasis in the presentation of correlation theory, both for random processes themselves and for their covariance. This is of importance in treatments of thermal noise in discrete dynamical systems (in continuous systems this noise falls into the category of random fields to be considered in Volume 3), and flicker effect. At the same time, the spectral approach complements the studies of the action of random processes on linear and nonlinear dynamical systems.

This allows the reader to get acquainted with types of random processes that in a sense are close to stationary ones. These include modulated processes, pulse processes with independent intervals, and quasi-stationary processes. This volume concludes with a description of spectral expansions of random processes, in particular periodically nonstationary ones.

The author of Volume 2 is S.M. Rytov. He is grateful to professors A.M. Yaglom, M.L. Levin, V.N. Tutubalin, and Ya.I. Khurghin for their valuable suggestions and comments that added immensely to the quality and arrangement of material in some chapters.

Moscow, March 1988
S. M. Rytov

Contents

1. Fundamentals of Correlation Theory ... 1
1.1 Complex Random Functions. Analytical Signal ... 1
1.2 Random Function Behavior and Properties of the Covariance ... 7
1.3 Spectral Representation of Random Functions ... 11
1.4 Stationary Random Functions ... 14
1.5 Examples of Spectral Representations of Stationary Functions ... 22
1.6 Exercises ... 34

2. Applications of Correlation Theory ... 42
2.1 White Noise and Black-Body Radiation ... 42
2.2 Modulated Random Processes ... 46
2.3 Spectrum of Oscillations with Fluctuating Frequency ... 59
2.4 Spectra of Pulse Processes with Independent Intervals ... 65
2.5 Correlation Theory of Coherence ... 75
2.6 Nonstationary Interference. Source Correlation ... 87
2.7 Statistical Properties of Polarization of Modulated Oscillations ... 95
2.8 Exercises ... 103

3. Spectral Theory of Random Actions on Dynamic Systems ... 110
3.1 Random Actions on Harmonic Systems ... 110
3.2 Random Actions on Memoryless Nonlinear Systems ... 120
3.3 Noise Measurement. Radiometers ... 130
3.4 Correlation Theory of Fluctuations in the Thomson Oscillator ... 139
3.5 Thermal Noise in Quasi-Stationary Networks. The Fluctuation-Dissipation Theorem ... 153
3.6 Exercises ... 167

4. Certain Kinds of Nonstationary Processes ... 177
4.1 The Flicker Effect ... 177
4.2 Random Functions with Stationary Increments. The Structure Function ... 181

4.3 Spectra of Nonstationary Processes. Quasi-Stationary
     Processes 189
4.4 Filtration of Nonstationary Processes. The Mean Power
     Spectrum 197
4.5 Periodically Nonstationary Processes 207
4.6 Exercises 218

References 223

Subject Index 227

Errata to Vol. 1 233

1. Fundamentals of Correlation Theory

We have already touched upon correlation theory in Sect. I.4.5*. The reader will recall that the theory describes a random function in a fairly incomplete way – just by using moments of the first (means) and second (covariances and mean-square values) order. For reasons to be clarified in Sect. 1.4, the most interesting and advanced form of correlation theory is that applied to *stationary* processes. It is these processes that form the main subject of our further discussion, although we also look at some types of nonstationary processes.

It will also be recalled that an additional attraction of correlation theory is that we very often have to deal with *normal* (Gaussian) processes, for which all the n-variate distributions w_n are defined completely, if the first and second moments are known.

The chapter begins with a generalization of the theory of random processes to the case of their complex representation (Sect. 1.1). Section 1.2 covers the key concepts and relationships of correlation theory, Sects. 1.3–5 deal with the most important issues of *spectral representations* of random functions and their covariances, as well as the interrelation between these two representations.

1.1 Complex Random Functions. Analytical Signal

Before we set out to discuss correlation theory and its applications it might be instructive to make one generalization and to introduce *complex* random functions.

A complex random function $\zeta(t)$ is generally represented by the following linear combination of two real random functions $\xi(t)$ and $\eta(t)$

$$\zeta(t) = \xi(t) + i\eta(t) \quad ,$$

defined by joint distribution functions of ξ and η. Clearly,

$$\overline{\zeta(t)} = \overline{\xi(t)} + i\overline{\eta(t)} \quad .$$

As for second moments, we usually need to employ those which yield real mean square values. Thus the mixed second moment is

---

* (Labels starting with a Roman number refer to the respective volume of the Principles, e.g. (I.2.1) stands for Eq. (2.1) of Vol 1.)

$$B(t,t') = \langle \zeta(t)\zeta^*(t') \rangle = \overline{\xi\xi'} + \overline{\eta\eta'} + \mathrm{i}(\overline{\eta'\xi} - \overline{\eta\xi'}) \quad , \tag{1.1}$$

where ξ' and η' are values of ξ and η at a time t'. Specifically,

$$B(t,t) = \overline{|\zeta(t)|^2} = \overline{\xi^2(t)} + \overline{\eta^2(t)},$$

which is a real and positive quantity. In much the same way we define the covariance

$$\psi(t,t') = \overline{[\zeta(t) - \overline{\zeta(t)}][\zeta^*(t') - \overline{\zeta^*(t')}]} = B(t,t') - \overline{\zeta(t)}\;\overline{\zeta^*(t')} \quad .$$

In particular, the variance $\zeta(t)$ will also be real and positive

$$\psi(t,t) = \overline{|\zeta(t) - \overline{\zeta(t)}|^2} = \overline{|\zeta(t)|^2} - \overline{|\zeta(t)|}^2 \equiv D[\zeta(t)] \quad .$$

In a number of problems, however, it is convenient to consider in addition other second moments derived without using complex-conjugate random functions. These moments will be termed "second" (second mixed moment, second covariance) and denoted by a tilde over the respective symbol

$$\tilde{B}(t,t') = \overline{\zeta(t)\zeta(t')} = \overline{\xi\xi'} - \overline{\eta\eta'} + \mathrm{i}(\overline{\xi'\eta} + \overline{\xi\eta'}) \quad ,$$
$$\tilde{\psi}(t,t') = \tilde{B}(t,t') - \overline{\zeta(t)} \cdot \overline{\zeta(t')} \quad . \tag{1.2}$$

In particular, at $t' = t$

$$\tilde{B}(t,t) = \overline{\zeta^2(t)} = \overline{\xi^2(t)} - \overline{\eta^2(t)} + 2\mathrm{i}\overline{\xi(t)\eta(t)} \quad .$$

$$\tilde{\psi}(t,t) = \overline{\zeta^2(t)} - \overline{\zeta(t)}^2 \quad .$$

Of course, the two complex functions $B(t,t')$ and $\tilde{B}(t,t')$ completely define all the second moments of real functions $\xi(t)$ and $\eta(t)$

$$\overline{\xi\xi'} = \tfrac{1}{2}\mathrm{Re}\{B + \tilde{B}\} \quad , \quad \overline{\eta\eta'} = \tfrac{1}{2}\mathrm{Re}\{B - \tilde{B}\} \quad ,$$
$$\overline{\xi'\eta} = \tfrac{1}{2}\mathrm{Im}\{B + \tilde{B}\} \quad , \quad \overline{\xi\eta'} = \tfrac{1}{2}\mathrm{Im}\{\tilde{B} - B\} \quad . \tag{1.3}$$

We will consider only those random functions whose moment $B(t,t')$ exists and is continuous at all $t' = t^1$, and hence exists and is continuous for all t and t'. The mean square of the modulus $\overline{|\zeta|^2}$ is finite; therefore the modulus of the mean is finite as well, since $D[\zeta] > 0$, i.e., $|\overline{\zeta}|^2 < \overline{|\zeta|^2}$.

---

[1] This condition distinguishes the class of so-called *random functions of the second order* that are everywhere continuous in the mean square (see below).

In the special case of the stationary function we will interpret stationarity in its *wide sense* (Sect. I.4.5). In other words, we take it to include: (1) the constancy of the mean $\overline{\zeta}$ (which enables us in all cases to consider $\zeta - \overline{\zeta}$ instead of ζ, i.e., to take $\overline{\zeta} = 0$), (2) the stationarity of B (and hence of the covariance ψ), i.e., their dependence on $t - t' = \tau$ alone, and (3) the continuity of $B(\tau)$ at $\tau = 0$.

Let, for instance, $\zeta(t) = cf(t)$, where $f(t)$ is a deterministic complex function, and c is a random complex variable. Condition (1) requires $\overline{c} = 0$. Then, $\overline{\zeta(t)} = 0$, and $B = \psi = \overline{|c|^2} f(t) f^*(t - \tau)$.

Conditions (2) and (3) are only met for the function $f(t) = \exp(\mathrm{i}\omega t)$ when

$$B(\tau) = \psi(\tau) = \overline{|c|^2} \mathrm{e}^{\mathrm{i}\omega\tau} \quad .$$

For $\overline{c} \neq 0$, we have

$$\overline{\zeta(t)} = \overline{c}\,\mathrm{e}^{\mathrm{i}\omega t} \quad , \qquad B(\tau) = \overline{|c|^2}\,\mathrm{e}^{\mathrm{i}\omega\tau} \quad , \qquad \psi(\tau) = (\overline{|c|^2} - |\overline{c}|^2)\mathrm{e}^{\mathrm{i}\omega\tau} \quad ,$$

i.e., conditions (2) and (3) are satisfied, but (1) is not. It follows thus that (1) is an independent condition.

It is to be noted that even if all the three conditions are met, second moments are, generally speaking, nonstationary (they depend not only on the shift τ but also on t). In this example, where $\zeta(t) = c\exp(\mathrm{i}\omega t)$ and $\overline{c} = 0$, we have

$$\tilde{B} = \tilde{\psi} = \overline{c^2}\exp[\mathrm{i}\omega(2t - \tau)] \quad .$$

Real processes are described by real functions of t, but it is common knowledge that complex variables are often more convenient to deal with $-$ a fact widely used in many branches of physics, including the theory of oscillations and waves. But when we replace a real process $\xi(t)$ by the complex function $\zeta(t) = \xi(t) + \mathrm{i}\eta(t)$, through the addition of an *arbitrary* imaginary part $\eta(t)$, we "double the information" in an unjustified manner. This can only be avoided by imposing a one-to-one relationship between $\eta(t)$ and $\xi(t)$. A means of establishing such a relationship that offers a number of advantages which will be seen below has been suggested by *Gabor* [1.1]. At this point we formulate this technique just as a recipe.

We require, above all, that the continuation of $\xi(t)$ to the complex plane $\tau = t + \mathrm{i}\alpha$ give an analytical function $\xi(\tau)$ that passes into $\xi(t)$ as $\alpha \to 0$. We now select a real function $\eta(t)$ such that the complex function $\zeta(t)$ is *regular and analytic in the upper half-plane of the complex variable τ and that at any t its modulus $|\zeta|$ tends to zero sufficiently fast as $\alpha \to +\infty$*. These are stringent requirements which result in $\xi(t)$ and $\eta(t)$ becoming a pair of Hilbert transforms, and hence uniquely interrelated. Indeed, by the Cauchy formula, we have

$$\zeta(\tau) = \frac{1}{2\pi\mathrm{i}} \oint_{\Gamma} \frac{\zeta(\theta)}{\theta - \tau} d\theta \quad ,$$

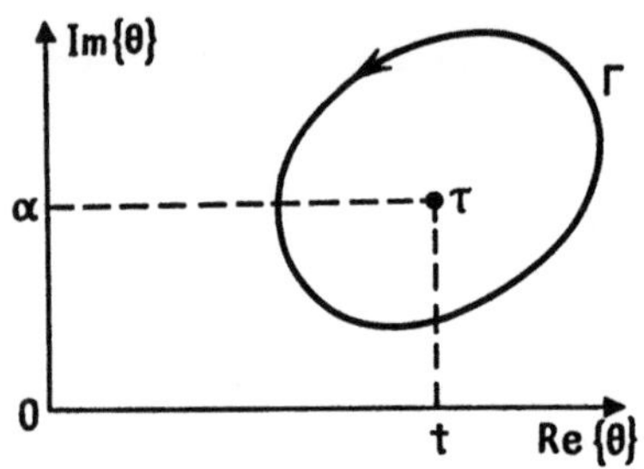

Fig. 1.1. An illustration to the derivation of the Hilbert transformation

where the pole τ lies in the upper half-plane ($\tau = t + i\alpha$, $\alpha > 0$), and the path Γ encompasses this pole (Fig. 1.1). Owing to the first requirement on $\zeta(\theta)$, the contour Γ can be stretched so as to include the whole of the upper half-plane. At the same time, as $\alpha \to \infty$, $|\zeta|$ must fall off sufficiently fast for the integral over the infinite semi-circle to give no contribution. What remains then is the integral along the real axis

$$\zeta(\tau) = \frac{1}{2\pi i} \int_{-\infty}^{+\infty} \frac{\zeta(\theta)}{\theta - \tau} d\theta \quad . \tag{1.4}$$

But

$$\frac{1}{\theta - \tau} = \frac{1}{\theta - t - i\alpha} = \frac{\theta - t}{(\theta - t)^2 + \alpha^2} + i\frac{\alpha}{(\theta - t)^2 + \alpha^2} \quad ,$$

so that in the limit $\alpha \to 0$, the real part of this expression yields the principal value of the integral of $\zeta(\theta)/(\theta - t)$, and the imaginary part becomes the delta-function $\pi\delta(\theta - t)$. Thus, at $\alpha = 0$, (1.4) gives

$$\zeta(t) = \frac{1}{2\pi i} \left[\fint_{-\infty}^{+\infty} \frac{\zeta(\theta)}{\theta - t} d\theta + \pi i \zeta(t) \right] \quad ,$$

hence

$$\zeta(t) = \frac{1}{\pi i} \fint_{-\infty}^{+\infty} \frac{\zeta(\theta)}{\theta - t} d\theta \quad .$$

A dash on the integral sign means that the integral is taken in the sense of the Cauchy principal value. Substituting $\zeta = \xi + i\eta$ shows that $\xi(t)$ and $\eta(t)$ are related by the Hilbert transform

$$\eta(t) = -\frac{1}{\pi} \fint_{-\infty}^{+\infty} \frac{\xi(\theta)}{\theta - t} d\theta \quad , \quad \xi(t) = \frac{1}{\pi} \fint_{-\infty}^{+\infty} \frac{\eta(\theta)}{\theta - t} d\theta \quad . \tag{1.5}$$

It is esily seen that these transforms can also be written otherwise:

$$\eta(t) = -\frac{1}{\pi} \int\limits_{0}^{\infty} \frac{\xi(t+\theta) - \xi(t-\theta)}{\theta} d\theta \quad ,$$

$$\xi(t) = \frac{1}{\pi} \int\limits_{0}^{\infty} \frac{\eta(t+\theta) - \eta(t-\theta)}{\theta} d\theta \quad . \tag{1.6}$$

The complex function $\zeta(t)$ obtained in this way from the real function $\xi(t)$ is said to be the *analytical signal*, the term "analytical" coming from the fact that $\zeta(t)$ is analytical in the upper half-plane of the complex t[2]. The notion of an analytical signal is widely used in information theory with which this book does not concern, and in coherence theory (to be touched upon in Sect. 1.9 and in Vol. 4). In the classical (nonquantum) theory of coherence of wave fields the use of the analytical signal only provides the formal advantages inherent in dealing with complex quantities, although the signal contains no information besides that contained in $\xi(t)$. But in quantum electrodynamics it also acquires a physical meaning as it describes eigenfunctions of the photon annihilation operator.

The derivation of the analytical signal does not, of course, rely on assumptions as to whether the original real function $\xi(t)$ is deterministic or random. Turning now to the latter case, it is easily seen that the one-to-one correspondence between $\eta(t)$ and $\xi(t)$ also predetermines the fact that the stochastic behavior of $\eta(t)$ is completely governed by the statistics of $\xi(t)$. Although a natural consequence, this is by no means trivial as the relation between $\eta(t)$ and $\xi(t)$ given by (1.5) is *nonlocal in* t. A rigorous treatment of the statistics of the analytical signal $\zeta(t)$ as compared with the statistics of the initial random process $\xi(t)$ is given in [1.2][3]. In discussing correlation theory we need only consider first and second moments.

The means $\overline{\xi(t)}$ and $\overline{\eta(t)}$ appear to be related by the same Hilbert transform (1.5). This suggests that for the stationary process $\xi(t)$ it is necessary that $\overline{\xi(t)} = \overline{\eta(t)} = 0$. Indeed, if $\overline{\xi(t)} = \overline{\xi} = \mathrm{const}$, it follows from the first relation in (1.5) [or (1.6)] that $\overline{\eta} = 0$. And then, by the second relation, $\overline{\xi} = 0$ as well.

Multiplying the first relation in (1.5) by $\eta(t')$, the second by $\xi(t')$, and averaging the results gives

$$B_\eta(t,t') \equiv \langle \eta(t)\eta(t') \rangle = -\frac{1}{\pi} \int\limits_{-\infty}^{+\infty} \frac{B_{\xi\eta}(\theta,t')}{\theta - t} d\theta \quad ,$$

---

[2] Of course, we can require that $\zeta(t)$ be analytical not in the upper, but in the lower half-plane of the complex t. This eventually comes down to deciding on the sign at i in the subsequent complex Fourier transforms: the function $\exp(i\omega t)$ decays exponentially into the upper half-plane, and $\exp(-i\omega t)$ into the lower one.

[3] In this work $\xi(t)$ is required to be quadratically integrable, which, strictly speaking, eliminates the stationary processes $\xi(t)$. The authors agree, however, that this requirement is rather too stringent. To all intents and purposes, it would be sufficient to have a spectral expansion of $\xi(t)$ in the mean square (Sect. 1.3).

$$B_\xi(t,t') \equiv \langle \xi(t)\xi(t') \rangle = \frac{1}{\pi} \int\limits_{-\infty}^{+\infty} \frac{B_{\xi\eta}(t',\theta)}{\theta - t} d\theta \quad , \tag{1.7}$$

where $B_{\xi\eta}(t,t') = \langle \xi(t)\eta(t') \rangle$. Multiplying the same equations by $\xi(t')$ and $\eta(t')$ respectively, and averaging, we obtain the *inverted* Hilbert transforms

$$B_{\xi\eta}(t,t') = \frac{1}{\pi} \int\limits_{-\infty}^{+\infty} \frac{B_\eta(\theta,t')}{\theta - t} d\theta \quad ,$$

$$B_{\xi\eta}(t',t) = -\frac{1}{\pi} \int\limits_{-\infty}^{+\infty} \frac{B_\xi(\theta,t')}{\theta - t} d\theta \quad . \tag{1.8}$$

Therefore, for second moments we have two pairs of Hilbert transforms.

We will now see what these relationships yield for the case of a (wide-sense) *stationary* process $\xi(t)$ with $B_\xi(t,t') = B_\xi(\tau) = B_\xi(-\tau)$, where $\tau = t - t'$. The second relation in (1.8) becomes

$$B_{\xi\eta}(t',t) = -\frac{1}{\pi} \int\limits_{-\infty}^{+\infty} \frac{B_\xi(\theta - t')}{\theta - t} d\theta = -\frac{1}{\pi} \int\limits_{-\infty}^{+\infty} \frac{B_\xi(\theta')}{\theta' - \tau} d\theta' = B_{\xi\eta}(\tau) \quad , \tag{1.9}$$

i.e., $\xi(t)$ and $\eta(t)$ are *stationarily related,* and it follows from (1.9) that $B_{\xi\eta}(\tau)$ is an odd function of τ

$$B_{\xi\eta}(-\tau) = -B_{\xi\eta}(\tau) \quad .$$

But then (1.7) suggests that B_η depends on $\tau = t - t'$ and is equal to B_ξ,

$$B_\eta(\tau) = B_\xi(\tau) \quad .$$

To sum up, if we have an original real process $\xi(t)$ then the conjugate (according to Hilbert) process $\eta(t)$ is also stationary and stationarily related to $\xi(t)$, so that

$$\overline{\xi} = \overline{\eta} = 0 \quad , \quad \overline{\xi\xi_\tau} = \overline{\eta\eta_\tau} \quad , \quad \overline{\xi\eta_\tau} = -\overline{\xi_\tau\eta} \quad , \tag{1.10}$$

where $\xi = \xi(t)$, $\xi_\tau = \xi(t + \tau)$. The same is true for η. This means, by (1.1,2), that for a stationary analytical signal

$$B_\zeta(\tau) = 2[B_\xi(\tau) + iB_{\xi\eta}(\tau)] \quad , \quad \tilde{B}_\zeta(\tau) = 0 \quad . \tag{1.11}$$

Since the real and imaginary parts of $B_\zeta(\tau)$ are related by the Hilbert transform, $B_\zeta(\tau)$ is also an analytical signal. We write $B_\zeta(\tau)$ as

$$B_\zeta(\tau) = |B_\zeta(\tau)|\exp[i\varphi(\tau)] \quad . \tag{1.12}$$

The analyticity and regularity of $B_\zeta(\tau)$ in the upper half-plane of the complex variable τ means that the function

$$\ln B_\zeta(\tau) = \ln |B_\zeta(\tau)| + i\varphi(\tau)$$

is also analytical in this half-plane, but at zeros of $|B_\zeta(\tau)|$ it can have logarithmic branch points. In the absence of such points in the upper half-plane, $\ln B_\zeta(\tau)$ is not only analytical but also regular, and hence, its real and imaginary parts are also related by the Hilbert transforms

$$\ln |B_\zeta(\tau)| = \frac{1}{\pi} \int\limits_{-\infty}^{+\infty} \frac{\varphi(\theta)}{\theta - \tau} d\theta \quad ,$$

$$\varphi(\tau) = -\frac{1}{\pi} \int \frac{\ln |B_\zeta(\tau)|}{\theta - \tau} d\theta \quad . \tag{1.13}$$

In this case the phase $\varphi(\tau)$ is thus uniquely defined by the modulus of $B_\zeta(\tau)$, and vice versa[4].

In the following, in dealing with complex random processes, we will use the analytical signal, where appropriate.

1.2 Random Function Behavior and Properties of the Covariance

The following reasoning is given for the mixed moment $B(t, t')$, although it is perfectly applicable to the covariance $\psi(t, t')$ as well.

It follows immediately from the definition of $B(t, t')$ that the Hermitian condition is met

$$B(t, t') = B^*(t', t) \quad . \tag{1.14}$$

Further, it can be seen that in the entire plane (t, t'), the moment $B(t, t')$ is *limited in modulus*. In fact, setting $B(t, t') = |B| \exp(i\varphi)$, we have, by (1.14),

$$\overline{|\zeta(t) + e^{i\alpha}\zeta(t')|^2} = \overline{|\zeta|^2} + \overline{|\zeta'|^2} + e^{i\alpha}B(t', t) + e^{-i\alpha}B(t, t')$$
$$= \overline{|\zeta|^2} + \overline{|\zeta'|^2} + 2|B| \cos(\alpha - \varphi) \geq 0 \quad ,$$

thus

$$|B| \leq \frac{\overline{|\zeta|^2} + \overline{|\zeta'|^2}}{2} \quad . \tag{1.15}$$

---

[4] The presence of zeros of $|B_\zeta(\tau)|$ in the upper half-plane is discussed in [1.3].

The properties (1.14, 15) follow from the more general statement that the moment B is a *positive definite function*. This means that, for any instants of time $t_1, t_2, \ldots, t_n$ and n arbitrary complex numbers $a_1, a_2, \ldots, a_n$, the sum

$$\sum_{j,k=1}^{n} B(t_j, t_k) a_j a_k^* \geq 0 \tag{1.16}$$

is real and nonnegative. The proof is extremely simple: (1.16) follows from the fact that

$$\overline{\left| \sum_{j=1}^{n} \zeta(t_j) a_j \right|^2} = \sum_{j,k=1}^{n} \overline{\zeta(t_j) \zeta^*(t_k)} a_j a_k^* \geq 0 \quad .$$

Clearly, the "second" mixed moment $\tilde{B}(t, t')$ is symmetric under the interchange $t \rightleftarrows t'$, but it is not positive definite.

It is also possible to prove the converse: any positive definite function $B(t, t')$ is the second moment of a random function of the type considered. The class of positive definite functions thus *coincides* with the class of moments $B(t, t')$ of random functions of the second order. Clearly, any linear combinations of the moments $B_i(t, t')$ with arbitrary positive coefficients are also positive definite, i.e., they exhibit the properties of the mixed moment B of a random function of the second order.

If $\zeta(t)$ is stationary, so that $B(t, t') = B(\tau)$ where $\tau = t - t'$, then (1.14–16) become

$$B(\tau) = B^*(-\tau) \tag{1.17}$$

(which follows directly from the invariance under time translation; see Sect. I.4.5),

$$|B(\tau)| \leq \overline{|\zeta|^2} = B(0) \quad , \tag{1.18}$$

$$\sum_{j,k=1}^{n} B(t_j - t_k) a_j a_k^* \geq 0 \quad . \tag{1.19}$$

Again we note that the above arguments also hold for the covariance $\psi(t, t')$. For the stationary function $\zeta(t)$, putting $\overline{\zeta} = 0$ allows us to identify $B(\tau)$ and $\psi(\tau)$.

We will now see that the assumption that $B(t, t')$ is continuous at any $t = t'$ suggests that $B(t, t')$ is uniformly continuous in the entire plane (t, t'). We have

$$\overline{\zeta(t + \tau)[\zeta^*(t - h) - \zeta^*(t)]} = B(t + \tau, t - h) - B(t + \tau, t) \quad .$$

Therefore, using the Cauchy-Buniakovski inequality

$$\overline{|ab^*|^2} \le \overline{|a|^2}\;\overline{|b|^2}$$

gives

$$|B(t+\tau,t-h) - B(t+\tau,t)|^2$$
$$= \overline{|\zeta(t+\tau)[\zeta^*(t-h)-\zeta^*(t)]|^2} \le \overline{|\zeta(t+\tau)|^2}\;\overline{|\zeta(t)-\zeta(t-h)|^2}$$
$$= B(t+\tau,t+\tau)\{B(t,t)+B(t-h,t-h)-B(t,t-h)-B(t-h,t)\} \quad .$$

Because $B(t,t')$ is continuous at all $t=t'$, the right-hand side vanishes as $h \to 0$ implying that the left-hand side tends to zero for any τ. In particular, if the random function $\zeta(t)$ is stationary, then the inequality takes the form

$$|B(\tau+h) - B(\tau)|^2 \le B(0)\{2B(0) - B(h) - B(-h)\}$$
$$= 2B(0) \cdot \mathrm{Re}\,\{B(0) - B(h)\} \quad ,$$

whence the mixed moment being continuous at zero ensures its continuity at all τ. Accordingly, out of the four real and even functions $B(\tau)$ shown in Fig. 1.2 only the last three *can* represent the second moment of a stationary process, whereas the first certainly can not.

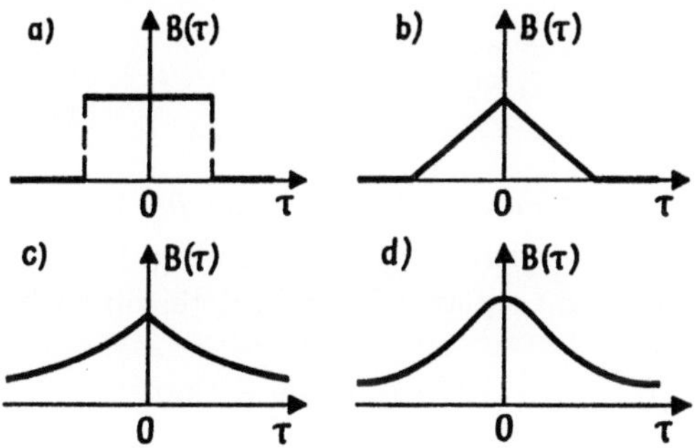

Fig. 1.2a–d. An discontinuous function of type (a) cannot be a second-order moment of a stationary process; continuous ones can

Certain conditions imposed on the moment $B(t,t')$ represent a number of properties of the random function $\zeta(t)$. So, for instance, the equality

$$\overline{|\zeta(t) - \zeta(t')|^2} = B(t,t) + B(t',t') - B(t,t') - B(t',t)$$

suggests that the continuity of $B(t,t')$ for $t' \to t$ means that $\zeta(t)$ is continuous in the mean square

$$\lim_{t' \to t} \overline{|\zeta(t) - \zeta(t')|^2} = 0 \quad , \quad \text{or} \quad \underset{t' \to t}{\mathrm{l.i.m.}}\ \zeta(t') = \zeta(t)$$

(see also Vol. I, p. 99).

It can also be shown quite simply that the derivative $\dot{\zeta}(t)$ exists and is a random function of the second order continuous in the mean square only if the mixed derivatie $\frac{\partial^2 B(t,t')}{\partial t\,\partial t'}$ exists and is continuous for all $t' = t$. It is, in fact, trivial to verify that here

$$\lim_{h,h' \to 0} \overline{\left| \frac{\zeta(t+h) - \zeta(t)}{h} - \frac{\zeta(t+h') - \zeta(t)}{h'} \right|^2} = 0 \quad ,$$

i.e., there exists a limit in the mean square of the ratio $[\zeta(t+h) - \zeta(t)]/h$ which does not depend on the path to $h = 0$.

We now wish to find the moment $B_1(t, t')$ of the random function $\dot\zeta(t)$. We have

$$
\begin{aligned}
B_1(t, t') &\equiv \overline{\dot\zeta(t)\dot\zeta^*(t')} \\
&= \lim_{h \to 0} \left\{ \overline{\frac{\zeta(t+h) - \zeta(t)}{h} \frac{\zeta^*(t'+h) - \zeta^*(t')}{h}} \right\} \\
&= \frac{\partial^2 B(t, t')}{\partial t\,\partial t'} \quad .
\end{aligned}
\tag{1.20}
$$

In particular, for the stationary random function $\zeta(t)$, a sufficient condition for the existence of $\dot\zeta(t)$ is the existence of the derivative $d^2 B(\tau)/d\tau^2$. The mixed moment of $\dot\zeta(t)$ will be

$$B_1(\tau) = \overline{\dot\zeta(t+\tau)\dot\zeta^*(t)} = -\frac{d^2 B(\tau)}{d\tau^2} \quad . \tag{1.21}$$

Therefore, the stationary random functions $\zeta(t)$ with the moments $B(\tau)$ shown in Fig. 1.2b, c have no derivative, while those with $B(\tau)$ shown in Fig. 1.2d do. The continuity of the derivative (in the mean square) is again determined by the continuity of $B_1(\tau)$ at zero.

In the same way as in finding $B_1(t, t')$, we can readily show that

$$\overline{\dot\zeta(t)\zeta^*(t)} = \frac{\partial B(t, t')}{\partial t'} \quad , \quad \overline{\dot\zeta(t)\zeta^*(t')} = \frac{\partial B(t, t')}{\partial t} \quad , \tag{1.22}$$

and for the stationary $\zeta(t)$

$$\overline{\dot\zeta(t+\tau)\zeta^*(t)} = -\overline{\dot\zeta(t+\tau)\zeta^*(t)} = -\frac{dB(\tau)}{d\tau} \quad . \tag{1.23}$$

Of course,

$$\overline{\dot\zeta(t+\tau)\zeta^*(t)} + \overline{\zeta(t+\tau)\dot\zeta^*(t)} = \frac{d}{dt}\overline{\zeta(t+\tau)\zeta^*(t)} = 0 \quad ,$$

because $\overline{\zeta(t+\tau)\zeta^*(t)} = B(\tau)$, i.e., B is independent of t.

Lastly, the necessary and sufficient condition for the existence in the mean square of the integral

$$I = \int\limits_a^b f(t)\xi(t)dt \quad ,$$

(where $f(t)$ is a deterministic function, and the limits can also be infinite), is the finiteness of the double integral

$$\int\limits_a^b \int\limits_a^b f(t)f^*(t')B(t,t')dt\,dt' \quad ,$$

which, if it exists, is the quantity $\overline{|I|^2}$. Clearly, the distribution function for the random variable I depends on a and b as parameters, and in addition is a functional of $f(t)$.

1.3 Spectral Representation of Random Functions

Central to correlation theory is the problem of spectral representation, both of random functions themselves and of their second moments, i.e., of mean bilinear variables. These latter are especially important because in many cases they carry the sense of an energy and in all cases serve as a simple measure of the intensity of random variations. The above, of course, in no way makes it clear why it is in general advantageous to use any expansions nor why, from among the various possible expansions

$$\zeta(t) = \int \varphi(t,\omega)dC(\omega) \quad ,$$

we should prefer the one in the functions $\varphi(t,\omega) = \exp(i\omega t)$. The following answers to both questions contain nothing specific for random functions.

The representation of an external action on a dynamic system, and its response to it, as a sum of some "elementary" summands becomes justified when the system is *linear*, i.e., when it satisfies the superposition principle (the response to the total force equals the sum of responses to each of the summands). This is true of both deterministic and random actions. The choice of "elementary" summands to represent the functions considered is also predetermined by the properties of the dynamic system. Quite common among linear systems are devices with *constant* parameters. These are often termed *harmonic* systems (in particular, harmonic filters) because of their basic property that an action $\exp(i\omega t)$ causes the *steady-state* (also harmonic) response $k(i\omega)\exp(i\omega t)$ with the same frequency ω. Just for that reason the functions $\exp(i\omega t)$ with a continuous or discrete set of frequencies ω are singled out for such devices from the various other complete systems of "elementary" functions. The special role of Fourier expansions thus stems from the wide use and special importance of linear dynamic systems with constant parameters, a fact which applies equally to deterministic and to random processes. That is why we will also concentrate our attention on the representation

$$\zeta(t) = \int\limits_{-\infty}^{+\infty} e^{i\omega t} dC(\omega) \quad , \tag{1.24}$$

where $dC(\omega)$ is a finite or infinitesimal (proportional to $d\omega$) increment of the function $C(\omega)$ in the interval $(\omega, \omega + d\omega)$:

$$dC(\omega) = C(\omega + d\omega) - C(\omega) \quad .$$

The Fourier-Stieltjes integral (1.24) covers the cases of continuous, discrete, and mixed spectra. In the purely discrete case we can write it in the more usual form of the generalized Fourier series

$$\zeta(t) = \sum_n c_n \exp(i\omega_n t) \quad . \tag{1.25}$$

For a purely continuous spectrum we can, in a certain formal sense to be discussed below, introduce the complex amplitude density

$$dC(\omega) = c(\omega)d\omega \quad , \tag{1.26}$$

so that (1.24) will take the form of the conventional Fourier integral

$$\zeta(t) = \int\limits_{-\infty}^{+\infty} e^{i\omega t} c(\omega)d\omega \quad . \tag{1.27}$$

Again, we can formally include in (1.27) the case of the discrete spectrum (1.25), assuming that

$$c(\omega) = \sum_n c_n \delta(\omega - \omega_n) \quad . \tag{1.28}$$

Note that for a real function $\zeta(t)$ the following condition must be met:

$$dC(\omega) = dC^*(-\omega) \quad . \tag{1.29}$$

Here $dC(-\omega) = C(-\omega + d\omega) - C(-\omega)$.

If $\zeta(t)$ is a *random* function, then the existence of the integral (1.24) should be taken to imply some kind of probabilistic convergence. The random functions that can be represented as (1.24) are referred to as *harmonizable*.

Within the framework of correlation theory based on moments of the first two orders, it is natural and prudent to treat the integral (1.24) as converging in the mean square. The necessary and sufficient condition for the random function $\zeta(t)$ to be harmonizable is then the existence, for any t and t', of the double Fourier-Stieltjes integral

$$B(t,t') = \iint\limits_{-\infty}^{+\infty} \exp\left[i(\omega t - \omega' t')\right] d^2 \Gamma(\omega, \omega') \quad , \tag{1.30}$$

which is the second moment $B(t, t') = \langle \zeta(t)\zeta^*(t') \rangle$ of the function $\zeta(t)$. Put another way, if the integral (1.30) exists, then a random complex function $C(\omega)$ exists such that the integral (1.24) converges in the mean square to $\zeta(t)$, the two-dimensional increment of the function $\Gamma(\omega, \omega')$ being given by

$$d^2\Gamma(\omega, \omega') = \langle dC(\omega)dC^*(\omega') \rangle \quad . \tag{1.31}$$

In particular, at $t' = t$ we have

$$\langle |\zeta(t)|^2 \rangle = B(t, t) = \iint\limits_{-\infty}^{+\infty} \exp\left[i(\omega - \omega')t\right] d^2\Gamma(\omega, \omega') \quad . \tag{1.32}$$

It should be noted that the condition that (1.30) be finite tells us nothing about the mean of the function $\zeta(t)$. In particular this condition does not presuppose that $\langle \zeta(t) \rangle$ is stationary (constant). That the expansion (1.24) subject to (1.30) is unique, i.e., that the relationship between the statistical properties of $\zeta(t)$ and $C(\omega)$ is unique, can only be established using *filtration* to *single out* some or other areas of the frequency plane (ω, ω'). But filtration will be discussed later (Sects. 3.1 and 4.4).

The function

$$\Gamma(\omega, \omega') = \int\limits_{-\infty}^{\omega} \int\limits_{-\infty}^{\omega'} \langle dC(\omega)dC^*(\omega') \rangle \tag{1.33}$$

will be referred to as the complex *spectral "mass"* of the process $\zeta(t)$. The "mass" may be concentrated at individual points in the plane (ω, ω'), or distributed along lines (with a continuous linear density) or, finally, be distributed over a plane with a continuous surface density. The two-dimensional increment (1.31), however, always has the property that

$$d^2\Gamma(\omega, \omega') = d^2\Gamma^*(\omega', \omega) \quad ,$$

i.e., at points located symmetrically about the bisector $\omega = \omega'$, the values of the densities are the complex conjugate of one another, being real on the bisector itself. Thus, integrals of the form

$$\iint \exp\left[i(\omega - \omega')t\right] d^2\Gamma(\omega, \omega') \quad ,$$

extended to include any region (ω, ω') symmetrical about the bisector $\omega = \omega'$ [specifically, integrals over the entire plane (ω, ω')], will be real. Since the integral (1.30) exists, in particular, at $t' = t = 0$, the total complex "mass" (1.33), i.e., the quantity $\Gamma(\infty, \infty) = B(0, 0) = \langle |\zeta(0)|^2 \rangle$, will be real and finite.

The fact that the "mass" (1.33) is distributed over the entire plane (ω, ω') due to the correlation of the amplitudes $dC(\omega)$ at different points $\omega \neq \omega'$ suggests that it is impossible to represent the second moments (1.30) and (1.32) in the form of *single* integrals with respect to frequency. In particular, if these moments have the sense of an energy, e.g., if $\langle|\overline{\zeta(t)}|^2\rangle$ is the mean instantaneous power of the random process $\zeta(t)$, then the above statement implies that a contribution to this power in the frequency interval $(\omega, \omega + d\omega)$ is made not only by the harmonic component $\exp(i\omega t)dC(\omega)$ of the process $\zeta(t)$, but also by all the other harmonic components, since they are correlated with $dC(\omega)$. Under these conditions, it is said that the *mean energy* (or more generally, *mean bilinear quantities*) *are not localizable in frequency*. Is the opposite possible?

As has already been stressed, interest in the harmonic representations (1.24) stems from the fact that for a linear system with constant parameters, they enable an appropriate response (output process) to be constructed immediately from a given external action (input process). In the process, each harmonic $\exp(i\omega t)dC(\omega)$ is transformed independently, i.e., a dynamic system (filter) with the transfer function $k(i\omega)$ transforms this harmonic, independently of other harmonics, into the oscillation $k(i\omega)\exp(i\omega t)dC(\omega)$. As a result, the process (1.24) at the system input gives rise to the output process

$$Z(t) = \int\limits_{-\infty}^{+\infty} k(i\omega)e^{i\omega t}dC(\omega) \quad .$$

It is well known, however, that the process $Z(t)$ can also be constructed differently or, so to speak, directly if we use the *temporal* rather than spectral approach. In that case, a dynamic system such as that under consideration is described not by the transfer function $k(i\omega)$, but by the response $H(t)$ to the impulse $\delta(t)$. The output process $Z(t)$ will then be expressed in terms of the Duhamel integral (Sect. I.6.5), and the transient phenomena will therefore also be covered quite simply.

To summarize, if the spectral representation has the constraint that only linear quantities are localized in frequency, then it is just one of the possible and equivalent representations and has no additional physical arguments in its favor. The situation changes and the spectral approach comes to the fore if the harmonics $\exp(i\omega t)dC(\omega)$ can also be individualized by their contribution to the "energy" (mean bilinear) quantities related to $\zeta(t)$ and $Z(t)$. This is exactly the case with *stationary* random functions, which are to be discussed in the following section.

1.4 Stationary Random Functions

To what constraints are the random complex spectral amplitude $C(\omega)$ and the distribution of the complex "mass" $\Gamma(\omega, \omega')$, subjected if the random function $\zeta(t)$ is wide-sense stationary?

The requirement $\overline{\zeta(t)} = \overline{\zeta} = \text{const}$ implies, according to (1.24), that at all $\omega \neq 0$

$$\overline{dC(\omega)} = 0 \quad , \tag{1.34}$$

and then

$$\overline{\zeta} = \overline{dC(0)} \quad .$$

We can always eliminate the constant mean from the function $\zeta(t) - \overline{\zeta}$, which indicates that (1.34) covers all the values of ω without exception. We further assume that for the stationary functions $\zeta(t)$ under consideration the mean $\overline{\zeta} = 0$, and hence the mixed moment coincides with the covariance ψ.

As is seen from (1.30), the stationarity condition for the covariance, whereby the latter is dependent only on the difference $\tau = t - t'$, can be fulfilled, only if

$$d^2\Gamma(\omega,\omega') = \langle dC(\omega)dC^*(\omega') \rangle = 0 \quad \text{at} \quad \omega \neq \omega' \quad ,$$

i.e., if the "mass" $\Gamma(\omega,\omega')$ is distributed only over the bisector $\omega = \omega'$. The increment $d^2\Gamma(\omega,\omega')$ is then always real and nonnegative. In terms of the delta-function, the above statement can be written as

$$d^2\Gamma(\omega,\omega') = \langle dC(\omega)dC^*(\omega') \rangle = \delta(\omega - \omega')d\omega' dG(\omega) \quad , \tag{1.35}$$

the real increment $dG(\omega)$ being nonnegative, i.e.,

$$dG(\omega) \geq 0 \quad .$$

Accordingly, *if $\zeta(t)$ is a stationary function, then $C(\omega)$ is a function with uncorrelated increments* (Sect. I.6.2)[5].

Substituting (1.35) into (1.30) gives

$$\iint\limits_{-\infty}^{+\infty} e^{i(\omega t - \omega' t')}\langle dC(\omega)dC^*(\omega')\rangle = \int\limits_{-\infty}^{+\infty} dG(\omega) \int\limits_{-\infty}^{+\infty} e^{i(\omega t - \omega' t')}\delta(\omega - \omega')d\omega'$$

$$= \int\limits_{-\infty}^{+\infty} e^{i\omega(t-t')}dG(\omega) = \psi(t - t') \quad , \tag{1.36}$$

i.e., *the covariance for a wide-sense stationary process can be represented in terms of the single Fourier-Stieltjes integral (1.36), where $G(\omega)$ is a nondecreasing real function of ω.* In addition, as

$$\int\limits_{-\infty}^{+\infty} dG(\omega) = \psi(0) = \langle |\zeta|^2 \rangle \quad , \tag{1.37}$$

and the second-order random functions are subject to the condition that $\langle |\zeta|^2 \rangle$ be finite, the function $G(\omega)$ must be *bounded:* $G(\omega) \leq G(\infty)$.

---

[5] As for the increments $dC(\omega)$, we do not need to resort to moments of higher order than two. The requirement that they be uncorrelated (but not necessarily independent) will be sufficient.

This fundamental theorem is due to *Khintchine* [1.4] who gave a proof in 1934 for (wide-sense) stationary random functions satisfying the condition that $\psi(\tau)$ be continuous at zero. The treatment used not the spectral representation (1.24) of the random function $\zeta(t)$ itself (whose applicability was established by *Kolmogorov* [1.5] at a later date), but the theorem of harmonic analysis proved earlier by *Bochner* [Ref. 1.6, p. 76]. The theorem states that any positive definite function $\psi(\tau)$[6] may be represented as

$$\psi(\tau) = \int\limits_{-\infty}^{+\infty} e^{i\omega\tau}\, dG(\omega) \quad ,$$

where $G(\omega)$ is real, nondecreasing and bounded. This theorm, established without any connection to the theory of random functions, immediately leads to the Khintchine theorem, if we take into account that the class of positive definite functions coincides with the class of covariances of stationary random functions that are continuous at zero.

It should be stressed once more that the stationarity of the random function $\zeta(t)$ imposes *two* requirements on the distribution of the complex "mass" $\Gamma(\omega,\omega')$ in the plane (ω,ω'). First, the mass must be concentrated solely on the bisector $\omega' = \omega$. Second, the total mass $G(\infty)$ must be finite, i.e., its linear density $g(\omega)$ must be integrable over the entire ω axis from $-\infty$ to $+\infty$. If any of these conditions is violated, $\zeta(t)$ will be nonstationary.

Thus, if the increments $dC(\omega)$ are delta-correlated in the spectral representation of a stationary random function, then the *mean bilinear quantities will be localizable in frequency*. In order to clarify this statement and illustrate the practical technique of going over from the spectral expansion $\zeta(t)$ to the spectral expansion $\psi(\tau)$, we will for the moment ignore the above-mentioned (but necessary) mathematical conditions and proceed to perform the transition in some special cases, drawing heavily on the delta-function technique.

Consider a purely continuous spectrum. We know that a continuous function with uncorrelated increments $C(\omega)$ is nowhere differentiable. In Sect. I.6.2 we considered at length the questions of why and in what sense we may practically employ the delta-correlated derivative of such a function, in this case – the complex *amplitude density* $c(\omega) = dC(\omega)/d\omega$. Of course we did introduce $c(\omega)$ not with the aim to impair the mathematical rigor. This quantity is actually useful since in physical problems, when going over from stochastic differential equations for $\zeta(t)$ to its spectral representation, it would be extremely unusual and inconvenient to make use of the increments $dC(\omega)$. All such equations are generally written for the amplitude spectral densities $c(\omega)$ and, as we have seen, pose no problem if care is exercised.

Thus, if $dC(\omega) = c(\omega)d\omega$ and (which is now quite correct) $dG(\omega) = g(\omega)d\omega$, then it follows immediately from (1.35) that

---

[6] That is, a function that is continuous and bounded over the entire τ axis and is subject to the condition (1.19).

$$\langle c(\omega)c^*(\omega')\rangle = g(\omega)\delta(\omega - \omega') \quad . \tag{1.38}$$

It may sometimes be convenient to write $g(\omega) = \overline{|c(\omega)|^2}$. This, of course, is just *notation*, for the function $c(\omega)$ has no finite mean square of the modulus. When using this notation we should remember that $\overline{|c(\omega)|^2}$ is a factor of the delta-function.

From (1.27) and with the aid of the covariance (1.38), by formally multiplying and averaging, we easily obtain the result

$$\psi(\tau) = \langle \zeta(t+\tau)\zeta^*(t)\rangle = \int\!\!\!\int\limits_{-\infty}^{+\infty} e^{i\omega(t+\tau)-i\omega't}\langle c(\omega)c^*(\omega')\rangle d\omega\, d\omega'$$

$$= \int\limits_{-\infty}^{+\infty} g(\omega)d\omega \int\limits_{-\infty}^{+\infty} e^{i\omega(t+\tau)-i\omega't}\delta(\omega - \omega')d\omega' = \int\limits_{-\infty}^{+\infty} e^{i\omega\tau}g(\omega)d\omega \quad, \tag{1.39}$$

i.e., a conventional Fourier integral for $\psi(\tau)$. Its inversion gives

$$g(\omega) = \frac{1}{2\pi}\int\limits_{-\infty}^{+\infty}\psi(\tau)e^{-i\omega t}d\tau \quad . \tag{1.40}$$

Of course, we can reverse the procedure and derive the covariance (1.38) for the amplitude densities $c(\omega)$ by requiring $\langle \zeta(t+\tau)\zeta^*(t)\rangle$ to be dependent only on τ. It is precisely this approach that we will now use in another special case – that of the purely discrete spectrum given by (1.25):

$$\zeta(t) = \sum_n c_n e^{i\omega_n t} \quad .$$

Multiplying and averaging leads to

$$\langle \zeta(t+\tau)\zeta^*(t)\rangle = \sum_{m,n} \overline{c_n c_m^*}\exp\left[i\omega_n(t+\tau) - i\omega_m t\right] \quad .$$

It is obvious that the double Fourier series on the right can only vary with τ if the random coefficients c_n are mutually uncorrelated

$$\overline{c_n c_m^*} = \overline{|c_n|^2}\delta_{mn} \equiv g_n\delta_{mn} \quad . \tag{1.41}$$

Then

$$\psi(\tau) = \sum_n g_n e^{i\omega_n \tau} \quad, \tag{1.42}$$

i.e., $\psi(\tau)$ is an *almost periodic* function[7] with real and positive coefficients g_n,

---

[7] Almost periodic functions occupy in a sense an intermediate position between periodic functions, which can be represented by a conventional Fourier series with frequencies that are miltiples of the basic frequency $\omega_0 = 2\pi/T_0$ (T_0 is the period), and nonperiodic functions, which can be represented by the Fourier integral. The harmonic series (1.42) with arbitrary ω_n (a *generalized* Fourier series) represents a function that, after a time T, recurs *approximately and all the more accurately, the larger T is*. In other words, we can always take T to be so large that the function will recur with predetermined accuracy.

$$g_n = \lim_{T \to \infty} \frac{1}{2T} \int\limits_{-T}^{+T} \psi(\tau) e^{-i\omega_n \tau} d\tau \quad . \tag{1.43}$$

Clearly, (1.39) and (1.42) are special cases of the general theorem (1.36). The conditions (1.34) and (1.37) are written in these special cases in the form (recalling that $\overline{\zeta} = 0$)

$$\overline{c(\omega)} = 0 \quad , \quad \overline{|\zeta|^2} = \psi(0) = \int\limits_{-\infty}^{+\infty} g(\omega) d\omega \quad ,$$

$$\overline{c}_n = 0 \quad , \quad \overline{|\zeta|^2} = \psi(0) = \sum_n g_n \quad .$$

The Fourier-Stieltjes integral (1.36) can, of course, be replaced by the conventional Fourier integral both for a mixed and for a purely discrete spectrum, the latter being described in terms of the spectral density

$$g(\omega) = \sum_n g_n \delta(\omega - \omega_n) \quad .$$

What is more, we can in all cases also utilize the inversion relation (1.40). It is sufficient to make use of the Fourier expansion for the delta-function when substituting the covariance (1.42) into (1.40).

A word about the terminology is now in order. In the literature, $G(\omega)$ and $g(\omega)$ are often described using energy-related terminology. Thus, $G(\omega)$ is termed the spectral distribution function for the mean instantaneous *power* $\overline{|\zeta|^2}$, spectral *power* in the interval $(-\infty, \omega)$, and so forth. Accordingly, $g(\omega)$ is termed the spectral *power* in a unit interval of ω. This is all very well, if $\overline{|\zeta|^2}$ really has the sense of an energy, representing, for example, the squared random current or voltage. But if $\zeta(t)$ is, say, the randomly varying refractive index what "power" is then involved? We prefer a more neutral terminology. When speaking about "intensity of fluctuations", we usually do not interpret "intensity" in energy terms. We will thus use this term and refer to $G(\omega)$ as the *spectral intensity* of the random function $\zeta(t)$ (in the interval from $-\infty$ to ω), and to $g(\omega)$ as the *spectral density* of the process. As noted above, we can only prove that $dG(\omega) = g(\omega)d\omega$ is the spectral density in the interval $(\omega, \omega + d\omega)$ by using the harmonic filtration (Sect. 3.1).

If, for some reason or other, we do not wish to commit ourselves to the assumption that $\overline{\zeta} = 0$, for the stationary process under consideration, then the mixed moment $B(\tau)$ and the covariance $\psi(\tau)$ no longer coincide,

$$B(\tau) = \psi(\tau) + |\overline{\zeta}|^2 \quad . \tag{1.44}$$

As a result, the spectra $B(\tau)$ and $\psi(\tau)$ will also be different. If, as in (1.39), we write

$$B(\tau) = \int\limits_{-\infty}^{+\infty} g_B(\omega) e^{i\omega\tau}\, d\omega \qquad (1.45)$$

and take into account that the constant quantity $|\bar\zeta|^2$ can always be represented in the form

$$|\bar\zeta|^2 = |\bar\zeta|^2 \int\limits_{-\infty}^{+\infty} \delta(\omega) e^{i\omega\tau}\, d\omega \quad ,$$

we get

$$g_B(\omega) = g(\omega) + |\bar\zeta|^2 \delta(\omega) \quad . \qquad (1.46)$$

The spectral densities $g_B(\omega)$ and $g(\omega)$ thus coincide at all ω save for the point $\omega = 0$, where the spectrum of the mixed moment $B(\tau)$ has a discrete line with the integral intensity $|\bar\zeta|^2$.

If $\zeta(t) \equiv \xi(t)$ is real, then $\psi(\tau)$ is an even function, thus suggesting that the spectral density $g(\omega)$, too, is an even function of ω. In that case, (1.39) can be rewritten as

$$\psi(\tau) = \int\limits_{-\infty}^{+\infty} g(\omega)\cos\omega\tau\, d\omega = \int\limits_{0}^{\infty} g_+(\omega)\cos\omega\tau\, d\omega \quad , \qquad (1.47)$$

where

$$g_+(\omega) = 2g(\omega) \qquad (1.48)$$

is the spectral density *in positive frequencies*. Formula (1.40) will then give

$$g_+(\omega) = \frac{2}{\pi} \int\limits_{0}^{\infty} \psi(\tau)\cos\omega\tau\, d\tau \quad . \qquad (1.49)$$

The fact that the covariance $\psi(\tau)$ is positive definite suggests that the spectral density $g(\omega)$ is nonnegative, and *vice versa*. Occasionally, therefore, in seeking an answer to the question of whether a function $\psi(\tau)$ subject to (1.17, 18) represents the covariance of a wide-sense stationary random process $\zeta(t)$, it is more convenient simply to check if the condition $g(\omega) \geq 0$ is fulfilled. If, say, $\psi(\tau)$ has the configuration of a rectangle (Fig. 1.2a) of width 2ϑ, then, by (1.40),

$$g(\omega) = \frac{\psi(0)}{2\pi} \int\limits_{-\vartheta}^{\vartheta} e^{-i\omega\tau}\, d\tau = \frac{\psi(0)}{\pi\omega} \sin\omega\vartheta \quad ,$$

i.e., the condition $g(\omega) \geq 0$, is violated. For the same reason it is impossible for the covariance to be $+\psi(0)$ or $-\psi(0)$ at some *finite intervals* of τ (even if symmetrical about $\tau = 0$).

We now turn to the spectral representations of the complex function $\zeta(t) = \xi(t) + i\eta(t)$ and its covariance, given that $\zeta(t)$ is an analytical signal (Sect. 1.1). We write the spectral expansions of both real functions $\xi(t)$ and $\eta(t)$

$$\xi(t) = \int_{-\infty}^{+\infty} c(\omega)e^{i\omega t}d\omega \quad \text{and} \quad \eta(t) = \int_{-\infty}^{+\infty} \tilde{c}(\omega)e^{i\omega t}d\omega$$

where we have

$$c(-\omega) = c^*(\omega) \quad , \quad \tilde{c}(-\omega) = \tilde{c}^*(\omega)$$

because ξ and η are real. In other words, in the spectral expansion of a real function the region $\omega<0$ yields no further information than that contained in the region $\omega>0$. Owing to the one-to-one correspondence between $\xi(t)$ and $\eta(t)$, the spectral amplitude densities $c(\omega)$ and $\tilde{c}(\omega)$ of the analytical signal are also uniquely interrelated.

Proceeding directly from (1.6) shows that the Hilbert transformation changes $\cos \omega t$ into $\pm \sin \omega t$, and $\sin \omega t$ into $\mp \cos \omega t$ $(\omega \gtrless 0)$. The exponential function $\exp(i\omega t)$ thus transforms into $-i\exp(i\omega t)\,\text{sgn}\,\omega$, where

$$\text{sgn}\,\omega = \begin{cases} +1 & \text{for} \quad \omega>0 \quad , \\ 0 & \text{for} \quad \omega = 0 \quad , \\ -1 & \text{for} \quad \omega<0 \quad . \end{cases}$$

Consequently, the spectral amplitude density $\tilde{c}(\omega)$ of the process $\eta(t)$ which is according to Hilbert conjugate to $\xi(t)$, will be

$$\tilde{c}(\omega) = -ic(\omega)\,\text{sgn}\,\omega \quad . \tag{1.50}$$

As a result, the spectral representation of an analytical signal has no negative frequencies

$$\zeta(t) = \xi(t) + i\eta(t) = \int_{-\infty}^{+\infty} [c(\omega) + i\tilde{c}(\omega)]e^{i\omega t}d\omega$$

$$= \int_{-\infty}^{+\infty} c(\omega)[1 + \text{sgn}\,\omega]e^{i\omega t}d\omega = 2\int_{0}^{\infty} c(\omega)e^{i\omega t}d\omega \quad . \tag{1.51}$$

It is therefore also possible to define the analytical signal as a complex process whose spectrum is nonzero only on the positive ω semiaxis.

If $\xi(t)$ is a stationary random process with $\overline{\xi(t)} = 0$ [so that $\overline{c(\omega)} = 0$] and

$$\langle c(\omega)c^*(\omega')\rangle = g(\omega)\delta(\omega - \omega') \quad ,$$

then, by (1.50), $\overline{\tilde{c}(\omega)} = 0$ and, accordingly, $\overline{\eta(t)} = 0$, and the covariance $\tilde{c}(\omega)$ is

$$\langle \tilde{c}(\omega)\tilde{c}^*(\omega') \rangle = g(\omega)\delta(\omega - \omega')\operatorname{sgn}^2(\omega) \quad .$$

Thus, the process $\eta(t)$ is also stationary and has the same spectral density $g(\omega)$ as the process $\xi(t)$ (with the exception of the point $\omega = 0$, where $\operatorname{sgn}^2\omega = 0$), and hence has the same covariance

$$B_\xi(\tau) = B_\eta(\tau) = \int\limits_{-\infty}^{+\infty} g(\omega)e^{i\omega\tau}d\omega = 2\int\limits_{0}^{\infty} g(\omega)\cos\omega\tau\,d\omega \quad . \tag{1.52}$$

Using (1.50), we further obtain

$$B_{\xi\eta}(\tau) = -i\int\limits_{-\infty}^{+\infty} g(\omega)e^{i\omega\tau}\operatorname{sgn}\omega\,d\omega = 2\int\limits_{0}^{\infty} g(\omega)\sin\omega\tau\,d\omega \quad , \tag{1.53}$$

whence we immediately see that $B_{\xi\eta}$ is odd in τ, and in particular that $\xi(t)$ and $\eta(t)$ are noncorrelated at a given time t ($\tau = 0$). This absence of correlation does not, of course, imply statistical independence, since in the general case ξ and η are not normal processes. Thus, the spectral treatment naturally includes the results obtained earlier in Sect. 1.1.

It follows from (1.11) and (1.52, 53) that the covariance for the stationary analytical signal $\zeta(t)$ is

$$B_\zeta(\tau) = 4\int\limits_{0}^{\infty} g(\omega)e^{i\omega\tau}d\omega = 2\int\limits_{0}^{\infty} g_+(\omega)e^{i\omega\tau}d\omega \quad . \tag{1.54}$$

In other words, the spectral density $\zeta(t)$ is

$$g_\zeta(\omega) = \begin{cases} 4g(\omega) & \text{for} \quad \omega > 0 \quad , \\ 0 & \text{for} \quad \omega < 0 \quad . \end{cases}$$

The inversion of the Fourier transforms (1.52, 53) gives the following expressions for the spectral density $g(\omega)$:

$$g(\omega) = \frac{1}{\pi}\int\limits_{0}^{\infty} B_\xi(\tau)\cos\omega\tau\,d\tau = \frac{1}{\pi}\int\limits_{0}^{\infty} B_{\xi\eta}(\tau)\sin\omega\tau\,d\tau \tag{1.55}$$

(the latter expression holds at $\omega > 0$).

Finally, consider the following question. Let a real stationary process $\xi(t)$ have a spectral density $g(\omega)$ (finite at zero) that can be represented in the form

$$\frac{dG}{d\omega} = g(\omega) = g(0) + g''(0)\frac{\omega^2}{2} + \ldots = g(0) + \omega^2 g_1(\omega^2) \quad ,$$

where $g_1(0)$ is finite. Under what condition will the *accumulation in time* t, i.e., the operation

$$\eta(t) = \int_0^t \xi(t)dt \quad , \qquad (1.56)$$

give a process that tends to be stationary with increasing t?

Using the spectral expansion of $\xi(t)$, we obtain

$$\eta(t) = \int_{-\infty}^{+\infty} dC(\omega) \int_0^t e^{i\omega t} dt = \int_{-\infty}^{+\infty} \frac{e^{i\omega t} - 1}{i\omega} dC(\omega) \quad ,$$

and hence, by (1.35),

$$\overline{\eta^2(t)} = \int_{-\infty}^{+\infty} \left(\frac{\sin (\omega t/2)}{\omega/2} \right)^2 g(\omega)d\omega$$

$$= g(0) \int_{-\infty}^{+\infty} \left(\frac{\sin (\omega t/2)}{\omega/2} \right)^2 d\omega + 2 \int_{-\infty}^{+\infty} g_1(\omega^2)(1 - \cos \omega t)d\omega \quad .$$

The first term is $2\pi g(0)t$. We can take the integral with $\cos \omega t$ around the residues of poles of $g_1(\omega^2)$, having first passed from $\cos \omega t$ to $\exp(i\omega t)$ and closed the integration path in the upper half-plane. We thus obtain

$$\overline{\eta^2(t)} = 2\pi g(0)t + \text{const} + \text{exponentially decaying terms} \quad .$$

It follows that the necessary condition for $\eta(t)$ to be stationary at $t \to \infty$ is

$$g(0) = 0. \qquad (1.57)$$

Otherwise, $\overline{\eta^2}$ will grow in a diffusional manner. Thus, if the path $s(t) = \int_0^t v(t)dt$ covered by a body is a stationary process, then the spectral density of the velocity $v(t)$ must vanish at $\omega = 0$. The same applies to the magnetic flux $\Phi(t) = -c \int_0^t \mathcal{E}(t)dt$ due to a random stationary e.m.f. $\mathcal{E}(t)$, e.g., when dealing with a so-called magnetic noise.

1.5 Examples of Spectral Representations of Stationary Functions

Turning to examples illustrating the Khintchine theorem, we begin with some general remarks.

The spectral intensity $G(\omega)$, just like its increment in any frequency interval, only depends on moduli of the spectral amplitudes $dC(\omega)$, and not on their arguments. Thus, the random stationary functions

$$\zeta(t) = \int\limits_{-\infty}^{+\infty} e^{i\omega t} dC(\omega) \quad , \quad \zeta_1(t) = \int\limits_{-\infty}^{+\infty} e^{i\omega t} e^{i\theta(\omega)} dC(\omega) \quad ,$$

[where $\theta(\omega)$ is a deterministic function or a random function independent of $C(\omega)$] which are generally completely different, possess the same spectral intensity $G(\omega)$, and hence the same covariance $\psi(\tau)$. This follows directly from the fact that the increments $dC(\omega)$ are delta-correlated, cf. (1.35). In particular, in the case of a continuous spectrum the substitution of $c_1(\omega) = c(\omega) \exp\left[i\theta(\omega)\right]$ for $c(\omega)$ does not change the spectral density $g(\omega)$, as it follows from (1.38) that

$$\langle c_1(\omega) c_1^*(\omega') \rangle = \langle c(\omega) c^*(\omega') \rangle \exp\left\{ i[\theta(\omega) - \theta(\omega')] \right\} = g(\omega)\delta(\omega - \omega') \quad .$$

The picture is the same in the case of a discrete spectrum. Subsituting $c_{1n} = c_n \exp(i\theta_n)$ with arbitrary phases θ_n (which are also either deterministic or random, but independent of c_n) for c_n does not, by virtue of (1.41), change the intensities g_n

$$\overline{c_{1n} c_{1m}^*} = \overline{c_n c_m^*} e^{i(\theta_n - \theta_m)} = g\delta_{mn} \quad .$$

Precisely because of this lack of sensitivity to phases, the covariance is the same for widely varying random processes. For example, the process (I.3.1) in the case of exponential pulses (Fig. I.3.5), the process (I.3.47) with an exponentially distributed duration of square pulses (Fig. I.3.3), the generalized telegraph signal (Fig. I.3.4), and the amplitude fluctuations in the Thomson self-oscillatory system $-$ all have the same (exponential) covariance [expressions (I.3.50), (I.3.48), (I.3.49), and (I.5.126), respectively]. It should be noted that the phase insensitivity means, in particular, that a delay line with arbitrary dispersion but uniform amplitude-frequency characteristic does not change the covariance of an input process.

The fact that the covariance $\psi(\tau)$ and spectral density $g(\omega)$ are Fourier conjugate, and hence uniquely define each other, is widely used in specific problems. In some cases it is easier to calculate $\psi(\tau)$ and then find the spectrum using the relations (1.40), (1.43) or (1.49); and conversely, in other problems it is easier to find first the spectrum and then to proceed to calculate the covariance with the help of (1.39), (1.42), or (1.47).

We also note that, given a continuous spectrum, the functions $\psi(\tau)$ and $g(\omega)$, related by a Fourier transform, obey the well-known uncertainty relation. Physically, it means that a narrow spectrum suggests a wide, or prolonged, correlation and notable "orderliness" of the process; and conversely, a wide spectrum gives a short correlation and more chaos. If the integrals

$$\overline{\omega} = \frac{1}{\psi(0)} \int\limits_{-\infty}^{+\infty} \omega g(\omega) d\omega \quad , \quad (\Delta\omega)^2 = \frac{1}{\psi(0)} \int\limits_{-\infty}^{+\infty} (\omega - \overline{\omega})^2 g(\omega) d\omega \quad ,$$

$$\bar{\tau} = \frac{1}{2\pi g(0)} \int\limits_{-\infty}^{+\infty} \tau \psi(\tau)d\tau \quad , \quad (\Delta\tau)^2 = \frac{1}{2\pi g(0)} \int\limits_{-\infty}^{+\infty} (\tau - \bar{\tau})^2 \psi(\tau)d\tau \quad ,$$

exist, then the uncertainty relation lends itself to a quantitative description: in the theory of Fourier integrals it is proved that the uncertainties $\Delta\omega$ and $\Delta\tau$ satisfy the inequality

$$(\Delta\omega)^2(\Delta\tau)^2 \geq \tfrac{1}{4} \quad . \tag{1.58}$$

The minimizing function $\psi(\tau)$, which turns (1.58) into an equality, is a Gaussian function[8]

$$\psi(\tau) = \psi(0)e^{-\tau^2/\tau_c^2} \quad . \tag{1.59}$$

The Gaussian correlation law (1.59) is an ivnariant law in the sense that the appropriate spectral density is also expressed by a Gaussian curve

$$g(\omega) = \frac{\tau_c\psi(0)}{2\sqrt{\pi}}e^{-\omega^2\tau_c^2/4} \quad . \tag{1.60}$$

The widths $\delta\tau$ and $\delta\omega$, determined at the level $1/e$ of the maximum values of $\psi(0)$ and $g(0)$, are $\delta\tau = 2\tau_c$ and $\delta\omega = 4/\tau_c$, so that $\delta\omega \cdot \delta\tau = 8$. Another example of the invariant correlation law is the inverse hyperbolic cosine

$$\psi(\tau) = \psi(0)/\cosh\left(\tau/\tau_c\right) \quad .$$

In that case

$$g(\omega) = \tau_c\psi(0)/2\cosh\left(\pi\omega\tau_c/2\right) \quad .$$

We now use the Khintchine theorem to calculate the spectra of a number of stationary processes considered earlier, whose covariances we have already found.

It was shown in Sect. I.3.4 that for a stationary pulse Poisson process

$$\xi(t) = \sum_{\nu} F(t - t_\nu, \boldsymbol{a}_\nu) \tag{1.61}$$

with a function $F(t)$ describing pulse shape, which decreases sufficiently fast, the covariance is

---

[8] *Mayer* and *Leontovich* [1.7] proved an analogous inequality for the real Fourier integral, covering only positive frequencies. In that case the minimizing function and the lower boundary of $(\Delta\omega)^2(\Delta\tau)^2$ are changed. The boundary lies between $(\pi - 2)/4\pi^2 = 0.0289\ldots$ and $1/12\pi = 0.0265\ldots$.

$$\psi(\tau) = n_1 \int w_a(a)da \int\limits_{-\infty}^{+\infty} F(\theta + \tau, a)F(\theta, a)d\theta \quad . \tag{1.62}$$

Here $w_a(a)$ is the probability density for a multidimensional random parameter $a = \{a_1, a_2, \ldots, a_m\}$. The dependence of the deterministic function F on this parameter enabled us to take into account random variations in the pulse *shape*[9]. We may now abandon this specific way of defining a random function and replace it by a general one, i.e., we simply treat the pulse shape as a complex *random function*[10], so that the process as a whole is complex

$$\zeta(t) = \sum_\nu F_\nu(t - t_\nu) \quad . \tag{1.63}$$

Assuming that $F_\nu(t)$ are statistically independent of the Poisson times $\{t_\nu\}$ and of $F_\mu(t)$ at $\mu \neq \nu$, where for any ν

$$\langle F_\nu(t) \rangle \equiv F(t) \quad , \quad \langle F_\nu(t + \tau)F_\nu^*(t) \rangle = B_F(t, \tau)$$

it is easy to perform calculations, similar to those carried out in Sect. I.3.4 to arrive at a nearly self-evident generalization of the results obtained. So, the mean of the process (1.63)

$$\overline{\zeta(t)} = \sum_\nu \overline{F_\nu(t - t_\nu)}$$

and its covariance

$$\psi(\tau) = \overline{\zeta(t + \tau)\zeta^*(t)} - \overline{\zeta(t + \tau)} \cdot \overline{\zeta^*(t)} \quad ,$$

(where the bar implies the total averaging over both the distribution F_ν and the Poisson distribution for the number of pulses n), will be

$$\overline{\zeta(t)} = n_1 \int\limits_{-\infty}^{+\infty} F(\theta)d\theta \quad , \tag{1.64}$$

$$\psi(\tau) = n_1 \int\limits_{-\infty}^{+\infty} B_F(\theta, \tau)d\theta \quad . \tag{1.65}$$

The general case of the non-Poisson and nonstationary flow of times t_ν is considered in [1.10].

---

[9] It will be recalled that the components of the random variable a can also be intercorrelated. The influence of this correlation on the spectrum of the process is studied in [1.8], where the components of a are taken to be the "amplitude" of pulse, a, the pulse duration ϑ and the time τ between a given pulse and a successive one.

[10] Such a more general treatment of the pulse processes is given in a number of works. See, e.g., [1.9, 10].

The spectrum of the process (1.63) is derived from the inversion relation (1.40). But we may write the result immediately using the well-known theorem on the spectrum of the integral $\int_{-\infty}^{+\infty} x_1^*(\theta)x_1(\theta+\tau)d\theta$. The integral is not a convolution, as the integration variable θ enters into both multipliers with the same sign [the convolution extended to include complex functions has the form $\int_{-\infty}^{+\infty} x_1^*(\theta)x_2(\tau-\theta)d\theta$]. As the derivation is straightforward and its result will be required in what follows, we will give it here.

Let $\tilde{x}_1(\omega)$ and $\tilde{x}_2(\omega)$ be spectral amplitude densities of the functions $x_1(t)$ and $x_2(t)$, so that

$$x_j(t) = \int_{-\infty}^{+\infty} \tilde{x}_j(\omega)e^{i\omega t}d\omega \quad (j=1,2) \quad .$$

Then

$$I(\tau) \equiv \int_{-\infty}^{+\infty} x_1^*(t)x_2(t+\tau)dt$$

$$= \iint_{-\infty}^{+\infty} \tilde{x}_1^*(\omega)\tilde{x}_2(\omega')e^{i\omega'\tau}d\omega\,d\omega' \int_{-\infty}^{+\infty} e^{i(\omega'-\omega)t}dt$$

$$= \iint_{-\infty}^{+\infty} \tilde{x}_1^*(\omega)\tilde{x}_2(\omega')e^{i\omega'\tau}2\pi\delta(\omega'-\omega)d\omega\,d\omega'$$

$$= \int_{-\infty}^{+\infty} 2\pi\tilde{x}_1^*(\omega)\tilde{x}_2(\omega)e^{i\omega\tau}d\omega \quad .$$

The spectral amplitude density $I(\tau)$ will thus be[11]

$$\tilde{I}(\omega) = 2\pi\tilde{x}_1^*(\omega)\tilde{x}_2(\omega) \quad . \tag{1.66}$$

Applying (1.66) to (1.65), where $x_1(t) = x_2(t) = F(t)$, we find

$$\psi(\tau) = 2\pi n_1 \int_{-\infty}^{+\infty} \langle|\tilde{F}(\omega)|^2\rangle e^{i\omega\tau}d\omega \quad ,$$

i.e., the spectral density of the process (1.63) is

$$g(\omega) = 2\pi n_1 \langle|\tilde{F}(\omega)|^2\rangle \quad , \tag{1.67}$$

where $\tilde{F}(\omega)$ is the amplitude density of the random pulse shape $F(t)$. In the

---

[11] The same derivation for the convolution $x_1(t)$ and $x_2(t)$ would yield the amplitude density $2\pi\tilde{x}_1^*(-\omega)\tilde{x}_2(\omega)$, which for real $x_1(t)$ and $x_2(t)$ gives $2\pi\tilde{x}_1(\omega)\tilde{x}_2(\omega)$, as, for example, in the case of the characteristic function in the composition (convolution) of probability densities (Sect. I.3.2).

special case of (1.62), relation (1.67) takes the form

$$g(\omega) = 2\pi n_1 \int\limits_{-\infty}^{+\infty} w_a(a)|\tilde{F}(\omega, a)|^2 da \quad , \tag{1.68}$$

where $\tilde{F}(\omega, a)$ is now a deterministic function. Finally, if $F(t, a) = aF(t)$, i.e., if there is only one random parameter — the amplitude of pulses of the same deterministic shape $F(t)$, so that

$$\psi(\tau) = n_1\overline{|a|^2} \int\limits_{-\infty}^{+\infty} F^*(\theta)F(\theta + \tau)d\theta \quad , \qquad \text{then} \tag{1.69}$$

$$g(\omega) = 2\pi n_1\overline{|a|^2}|\tilde{F}(\omega)|^2 \quad . \tag{1.70}$$

In this latter case, the "thickness" of the noise n_1, and the random amplitudes a_ν solely affect the general level of the noise, whereas the *shape* of the spectrum is completely determined by the pulse shape[12]. This, of course, is inherent already in (1.69), which suggests that the behavior of the dependence of covariance on τ is only governed by the form of $F(t)$. This result permits, however, of another interpretation which is given below.

The spectrum of an individual pulse has an amplitude density $\tilde{F}(\omega)$ independent of the time of pulse emergence, t_ν. The latter enters into the spectral expansion of the pulse $F(t - t_\nu)$ only through the *phase* factor $\exp(-i\omega t_\nu)$. This implies that for a sequence of pulses distributed over the time axis, the spectrum will have the general envelope $|\tilde{F}(\omega)|$ "indented" as a result of interference occurring at each frequency ω with various additional phases ωt_ν. Since t_ν are random and uniformly distributed over any interval, the interference will *on average* be levelled out and the intensity will only contain the square of the envelope $|\tilde{F}(\omega)|$, which is the same for all pulses.

We will now give the expressions for the covariance (1.69) and the spectral density (1.70) for two special pulse shapes $F(t)$.

Consider the *square pulses* (Fig. 1.3)

$$F(t) = \begin{cases} 1 & \text{for} \quad 0 < t < \vartheta \quad , \\ 0 & \text{elsewhere} \quad . \end{cases} \tag{1.71}$$

The covariance is given by the triangle

$$\psi(\tau) = \begin{cases} n_1\overline{|a|^2}(\vartheta - |\tau|) & \text{for} \quad |\tau| < \vartheta \quad , \\ 0 & \text{for} \quad |\tau| > \vartheta \quad , \end{cases} \tag{1.72}$$

and the spectral density is

---

[12] It will be recalled that n_1, or rather $n_1\theta$ — the mean number of the pulses produced during a pulse — drastically influences the noise *distribution* (Sect. I.3.3).

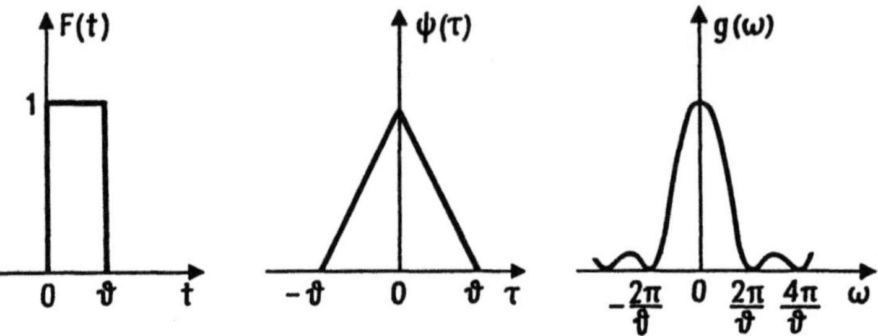

Fig. 1.3. Covariance $\psi(\tau)$ and spectral density $g(\omega)$ for a pulse Poisson process with square pulses $F(t)$.

$$g(\omega) = \frac{2}{\pi} n_1 \overline{|a|^2} \left(\frac{\sin\,(\omega\vartheta/2)}{\omega} \right)^2 \quad . \tag{1.73}$$

If the areas under the pulses are normalized to unity, for which purpose it is sufficient to set $a_\nu = 1/\vartheta$ (i.e., $\overline{|a|^2} = 1/\vartheta^2$), then

$$\psi(\tau) = \frac{n_1}{\vartheta} \left(1 - \frac{|\tau|}{\vartheta} \right) \quad , \quad g(\omega) = \frac{n_1}{2\pi} \left(\frac{\sin\,(\omega\vartheta/2)}{\omega\vartheta/2} \right)^2 \quad . \tag{1.74}$$

The covariance is now represented by a triangle with base 2ϑ and altitude n_1/ϑ, so that for any ϑ its area is n_1. As $\vartheta \to 0$, the triangle becomes an infinite peak at zero, i.e.

$$\psi(\tau) = n_1\delta(\tau) \quad . \tag{1.75}$$

The width of the main maximum $g(\omega)$ grows indefinitely with decreasing ϑ, the height $n_1/2\pi$ remaining unchanged at $\omega = 0$. In the limit $\vartheta \to 0$ we obtain the constant spectral density

$$g(\omega) = n_1/2\pi \quad . \tag{1.76}$$

It is easily seen that the limiting expressions (1.75, 76) formally satisfy the relationships (1.39, 40). Really,

$$\psi(\tau) = \int\limits_{-\infty}^{+\infty} \frac{n_1}{2\pi} e^{i\omega\tau} \, d\omega = n_1\delta(\tau) \quad ,$$

$$g(\omega) = \frac{1}{2\pi} \int\limits_{-\infty}^{+\infty} n_1\delta(\tau) e^{i\omega\tau} \, d\tau = \frac{n_1}{2\pi} \quad .$$

But a random process with the delta-correlation (1.75) no longer belongs to the class of stationary functions under consideration as it has no finite mean square of the modulus, so that there is no quantity

$$\overline{|\zeta|^2} = \psi(0) = \int\limits_{-\infty}^{+\infty} g(\omega)d\omega$$

for it.

We now look at the next *non-Poisson* stationary process: the values $\xi(t) = \mp a$ alternate with probabilities 1/2. The values hold for the same time interval ϑ. The start of the "zero" pulse is uniformly distributed over the interval $(-\vartheta, 0)$ (otherwise the process will not be stationary). Such a random function may be simulated by tossing a coin. We have here chaotic pulses of identical shape that do not overlap (Fig. 1.4) but follow one another, i.e., there are no gaps between pulses and their number per unit time is well specified ($n_1 = 1/\vartheta$). It can be easily seen, nonetheless, that the covariance is the same as for the Poisson process with square pulses (1.71), namely

$$\psi(\tau) = \begin{cases} a^2\left(\dfrac{1 - |\tau|}{\vartheta}\right) & \text{for} \quad |\tau| < \vartheta \;, \\ 0 & \text{for} \quad |\tau| \geq \vartheta \;. \end{cases}$$

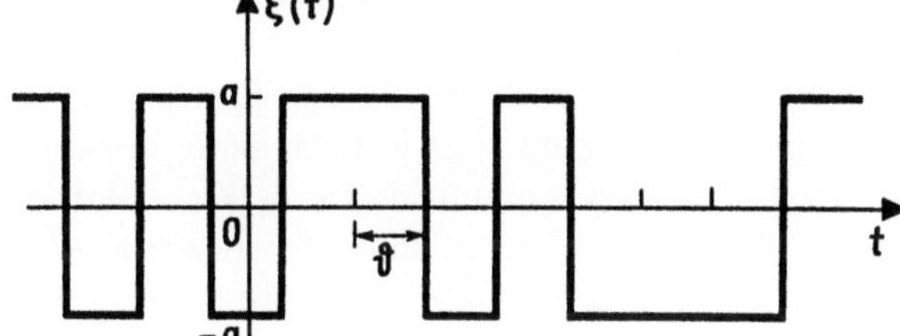

Fig. 1.4. Realization of a non-Poisson stationary process for which $\xi(t) = \pm a$ with probabilities 1/2; the pulses succeed one another without gaps and each lasts the time ϑ

This is just another illustration of the insensitivity of the covariance to phases of spectral amplitudes of the random function itself.

Let us now return to the Poisson process of the form

$$\zeta(t) = \sum_\nu a_\nu F(t - t_\nu) \;.$$

Consider the exponential pulses

$$f(t) = \begin{cases} e^{-t/\vartheta} & \text{for} \quad t \geq 0 \;, \\ 0 & \text{for} \quad t < 0 \;. \end{cases} \tag{1.77}$$

In this case, according to (1.69, 70), we have (Fig. 1.5)

$$\psi(\tau) = \frac{n_1\vartheta\overline{|a|^2}}{2}e^{-|\tau|/\vartheta} \;, \quad g(\omega) = \frac{n_1\vartheta^2\overline{|a|^2}}{2\pi(1 + \omega^2\vartheta^2)} \;. \tag{1.78}$$

Function (1.77) can represent the response of a RC-cell to a delta-impulse, so that the process $\zeta(t)$ with such a pulse shape describes the output voltage of

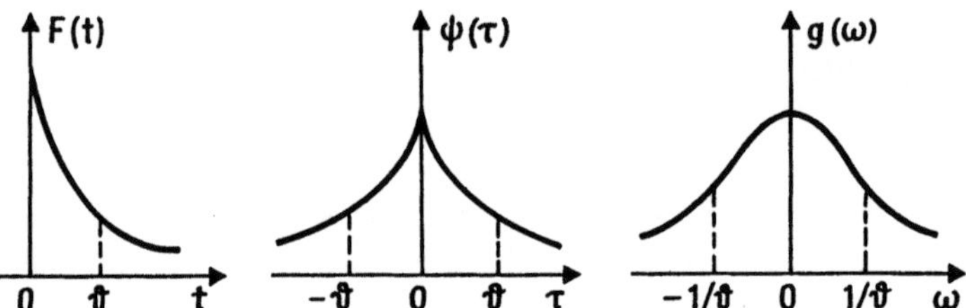

Fig. 1.5. Covariance $\psi(\tau)$ and spectral density $g(\omega)$ for a pulse Poisson process with exponential pulses $F(t)$

the RC-cell to which a chaotic (Poisson) sequence of delta-impulses is applied. It should, however, be recalled once more that one and the same covariance (and hence one and the same spectrum) may correspond to absolutely different random functions, a fact already stressed at the beginning of this section with reference to the exponential function $\psi(\tau)$.

This example also brings out the general connection between widths of $\psi(\tau)$ and $g(\omega)$. The width of $\psi(\tau)$ at the level $\psi(0)/e$ is 2ϑ, and the width of $g(\omega)$ at the level $g(0)/2$ is $2/\vartheta$, thus the product of these width measurements is 4 at any ϑ. If we fix the unit area of pulses by setting $a_\nu = 1/\vartheta$, then the limit $\vartheta \to 0$ yields again the expressions (1.75, 76).

The spectral form of the differentiability condition will be touched upon below (Sect. 3.1). But it is to be noted at this point that in both examples considered $\psi''(0)$ does not exist so that the existence condition for $\dot{\zeta}(t)$ is violated. In spectral expansions of $\psi(\tau)$ this becomes evident in that the densities $g(\omega)$ fall off slowly as $|\omega| \to \infty$, namely as $1/\omega^2$. If we formally calculate the spectral expansion of the second derivative of ψ

$$\psi''(\tau) = - \int_{-\infty}^{+\infty} \omega^2 g(\omega) e^{i\omega\tau} d\omega \quad ,$$

then, according to (1.73), we obtain

$$\psi''(\tau) = -\frac{n_1 \overline{|a|^2}}{\pi} \int_{-\infty}^{+\infty} (1 - \cos \omega\vartheta) e^{i\omega\tau} d\omega$$

$$= -n_1 \overline{|a|^2} \{2\delta(\tau) - \delta(\tau + \vartheta) - \delta(\tau - \vartheta)\} \quad ,$$

and, according to (1.78),

$$\psi''(\tau) = -\frac{n_1 \overline{|a|^2}}{2\pi} \int_{-\infty}^{+\infty} \frac{\omega^2 \vartheta^2}{1 + \omega^2\vartheta^2} e^{i\omega\tau} d\omega = -n_1 \overline{|a|^2} \left\{ \delta(\tau) - \frac{e^{-|\tau|/\vartheta}}{2\vartheta} \right\} \quad .$$

In both cases $\psi''(0)$ does not exist.

In all the examples just considered we found that the product $\Delta\omega \cdot \Delta\tau$ is a quantity of the order of unity. It should be remembered, however, that the uncertainty relation (1.58) is an *inequality*, so that the product may be much (even by several orders of magnitude) larger than unity. This occurs when we decompose into a spectrum a pulse of a *complicated* shape, e.g., one with a simple deterministic envelope but with inscribed oscillations varying according to a complicated law. These oscillations may be both a deterministic process (e.g., phase-shifted or frequency modulated in an intricate way), and a random process. In essence, it is the strong inequality $\Delta\omega \cdot \Delta\tau \gg 1$ that is here the criterion for the intricacy of the pulse shape $F(t)$.

We illustrate the above by referring to a pulse with a *Gaussian envelope* of width $\sqrt{2}\vartheta$ at $1/e$ of its maximum, and with an internal oscillation in the form of a *stationary* random process $\xi(t)$, where $\bar{\xi} = 0$ and the covariance is given by (1.59).

Owing to the random nature of the process $\xi(t)$, the pulse shape

$$F(t) = e^{-2t^2/\vartheta^2}\xi(t)$$

is on the whole random, as is its spectral amplitude

$$\tilde{F}(\omega) = \frac{1}{2\pi}\int\limits_{-\infty}^{+\infty} F(t)e^{i\omega t}dt \quad .$$

Let us find, therefore, the mean square of the spectral amplitude modulus

$$\langle|\tilde{F}(\omega)|^2\rangle = \frac{1}{4\pi^2}\iint\limits_{-\infty}^{+\infty} \langle F(t)F(t')\rangle \exp\left[i\omega(t'-t)\right]dt\,dt'$$

$$= \frac{1}{4\pi^2}\iint\limits_{-\infty}^{+\infty} \exp\left[-2(t^2+t'^2)/\vartheta^2\right]\langle\xi(t)\xi(t')\rangle e^{i\omega(t'-t)}dt\,dt' \quad .$$

Substituting (1.59), i.e., $\langle\xi(t)\xi(t')\rangle = \psi(t'-t) = \sigma^2\exp\left[-(t'-t)^2/\tau_c^2\right]$ and introducing the integration variables $\tau = t' - t$ and $\theta = (t'+t)/2$ gives

$$\langle|\tilde{F}(\omega)|^2\rangle = \frac{\sigma^2}{4\pi^2}\int\limits_{-\infty}^{+\infty} e^{-4\theta^2/\vartheta^2}d\theta \int\limits_{-\infty}^{+\infty} \exp\left[-(\vartheta^2+\tau_c^2)\tau^2/\tau_c^2\vartheta^2\right]e^{i\omega\tau}d\tau \quad .$$

If we take both integrals, we obtain

$$\langle|\tilde{F}(\omega)|^2\rangle = \frac{\sigma^2\vartheta^2\tau_c^2}{8\pi\sqrt{\vartheta^2+\tau_c^2}}\exp\left[-\frac{\vartheta^2\tau_c^2\omega^2}{4(\vartheta^2+\tau_c^2)}\right] \quad .$$

Thus, the spectrum is on average Gaussian, the width at $1/e$ of its maximum being

$$\Delta\omega = \frac{4}{\vartheta\tau_c}\sqrt{\vartheta^2 + \tau_c^2} \quad .$$

The product of $\Delta\omega$ by the $\Delta\tau$ of the whole pulse, i.e., of its envelope ($\Delta\tau = \sqrt{2}\vartheta$), will therefore be

$$\Delta\omega \cdot \Delta\tau = 4\sqrt{2}\sqrt{1 + \vartheta^2/\tau_c^2} \quad .$$

It is not large if the correlation time τ_c of the inscribed process $\xi(t)$ is of the same or a larger order of magnitude than ϑ. In that case, the envelope is nearly Gaussian, since $\xi(t)$ cannot change appreciably during ϑ. On the contrary, if $\tau_c \ll \vartheta$, then the pulse is markedly "indented" by the process $\xi(t)$ and the product

$$\Delta\omega \cdot \Delta\tau \approx 4\sqrt{2}\frac{\vartheta}{\tau_c} \gg 1$$

is large. The spectrum width $\Delta\omega = 4/\tau_c$ is governed here not by the envelope length, but by the correlation time τ_c of the random process $\xi(t)$. If $\tau_c \to 0$ [the process $\xi(t)$ is delta-correlated], then $\Delta\omega \cdot \Delta\tau \to \infty$ for any fixed ϑ.

In Ex. 1.7.6, we consider an example of a *deterministic* inscribed process for the same Gaussian envelope of the pulse, i.e., a case where the pulse shape $F(t)$ is completely deterministic. The "energy" spectrum of the pulse is described by the function $|\tilde{F}(\omega)|^2$, which is also deterministic. According to (1.66), this spectrum corresponds to the integral

$$\Psi_F(\tau) \equiv \int\limits_{-\infty}^{+\infty} F(t)F(t+\tau)dt = 2\pi \int\limits_{-\infty}^{+\infty} |\tilde{F}(\omega)|^2 e^{i\omega\tau}d\omega \quad . \tag{1.79}$$

We can therefore agree to refer to $\Psi_F(\tau)$ as the "covariance" of the deterministic pulse $F(t)$, and to the interval τ, where $\Psi_F(\tau)$ mainly lies, as the effective "correlation time" τ_{F_c} of the pulse. Since $\Psi_F(\tau)$ and $|\tilde{F}(\omega)|^2$ are connected by the complex Fourier transform, τ_{F_c} and $\Delta\omega$ satisfy the uncertainty relation $\Delta\omega \cdot \tau_{F_c} \geq 1$. Even if we have equality here, but $\tau_{F_c} \ll \vartheta$ (complicated pulse shape), the product of $\Delta\omega$ by the envelope width $\Delta\tau \approx \vartheta$ will be large.

Finally, we look at the natural nonmonochromaticity of the Thomson self-oscillation system. In Sect. (I.5.9) we considered the oscillation

$$x(t) = (\sqrt{p} + \varrho) \cos (t' + \varphi) \quad , \quad t' = \omega_0 t \quad ,$$

in this system for the case of so strong an excitation that the radius of the limiting cycle, $r_0 = \sqrt{p}$, is much larger than the standard deviation of amplitude fluctuations ($r_0 \gg \sqrt{\varrho^2}$). We obtained the following expression for the covariance [equation (I.5.129) which is rewritten here in terms of the dimensional time shift $\Delta t = \tau$]

$$\psi_x(\tau) = \frac{r_0^2}{2}\left(1 + \frac{\mathcal{D}}{h}e^{-h|\tau|}\right)e^{-\mathcal{D}|\tau|}\cos\omega_0\tau \quad . \tag{1.80}$$

We have now also introduced the dimensional increment $h = \mu p\omega_0$ and the phase diffusion coefficient $2\mathcal{D} = 2\mu D\omega_0$, so that

$$\overline{(\varphi_\tau - \varphi)^2} = 2\mathcal{D}\tau \quad . \tag{1.81}$$

Using (1.49), we can readily derive the spectral density of self-oscillations for positive frequencies

$$\begin{aligned}
g_+(\omega) &= \frac{2}{\pi}\int_0^\infty \psi_x(\tau)\cos\omega\tau\,d\tau \\
&= \frac{r_0^2}{2\pi}\int_0^\infty\left(1 + \frac{\mathcal{D}}{h}e^{-h\tau}\right)e^{-\mathcal{D}\tau}[\cos(\omega - \omega_0)\tau + \cos(\omega + \omega_0)\tau]d\tau \\
&= \Phi(\omega - \omega_0) + \Phi(\omega + \omega_0) \quad ,
\end{aligned}$$

where

$$\Phi(\alpha) = \frac{\mathcal{D}r_0^2}{2\pi}\left[\frac{1}{\alpha^2 + \mathcal{D}^2} + \frac{\mathcal{D} + h}{h}\frac{1}{\alpha^2 + (\mathcal{D} + h)^2}\right] \quad .$$

In actual practice $\omega_0 \gg h \gg \mathcal{D}$. This allows us to ignore $\mathcal{D}$ as compared with h and, moreover, to discard $\Phi(\omega + \omega_0)$ since the argument $\alpha = \omega + \omega_0$ is at all times no less than ω_0. As a result

$$g_+(\omega) \approx \Phi(\omega - \omega_0) \approx \frac{\mathcal{D}r_0^2}{2\pi}\left[\frac{1}{(\omega - \omega_0)^2 + \mathcal{D}^2} + \frac{1}{(\omega - \omega_0)^2 + h^2}\right] \quad . \tag{1.82}$$

Figure 1.6 is a plot of the expression derived, but is not drawn to scale because $\mathcal{D}$ and h differ enormously. The spectrum consists of two Lorentzians: an extremely sharp line due to the natural phase diffusion (its half-width at half-height is $\mathcal{D}$); and a very much weaker and wider background due to amplitude fluctuations (with half-width h). For practical purposes, this amplitude background in the neighborhood of the diffusion line may be ignored.

This theory of the natural (or fluctuation) spectral line width of the Thomson self-oscillatory system based on the solution of the Einstein-Fokker-Planck equation (Sect. I.5.8), was developed by *Bershtein* [1.11] who also verified it experimentally [1.12]. As noted earlier, the measurement of the natural non-monochromaticity of an oscillator at that time presented an exceptionally difficult problem, first because of the extreme narrowness of the line, and second because of slow and much more significant technical frequency drifts. Even if one assumed for a moment that the technical drifts had been completely

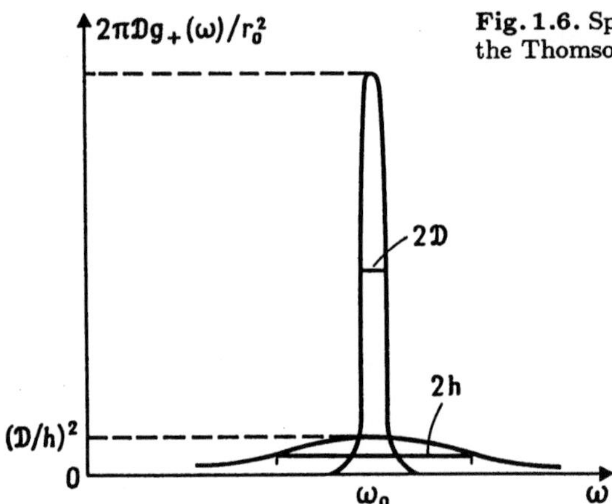

Fig. 1.6. Spectral density of natural fluctuations in the Thomson self-oscillatory system

removed (which, is a matter of skill and in principle possible since, by definition, technical drifts are due to the imperfection of the apparatus), one would still be unable to employ the conventional methods to measure such a small nonmonochromaticity.

Really, to measure a line width $\delta\omega \sim (10^{-10}\text{--}10^{-13})\omega_0$ using a *resonator* requires that its proper resonance curve be still narrower, i.e., its *quality* Q be higher than $\omega_0/\delta\omega \sim 10^{10} \div 10^{13}$. If we employed an *interferometer* to observe an interference pattern in the *conventional* way, the pattern would only be found to be blurred at path differences of the order of the coherent train length $L \sim 2\pi c/\delta\omega$ (see Sect. 2.5). This would imply using an interferometer of astronomical size, because

$$L \approx 2\pi c/\delta\omega = (10^{10} \div 10^{13})2\pi c/\omega_0 = (10^{10} \div 10^{13})\lambda_0 \quad .$$

Even at $\lambda_0 = 10\,\mathrm{cm}$, $L \sim 10^6 \div 10^9\,\mathrm{km}$ would be needed.

If the interferometer were sensitive to extremely small phase differences of the interfering oscillations $\Delta\varphi = 2\pi l/\lambda_0$, so that the detectable path difference l accounted for a negligible fraction of the coherent train length L (e.g., $\Delta\varphi = 0.01''$, thus yielding $l \approx 5\cdot 10^{-8}L$), then the situation would change appreciably. Of course, this would still leave us with the problem of how to get rid of the technical drifts, or at least to circumvent their influence. It is exactly these difficulties that have been tackled using an ingenious and elegant method suggested and realized by Bershtein, see Sect. 3.4.

1.6 Exercises

1.6.1. Derive an expression for the multivariate characteristic function of the complex normal process $\zeta(t) = \xi(t) + i\eta(t)$ (with $\langle \zeta(t) \rangle = 0$) in terms of its complex correlation matrices, proceeding from a $2n$-variate characteristic function of the real processes $\xi(t)$ and $\eta(t)$.

Solution. We denote the possible values of $\xi(t_k)$ by x_k, and $\eta(t_k)$ by x_{n+k} $(k = 1, 2, \ldots, n)$. The characteristic function of $2n$ Gaussian variates x_j $(j = 1, 2, \ldots, 2n)$ is

$$\varphi_{2n}(\boldsymbol{u}) = \exp\left\{-\frac{1}{2}\left\langle\left(\sum_{j=1}^{2n} x_j u_j\right)^2\right\rangle\right\} \quad .$$

The parameters u_j may also be represented as u_k and u_{n+k} $(k = 1, 2, \ldots, n)$ so that

$$\sum_{j=1}^{2n} x_j u_j = \sum_{k=1}^{n} (x_k u_k + x_{n+k} u_{n+k}) \quad .$$

We now make use of the complex possible values $z_k = x_k + i x_{n+k}$ of the variate $\zeta(t_k)$ and the complex parameters $U_k = u_k + i u_{n+k}$. Since

$$x_k = \tfrac{1}{2}(z_k + z_k^*) \quad , \quad x_{n+k} = \tfrac{1}{2i}(z_k - z_k^*) \quad ,$$
$$u_k = \tfrac{1}{2}(U_k + U_k^*) \quad , \quad u_{n+k} = \tfrac{1}{2i}(U_k - U_k^*) \quad ,$$

we get

$$\sum_{j=1}^{2n} x_j u_j = \frac{1}{2}\sum_{k=1}^{n} (z_k U_k^* + z_k^* U_k) \quad ,$$

hence

$$\left\langle\left(\sum_{j=1}^{2n} x_j u_j\right)^2\right\rangle$$

$$= \frac{1}{2}\sum_{k,l=1}^{2n} [b_{kl} U_k^* U_l + b_{kl}^* U_k U_l^* + \tilde{b}_{kl} U_k^* U_l^* + \tilde{b}_{kl}^* U_k U_l] \quad , \tag{1.83}$$

where b_{kl} and $\tilde{b}_{kl}$ stand for *halved* values of correlation matrix elements

$$b_{kl} = \tfrac{1}{2}\langle z_k z_l^*\rangle \quad , \quad \tilde{b}_{kl} = \tfrac{1}{2}\langle z_k z_l\rangle \quad .$$

Since the matrix B is Hermitian $(b_{kl} = b_{kl}^*)$, the first two terms on the right of (1.83) coincide and the characteristic function takes the form

$$\varphi_{2n}(\boldsymbol{u}) \equiv \varphi_n(\boldsymbol{U}, \boldsymbol{U}^*)$$

$$= \exp\left\{-\frac{1}{4}\sum_{k,l=1}^{n} [2b_{kl} U_k^* U_l + \tilde{b}_{kl} U_k^* U_l^* + \tilde{b}_{kl}^* U_k U_l]\right\} \quad . \tag{1.84}$$

If $\zeta(t)$ is a stationary analytical signal, then $\tilde{b}_{kl} = 0$ (Sect. 1.1), and

$$\varphi_n(\boldsymbol{U}, \boldsymbol{U}^*) = \exp\left\{-\frac{1}{2}\sum_{k,l=1}^{n} b_{kl} U_k^* U_l\right\} \quad . \tag{1.85}$$

The distribution corresponding to (1.85) is

$$
w_n(z_1, \ldots, z_n)d\boldsymbol{x}
$$

$$
= \frac{d\boldsymbol{x}}{(2\pi)^n} \int\limits_{-\infty}^{+\infty} \cdots \int \varphi_n(\boldsymbol{U}, \boldsymbol{U}^*) \exp\left\{ -\frac{\mathrm{i}}{2} \sum_{k=1}^{n} (z_k U_k^* + z_k^* U_k) \right\} d\boldsymbol{u}
$$

$$
= \frac{d\boldsymbol{x}}{(2\pi)^n |B|} \exp\left\{ -\frac{1}{2} \sum_{k,l=1}^{n} B_{kl}^{-1} z_k^* z_l \right\} \quad,
$$

where B_{kl}^{-1} are the elements of the matrix inverse to $B = (b_{kl})$, $d\boldsymbol{x} = dx_1 \ldots dx_{2n}$ and $d\boldsymbol{u} = du_1 \ldots du_{2n}$, $|B|$ being the determiant of the matrix B.

A fairly large body of literature is devoted to the complex normal distribution (see, e.g., [1.13] and references cited there).

1.6.2. Derive moments of arbitrary order for the stationary analytical signal using the characteristic function (1.85) derived in the previous problem

$$
\varphi(\boldsymbol{U}, \boldsymbol{U}^*) = e^{-S(\boldsymbol{U}, \boldsymbol{U}^*)} \quad, \quad S(\boldsymbol{U}, \boldsymbol{U}^*) = \frac{1}{2} \sum_{k,l} b_{kl} U_k^* U_l \quad.
$$

Solution. The general rule of calculating moments by differentiating the characteristic function remains the same

$$
\langle z_{k_1} \ldots z_{k_m} z_{l_1}^* \ldots z_{l_n}^* \rangle = \frac{\partial^{m+n} \varphi(\boldsymbol{U}, \boldsymbol{U}^*)}{\mathrm{i}^{m+n} \partial U_{k_1}^* \ldots \partial U_{k_m}^* \partial U_{l_1} \ldots \partial U_{l_n}} \bigg|_{U=0} \quad.
$$

But now it should be taken into account that the derivative $S(\boldsymbol{U}, \boldsymbol{U}^*)$ with respect to any U_n is dependent only on $\boldsymbol{U}^*$, and vice versa

$$
\frac{\partial S}{\partial U_n} = \frac{1}{2} \sum_k b_{kn} U_k^* \quad, \quad \frac{\partial S}{\partial U_n^*} = \frac{1}{2} \sum_l b_{nl} U_l \quad.
$$

Of the second derivatives S, therefore, only the mixed ones are nonzero

$$
\frac{\partial^2 S}{\partial U_m \partial U_n^*} = \frac{1}{2} b_{nm} = \frac{1}{4} \langle z_n z_m^* \rangle \quad.
$$

It follows that only those even moments which contain an equal number of multipliers z_k and z_l^* are nonzero. The result can be written as follows [1.14]:

$$
\langle z_{k_1} \ldots z_{k_m} z_{l_1}^* \ldots z_{l_n}^* \rangle = 0 \qquad \text{for} \quad m \neq n \quad,
$$

$$
\langle z_{k_1} \ldots z_{k_n} z_{l_1}^* \ldots z_{l_n}^* \rangle
$$
$$
= Z S_\pi \langle z_{k_1} z_{l\pi(1)}^* \rangle \langle z_{k_2} z_{l\pi(2)}^* \rangle \ldots \langle z_{k_n} z_{l\pi(n)}^* \rangle \text{ for } m = n \quad,
$$

where π is a permutation of integers $1, 2, \ldots, n$. The sum involves all these permutations, i.e., it contains $n!$ terms, e.g.,

$$
\langle z_1 z_2 z_3^* z_4^* \rangle = \langle z_1 z_3^* \rangle \langle z_2 z_4^* \rangle + \langle z_1 z_4^* \rangle \langle z_2 z_3^* \rangle \quad,
$$

$$\langle z_1 z_2 z_3 z_4^* z_5^* z_6^* \rangle = \langle z_1 z_4^* \rangle \langle z_2 z_5^* \rangle \langle z_3 z_6^* \rangle + \langle z_1 z_4^* \rangle \langle z_2 z_6^* \rangle \langle z_3 z_5^* \rangle$$
$$+ \langle z_1 z_5^* \rangle \langle z_2 z_4^* \rangle \langle z_3 z_6^* \rangle + \langle z_1 z_5^* \rangle \langle z_2 z_6^* \rangle \langle z_3 z_4^* \rangle$$
$$+ \langle z_1 z_6^* \rangle \langle z_2 z_4^* \rangle \langle z_3 z_5^* \rangle + \langle z_1 z_6^* \rangle \langle z_2 z_5^* \rangle \langle z_3 z_4^* \rangle \quad .$$

In particular,

$$\langle (z_1 z_2^*)^n \rangle = n! \langle z_1 z_2^* \rangle^n \quad , \quad \langle |z|^{2n} \rangle = n! \langle |z|^2 \rangle^n \quad .$$

1.6.3 Use the Khintchine theorem to show that the covariance $\psi(\tau)$ of a wide-sense stationary random process $\xi(t)$ obeys the inequality

$$2 \left[\int_0^T \psi(\tau) d\tau \right]^2 \le \sigma^2 \int_0^{2T} dt \int_0^t \psi(\tau) d\tau \quad ,$$

where $\sigma^2 = \psi(0) = D[\xi]$ [1.15].
Solution. Suppose, for simplicity, that $\overline{\xi} = 0$, so that

$$\psi(\tau) = \overline{\xi(t+\tau)\xi(t)} = \int_{-\infty}^{+\infty} \cos \omega\tau \, dG(\omega) \quad , \quad \sigma^2 = \int_{-\infty}^{+\infty} dG(\omega) \quad .$$

Thus,

$$\int_0^T \psi(\tau) d\tau = \int_{-\infty}^{+\infty} \frac{\sin \omega T}{\omega} dG(\omega) \quad ,$$

$$\int_0^{2T} dt \int_0^t \psi(\tau) d\tau = \int_{-\infty}^{+\infty} \frac{1}{\omega} dG(\omega) \int_0^{2T} \sin \omega t \, dt = 2 \int_{-\infty}^{+\infty} \left(\frac{\sin \omega T}{\omega} \right)^2 dG(\omega) \quad .$$

By virtue of the Schwarz inequality, we arrive at

$$2 \left[\int_0^T \psi(\tau) d\tau \right]^2 = 2 \left[\int_{-\infty}^{+\infty} \frac{\sin \omega T}{\omega} dG(\omega) \right]^2$$

$$\le 2 \int_{-\infty}^{+\infty} \left(\frac{\sin \omega T}{\omega} \right)^2 dG(\omega) \int_{-\infty}^{+\infty} dG(\omega) = \sigma^2 \int_0^{2T} dt \int_0^t \psi(\tau) d\tau \quad .$$

1.6.4. Using the technique employed to derive (1.42), find the stationarity conditions for the real random function

$$\xi(t) = \sum_n (a_n \cos \omega_n t + b_n \sin \omega_n t) \quad .$$

Solution.

$$\overline{a_n} = \overline{b_n} = 0 \quad , \quad \overline{a_n^2} = \overline{b_n^2} \equiv g_n \quad ,$$
$$\overline{a_n a_m} = \overline{b_n b_m} = g_n \delta_{mn} \quad , \quad \overline{a_n b_m} = 0 \quad .$$

Under these conditions

$$\psi(\tau) = \sum_n g_n \cos \omega_n \tau \quad \left(\sum_n g_n < \infty \right) \quad .$$

Consequently, the harmonic oscillation $A \cos (\omega t + \varphi_o)$, where φ_o is a constant and A is a random variable, cannot be a stationary process even at $\overline{A} = 0$.

1.6.5. Find the covariance of the process $Z(t)$ deduced from the stationary process $\zeta(t)$ by displaced averaging

$$Z(t) = \frac{1}{T} \int\limits_{t-T/2}^{t+T/2} \zeta(t) dt \quad .$$

Solution. We insert the spectral expansion of $\zeta(t)$ into the expression for $Z(t)$ to get

$$Z(t) = \int\limits_{-\infty}^{+\infty} dC(\omega) \frac{1}{T} \int\limits_{t-T/2}^{t+T/2} e^{i\omega t} dt = \int\limits_{-\infty}^{+\infty} \frac{\sin (\omega T/2)}{\omega T/2} e^{i\omega t} dC(\omega) \quad .$$

Using (1.35) and assuming $\overline{\zeta} = 0$, we find the covariance

$$\psi_Z(\tau) = \overline{Z(t+\tau)Z^*(t)} = \int\limits_{-\infty}^{+\infty} \left(\frac{\sin (\omega T/2)}{\omega T/2} \right)^2 e^{i\omega \tau} dG(\omega) \quad .$$

The first factor in the integrand suppresses the frequencies $|\omega| > 2\pi T$, i.e., the displaced average levels out fast as compared with T variations of $\zeta(t)$. This operation may be said to correspond to the filter action with a transfer function $k(i\omega)$ (see Sect. 3.1), for which

$$|k(i\omega)| = \left| \frac{\sin (\omega T/2)}{\omega T/2} \right| \quad .$$

1.6.6. Consider the deterministic pulse $F(t) = \exp(-2t^2/\vartheta^2)\xi(t)$ whose inscribed process $\xi(t)$ is a sinusoidal oscillation with a linearly varying frequency $\omega(t) = \omega_0 + \alpha t$, i.e., $\xi(t) = \cos (\omega_0 t + \alpha t^2/2)$. Find the covariance

$$\Psi_F(\tau) = \int\limits_{-\infty}^{+\infty} F(t)F(t+\tau) dt \quad .$$

Estimate the effective "correlation time" τ_{F_c} for the case where the deviations of the instantaneous frequency in time ϑ is small as compared with ω_0 ($\alpha\vartheta \ll \omega_0$), and ϑ is much longer than the average period $1/\omega_0$ ($\omega_0\vartheta \gg 1$).

Solution. Inserting $F(t) = \exp\left(-2t^2/\vartheta^2\right)\cos\left(\omega_0 t + \alpha t^2/2\right)$ into $\Psi_F(\tau)$ and substituting the integration varaible $x = t + \tau/2$ for t, we obtain $\psi_F(\tau)$ in the form

$$\Psi_F(\tau) = \frac{e^{-\tau^2/\vartheta^2}}{2} \int\limits_{-\infty}^{+\infty} e^{-4x^2/\vartheta^2} \left\{ \cos\left(\alpha\tau x + \omega_0\tau\right) \right.$$

$$\left. + \cos\left(\alpha x^2 + 2\omega_0 x + \frac{\alpha\tau^2}{4}\right) \right\} dx \quad .$$

We work out the integral using the relation

$$\int\limits_{-\infty}^{+\infty} e^{-ax^2} \cos\left(px^2 + 2qx + r\right) dx$$

$$= \frac{\sqrt{\pi}}{\sqrt[4]{a^2 + p^2}} \exp\left\{ -\frac{aq^2}{a^2 + p^2} \right\} \cos\left\{ \frac{1}{2}\arctan\frac{p}{a} + r - \frac{pq^2}{a^2 + p^2} \right\} \quad (a>0)$$

taken from [1.16] which gives

$$\Psi_F(\tau) = \frac{\sqrt{\pi}\vartheta}{4} \left\{ e^{-s\tau^2/\vartheta^2} \cos\omega_0\tau \right.$$

$$\left. + \frac{1}{\sqrt[4]{s}} \exp\left[-\left(\frac{\tau^2}{\vartheta^2} + \frac{\omega_0^2\vartheta^2}{4s}\right) \right] \cos\left(\frac{1}{2}\arctan\frac{\alpha\vartheta^2}{4} + \frac{\alpha\tau^2}{4} - \frac{\alpha\omega_0^2\vartheta^4}{16s} \right) \right\} \quad ,$$

where $s = 1 + \alpha^2\vartheta^4/16$.

If $\alpha\vartheta^2 \ll 4$, i.e., s is close to unity, then the two Gaussian factors have the same width ($\sim 2\vartheta$), but the second term is small owing to the factor $\exp\left(-\omega_0^2\vartheta^2/4\right)$, since we assume that $\omega_0\vartheta \gg 1$. If, however, $\alpha\vartheta^2 \gg 4$, so that $s \approx \alpha^2\vartheta^4/16 \gg 1$ as well, then in the first term the Gaussian factor has the width $\sim 2\vartheta/\sqrt{s}$, which is small as compared with that of its counterpart in the second term ($\sim 2\vartheta$). The condition that the second term be small within the width of the first term ($\tau \lesssim \vartheta/\sqrt{s}$) reduces to the requirement that

$$1 + \frac{\omega_0^2\vartheta^2}{4} \gg s, \quad \text{or} \quad \frac{\omega_0^2\vartheta^2}{4} \gg \frac{\alpha^2\vartheta^4}{16} \quad ,$$

which is fulfilled because of the assumption that the deviation is small ($\alpha\vartheta \ll \omega_0$). Thus, given the above conditions the function $\Psi_F(\tau)$ is described in the entire range τ to a sufficient accuracy by the first term of the relation derived. Consequently it has the effective "correlation time"

$$\tau_{F_c} = \frac{2\vartheta}{\sqrt{s}} = \frac{2\vartheta}{\sqrt{1 + \frac{\alpha^2\vartheta^2}{16}}} = \frac{2\vartheta\tau_c}{\sqrt{\vartheta^2 + \tau_c^2}} \quad ,$$

where $\tau_c = 4/\alpha\vartheta$ is the "correlation time" of the frequency-modulated process $\xi(t)$, inverse to the deviation $\alpha\vartheta$ of the instantaneous frequency $\omega(t) = \omega_0 + \alpha t$ within the width of the envelope ϑ. For a complicated pulse, $\tau_c = 4/\alpha\vartheta \ll \vartheta$ and $\tau_{F_c} \approx 2\tau_c$. Conversely, as α decreases and the process $\xi(t)$ tends to a harmonic oscillation for which the "correlation time" τ_c is infinite, the pulse becomes simple and $\tau_{F_c} \to 2\vartheta$.

1.6.7. In Sect. I.3.4 the pulse Poisson process with a random duration of pulses ϑ_ν was considered, and formula (I.3.46) obtained for the covariance. In the special case of the process (I.3.47)

$$\xi(t) = \sum_\nu a_\nu F\left(\frac{t - t_\nu}{\vartheta_\nu}\right) \quad ,$$

the formula takes the form

$$\psi_\xi(\tau) = n_1 \overline{a^2} \int\limits_{-\infty}^{+\infty} \vartheta w_\vartheta(\vartheta) d\vartheta \int\limits_{-\infty}^{+\infty} F(\theta) F\left(\theta + \frac{\tau}{\vartheta}\right) d\theta \quad .$$

Find the spectral density of this process using theorem (1.66).
Solution. By introducing the spectral amplitude density of the pulse in the dimensionless variable $\theta = t/\vartheta$

$$\tilde{F}(\alpha) = \frac{1}{2\pi} \int\limits_{-\infty}^{+\infty} F(\theta) e^{-i\alpha\theta} d\theta$$

we find, according to (1.66), that

$$\psi_\xi(\tau) = 2\pi n_1 \overline{a^2} \int\limits_0^\infty \vartheta w_\vartheta(\vartheta) d\vartheta \int\limits_{-\infty}^{+\infty} |\tilde{F}(\alpha)|^2 e^{i\alpha\tau/\theta} d\alpha \quad ,$$

or, in terms of the dimensional frequency $\omega = \alpha/\vartheta$,

$$\psi_\xi(\tau) = 2\pi n_1 \overline{a^2} \int\limits_0^\infty \vartheta^2 w_\vartheta(\vartheta) d\vartheta \int\limits_{-\infty}^{+\infty} |\tilde{F}(\omega\vartheta)|^2 e^{i\omega\tau} d\omega \quad .$$

The spectral density of the process (I.3.47) will thus be

$$g_\xi(\omega) = 2\pi n_1 \overline{a^2} \int\limits_0^\infty \vartheta^2 |\tilde{F}(\omega\vartheta)|^2 w_\vartheta(\vartheta) d\vartheta \quad . \tag{1.86}$$

1.6.8. In Sect. I.3.4 for square pulses with exponential distributions of duration ϑ_ν

$$w_\vartheta(\vartheta)d\vartheta = \mathrm{e}^{-\vartheta/\overline{\vartheta}}\frac{d\vartheta}{\overline{\vartheta}}$$

the exponential covariance was derived

$$\psi_\xi(\tau) = n_1\overline{a^2\vartheta}\,\mathrm{e}^{-|\tau|/\overline{\vartheta}} \quad ,$$

for which, by the Khintchine theorem (1.40), there corresponds the spectral density

$$g_\xi(\omega) = \frac{n_1\overline{a^2\vartheta^2}}{\pi(1+\omega^2\overline{\vartheta}^2)} \quad .$$

Verify that the same result can be obtained from (1.86).

Solution. For the square pulse with $F(\theta) = 1$ for θ in the interval $(0,1)$ we have

$$\tilde{F}(\alpha) = \frac{1}{2\pi}\int\limits_0^1 \mathrm{e}^{-i\alpha\theta}\,d\theta = \frac{\mathrm{e}^{-i\alpha/2}}{\pi\alpha}\,\sin\frac{\alpha}{2},$$

so that

$$|\tilde{F}(\omega\vartheta)|^2 = \left(\frac{\sin\,(\omega\vartheta/2)}{\omega\vartheta/2}\right)^2 \quad ,$$

and, by (1.86), we obtain

$$g_\xi(\omega) = \frac{2n_1\overline{a^2}}{\pi\omega^2\overline{\vartheta}}\int\limits_0^\infty \sin^2\frac{\omega\vartheta}{2}\cdot\mathrm{e}^{-\vartheta/\overline{\vartheta}}\,d\vartheta = \frac{n_1\overline{a^2\vartheta^2}}{\pi(1+\omega^2\overline{\vartheta}^2)} \quad .$$

2. Applications of Correlation Theory

The problems covered in the seven sections of this chapter do not refer to any single physical topic. On the contrary, they have been chosen to be as diverse as possible in order to illustrate the remarkable variety of situations to which correlation theory can be applied. The section headings are self-explanatory, making any further preliminary comments superfluous.

2.1 White Noise and Black-Body Radiation

The narrower the covariance of a stationary process, the wider is its spectrum. This was shown in Sect. 1.5 with reference to specific examples which demonstrated that, as $\psi(\tau)$ narrows indefinitely, the spectral density $g(\omega)$ tends to a constant value for all ω. The same result is readily obtained by simply assuming that $\psi(\tau) = 0$ for $|\tau| > \vartheta$, the form of $\psi(\tau)$ being otherwise arbitrary. In fact, for all the frequencies $|\omega| \ll 1/\vartheta$ we have

$$g(\omega) = \frac{1}{2\pi} \int\limits_{-\infty}^{+\infty} e^{-i\omega\tau} \psi(\tau) d\tau$$

$$= \frac{1}{2\pi} \int\limits_{-\vartheta}^{\vartheta} e^{-i\omega\tau} \psi(\tau) d\tau \approx \frac{1}{2\pi} \int\limits_{-\vartheta}^{\vartheta} \psi(\tau) d\tau = \text{const} \quad .$$

As long as ϑ is finite, the spectral density begins to fall off as $|\omega|$ tends to $1/\vartheta$. If, however, $\vartheta \to 0$, the frequency interval where $g = \text{const}$ widens undefinitely. The random function $\xi(t)$ may be stationary for arbitrarily small ϑ, but the limiting case of $\vartheta = 0$ (delta-correlation) is already a special one; as the variance of $\xi(t)$ becomes infinite, the integral

$$\int\limits_{-\infty}^{+\infty} g\, d\omega = \psi(0) = \overline{\xi^2} - \bar{\xi}^2$$

diverges.

Clearly, in all the cases where the spectral apparatus (filter) used provides *sufficiently sharp boundaries* of its pass-band, so that the latter completely lies in the region where $g = \text{const}$, the behavior of $g(\omega)$ away from the pass-band is of no significance and, in particular, the diminution of the spectrum will be of no consequence at frequencies $|\omega| \sim 1/\vartheta$ that are much higher than the upper band boundary. Figure 2.1 shows the behavior of the density at positive

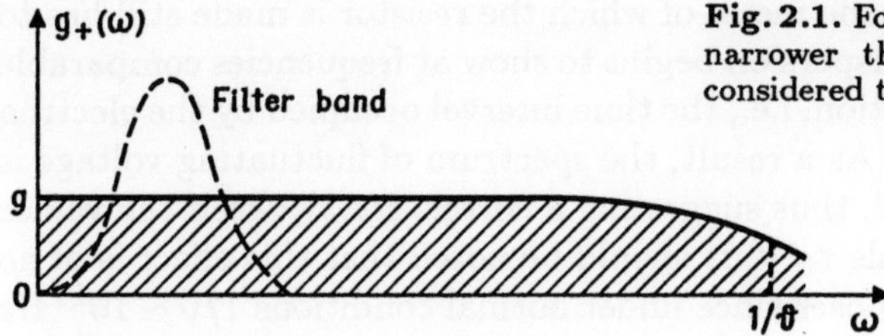

Fig. 2.1. For a filter whose pass-band is much narrower than $1/\vartheta$ the shot noise can be considered to be white

frequencies. In these cases the results are nearly unchanged, if we use a delta-correlated noise, with $g = \text{const}$ over the entire frequency range instead of the real noise with its small, but finite, correlation time ϑ[1].

An example of a noise that can nearly always be treated as delta-correlated is the shot noise of a tube. Its correlation time ϑ is governed by the electron transit time. If this quantitiy is of the order of 10^{-9} s, then the spectral density will be constant up to frequencies of hundreds of megahertzs. For these frequencies the shot current may be taken to be delta-correlated, i.e., it may be regarded as a chaotic (Poisson) sequence of *instantaneous* pulses with the integral value of each pulse equal to the electron charge e (Sect. I.3.3),

$$I(t) = \sum_\nu e\delta(t - t_\nu) \quad . \tag{2.1}$$

Then, by (1.69, 70),

$$\psi(\tau) = n_1 e^2 \delta(\tau) = e\overline{I}\delta(\tau) \quad , \quad g(\omega) = e\overline{I}/2\pi \quad . \tag{2.2}$$

Considering the correlation time ϑ, we have obtained the following variance of $I(t)$ [see (I.3.27)]:

$$\overline{(\Delta I)^2} = \psi(0) = e\overline{I}/\vartheta \quad .$$

As $\vartheta \to 0$ the variance becomes infinite, as was the case with the process (2.1).

A further example of noise that can be viewed as delta-correlated is *thermal* noise due to the thermal motion of microcharges in bodies. The theory of this noise will be discussed later (Sect. 3.5), but as we are interested at the moment in the uniformity of the spectrum, we should point out here that the spectral density of the thermal e.m.f. (electromotive force) produced by a section of an electric circuit is proportional to the real part of the impedance of this section. Thus, a conductor whose resistance R is constant within a frequency interval is a source of random e.m.f., with a uniform spectrum in this frequency interval.

If we forget for a moment the macroscopic factors controlling the variation of R with ω (reactive parameters of the circuit, skin-effect), the frequency

---

[1] The same applies to a more general case where the density $g(\omega)$, being constant within a pass-band, varies in an arbitrary manner on *either side* of the band, but without any exceedingly large increase or outbursts capable of making a notable contribution due to the "wings" of the pass-band of the filter.

dispersion of conductivity of the metal of which the resistor is made still has to be taken into account. This dispersion begins to show at frequencies comparable with $1/\vartheta$, where ϑ is the duration, i.e., the time interval occupied by the electron mean free path in the metal. As a result, the spectrum of fluctuating voltage is cut off at frequencies $\omega \sim 1/\vartheta$, thus suggesting a correlation between the values of the voltage in time intervals $\tau < \vartheta$. It should be noted that this effect is of no significance for practical purposes since under normal conditions $1/\vartheta \sim 10^{14}$ Hz for almost all metals. The frequency variation of R due to reactive parameters and skin-effect manifests itself at much lower frequencies.

In the literature delta-correlated noise was dubbed "white noise" and the term has stuck – perhaps by association with "white light" which, in many interference experiments, is alleged to exhibit complete incoherence (absence of correlation). But white light, i.e., an equilibrium (or black-body) thermal radiation, is not white noise, since it does not show constant spectral density. The latter is given by the Planck formula for the average density of electromagnetic energy

$$u_\omega = \frac{\hbar\omega^3}{\pi^2 c^3}[\exp(\hbar\omega/kT - 1)]^{-1} \quad \text{for} \quad \omega > 0 \;, \tag{2.3}$$

where $\hbar$ is Planck's constant divided by 2π, c is the velocity of light in a vacuum, k is Boltzmann's constant and T is the absolute temperature. Figure 2.2 shows the variation of u_ω with $x = \hbar\omega/kT$. The maximum of u_ω lies at $x_m \approx 2.82$.

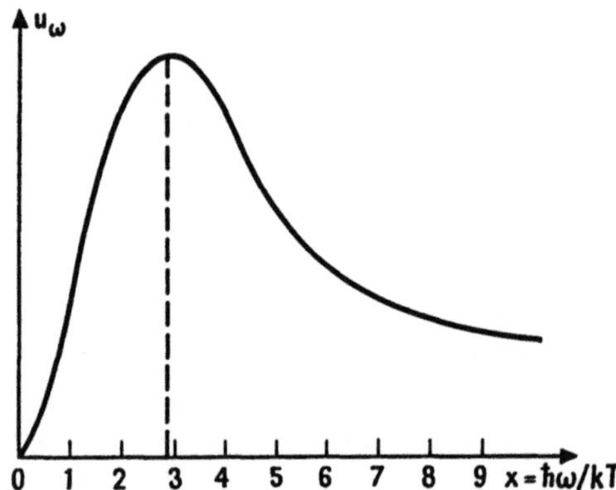

Fig. 2.2. Spectral density of black-body radiation (Planck's curve)

From the viewpoint of the theory of random processes, u_ω is the spectral density of fluctuating strengths $\boldsymbol{E}(t)$ and $\boldsymbol{H}(t)$ of the thermal radiation field. For each plane wave into which the field is decomposable, the relation $E(t) = H(t)$ holds with all directions being equiprobable. As a result, the electric and magnetic energies are equal and the components of $\boldsymbol{E}$ and $\boldsymbol{H}$ in any arbitrary direction feature the same covariances, but are not correlated mutually.

We will now find the correlation coefficient corresponding to the spectral density (2.3), i.e., the quantity

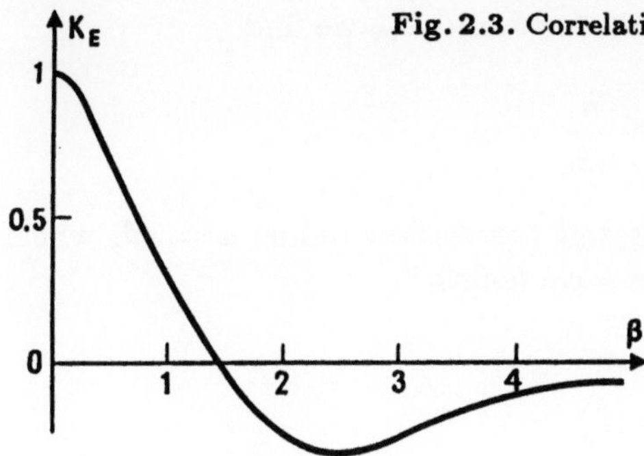

Fig. 2.3. Correlation coefficient of the black-body radiation

$$K_E(\tau) = \frac{\psi_E(\tau)}{\psi_E(0)} \quad , \quad \psi_E(\tau) = \frac{2}{\pi} \int\limits_0^\infty u_\omega \cos \omega\tau \, d\omega \quad .$$

Substituting (2.3) and taking the integral gives[2]

$$K_E(\tau) = 15\left(\frac{3}{\sinh^4\beta} - \frac{3}{\beta^4} + \frac{2}{\sinh^2\beta}\right) = -\frac{15}{2}\frac{d^3}{d\beta^3}L(\beta) \quad ,$$

$$\beta = \frac{\pi kT}{\hbar}\tau \quad , \tag{2.4}$$

where $L(\beta) = \coth\beta - 1/\beta$ is the Langevin function. Figure 2.3 gives the K_E vs β behavior, which clearly bears no resemblance to a delta-function. At $\beta \approx 1.37$, which corresponds to

$$\tau = \frac{1.37\hbar}{\pi kT} = 0.44\frac{\hbar}{kT} \equiv \vartheta \quad ,$$

the positive correlation becomes negative. It follows that at time shifts $\tau < \vartheta$ the values of the component $E_p(t)$ for any fixed direction p will often have the same sign at t and $t+\tau$, and the opposite signs for $\tau > \vartheta$. This can be explained as follows. The frequency ω_m, at which u_ω has its maximum, is

$$\omega_\mathrm{m} = x_\mathrm{m}\frac{kT}{\hbar} \approx 2.82\frac{kT}{\hbar} \quad .$$

Thus, waves with frequencies close to ω_m make the largest contribution, and at each point of space the alternation of the direction of E is mainly dictated by half-periods of these waves, the higher frequencies making a larger contribution to u_ω than the lower ones (Fig. 2.2). Clearly, the time ϑ should be of the same order as the half-period π/ω_m at the frequency ω_m, and should lie below this

---

[2] The relationship was first derived by M. L. Levin and the author in 1956, but not published. The same result was also obtained later by *Bourret* [2.1]; see also [2.2].

half-period. Really, from the expressions for ϑ and ω_{m} we find

$$\vartheta = \frac{1.37\hbar}{\pi kT} = \frac{1.37 \cdot 2.82}{\pi^2}\frac{\pi}{\omega_{\mathrm{m}}} = 0.39\frac{\pi}{\omega_{\mathrm{m}}} \quad .$$

The time ϑ may be likened to the *spatial* correlation radius $a = c\vartheta$, which, within a coefficient, coincides with the wave length

$$\lambda_{\mathrm{m}} = 0.4\pi\frac{\hbar c}{kT}$$

in the Wien displacement law (λ_{m} corresponds to a maximum u_λ of the electromagnetic energy density in terms of the wave length). From the expressions for ϑ and λ_{m} it follows that

$$a = 0.44\frac{\hbar c}{kT} = 0.35\lambda_{\mathrm{m}} \quad .$$

2.2 Modulated Random Processes

We will now discuss in more detail a special form of random processes − the so-called *modulated* random processes. If they are stationary (and consequently if there exists a spectral intensity), they may also be referred to as *narrow-band* or *quasi-monochromatic*. We dealt with such a process earlier, when discussing the spectrum of self-oscillations in the Thomson self-oscillatory system (Sect. 1.5).

Several preliminary remarks are in order here. They apply equally well to both deterministic and random modulated processes.

Modulated oscillations are often said to be harmonic oscillations with slowly varying amplitude and/or frequency. Strictly speaking, this is a misnomer. By definition, the amplitude and frequency of a harmonic oscillation are constant parameters for all t. However, this inexact term is only meaningful if the frequency resolution selectivity of the spectral apparatus used is sufficiently small. At the output of such an apparatus the response faithfully follows the frequency and amplitude variations at the input, i.e., it reproduces the modulation in a *quasi-stationary* way. At each instant of time the response is about the same as would be produced by a harmonic oscillation with the momentary values of frequency and amplitude. In this case, it is expedient to write the modulated oscillations in the quasi-harmonic form

$$\xi(t) = A(t) \cos \left[\omega_0 t + \varphi(t)\right] \quad , \tag{2.5}$$

where $A(t)$ and $\varphi(t)$ are time functions varying slowly as compared with oscillations of frequency ω_0. In other words, $A(t)$ and $\varphi(t)$ vary insignificantly during the high-frequency period $T_0 = 2\pi/\omega_0$. This is not to say that such a representation is also justified for a more selective spectral apparatus capable of *separating the spectral components* of the oscillation (2.5). To illustrate the point, we take the simple example of a harmonic amplitude modulation where $A(t) = 1 + k \cos \Omega t$ for $\Omega \ll \omega_0$, and for simplicity $\varphi(t) = \varphi_0 = 0$. Thus,

$$\xi(t) = (1 + k \cos \Omega t) \cos \omega_0 t$$

$$= \cos \omega_0 t + \frac{k}{2} \cos (\omega_0 - \Omega)t + \frac{k}{2} \cos (\omega_0 + \Omega)t \quad . \tag{2.6}$$

There has been much dispute in the literature as to the "reality" of the *side frequencies* $\omega_0 \mp \Omega$ or, for the general case, of side bands or Fourier expansion components. What was unclear was what occurred in reality: oscillations at *one* frequency ω_0 with a slowly varying amplitude, *or* oscillations at the *three* frequencies ω_0 and $\omega_0 \pm \Omega$. This discussion arose for the first time in the context of the important practical question of whether it would be necessary to allocate a frequency band of finite width to each radio station, or simply an arbitrarily narrow neighborhood around the carrier frequency ω_0. (Thus allowing the radio spectrum to be more densely populated.) As early as the beginning of the 1930s this confusion was clarified by *Mandelshtam* [Ref. 2.3, p. 35; Ref. 2.4, p. 62 and 436] , though this did not preclude further disputes in the literature at a later date [2.5].

Mandelshtam pointed out that the very question as to the "reality" of the left *or* right sides of (2.6) makes no sense, as the relation is an identity. In exactly the same way we could argue as to what is "real": $(a + b)^2$ or $a^2 + 2ab + b^2$. The question only makes sense as applied to the spectral properties (selectivity) of an apparatus detecting the oscillations (2.6), "reality" being understood as no more than the *advisability* of adopting either one of the two mathematically identical representations of $\xi(t)$. If the oscillation (2.6) affects, say, an oscillatory circuit with a resonance curve of half-width h, then for $h \gg \Omega$ (low selectivity) the circuit response will reproduce the modulation adequately, but for $h \ll \Omega$ (high selectivity) the circuit may single out three harmonic oscillations with frequencies ω_0 and $\omega_0 \pm \Omega$. In the first case the quasi-harmonic representation is real; in the second, the decomposition into three harmonic oscillations. Everything is thus predetermined by the relationship between the behavior of the oscillation (Ω) and that of the device (h). If the device is not included, the question is meaningless.

Representation (2.5), in which *one* function $\xi(t)$ is expressed through *two* functions $A(t)$ and $\varphi(t)$, is not unique. For it to be unique, it is necessary to define somehow either $A(t)$ or $\varphi(t)$. One possible way of arriving at a definition is based on the properties of the analytical signal (Sect. 1.1). It has no connection with the condition that $A(t)$ and $\varphi(t)$ be slowly varying and, at the same time, adds no information to that already contained in $\xi(t)$.

We now pass from the real oscillation $\xi(t)$ to the complex one

$$\zeta(t) = A(t) \exp \{i[\omega_0 t + \varphi(t)]\} \quad , \tag{2.7}$$

and require that $\zeta(t)$ be an analytical signal, i.e., that the imaginary part of (2.7),

$$\eta(t) = A(t) \sin \left[\omega_0 t + \varphi(t)\right] \quad , \tag{2.8}$$

be the Hilbert transform of the real part of $\xi(t)$, [see (1.5)]. Given the functions $\xi(t)$ and $\eta(t)$, we derive the unique expressions for $A(t)$ and $\varphi(t)$

$$A(t) = \sqrt{\xi^2(t) + \eta^2(t)} = |\zeta(t)| \quad ,$$

$$\omega_0 t + \varphi(t) = \arctan \frac{\eta(t)}{\xi(t)} = \arg \zeta(t) \quad . \tag{2.9}$$

In the general case, the $A(t)$ and $\varphi(t)$ formally deduced by the above procedure will not be slowly varying functions, and to refer to them as instantaneous amplitude and phase, respectively, will not lead to any physically justifiable result. On the contrary, if $A(t)$ and $\varphi(t)$ vary slowly, then we can attach some pictorial physical meaning to the notion of an *envelope* "stuffed" with "harmonic" oscillations with a slowly varying high *frequency* $\omega(t) = \omega_0 + \dot{\varphi}(t)$. In that case, definitions (2.9) essentially correspond to other possible definitions of the envelope and instantaneous frequency borrowed from treatments of the properties of the *harmonic* oscillation $\xi(t) = A \cos \omega_0 t$. So, for example, for this oscillation

$$A = \sqrt{\xi^2(t) + \frac{1}{\omega_0^2}\dot{\xi}^2(t)} \quad , \quad \tan \omega_0 t = -\frac{\dot{\xi}(t)}{\omega_0 \xi(t)} \quad .$$

Discarding in a first approximation the derivatives $\dot{A}(t)$ and $\dot{\varphi}(t)$ and differentiating (2.5) yields

$$\dot{\xi}(t) = -\omega_0 \eta(t) \quad ,$$

i.e., the Hilbert transform approximately reduces to differentiating the original function $\xi(t)$ with respect to $\omega_0 t$, see [2.6].

Oscillations (2.5) and (2.8) may be treated as the projections on the mutually orthogonal axes ξ and η of a vector of length $A(t)$ that rotates with angular velocity $\omega_0 + \dot{\varphi}(t)$. Introducing the complex amplitude

$$\boldsymbol{A}(t) = A(t)\mathrm{e}^{\mathrm{i}\varphi(t)} \quad ,$$

which is also called the modulating function, we can rewrite the analytical signal (2.7) in the form

$$\zeta(t) = \boldsymbol{A}(t)\mathrm{e}^{\mathrm{i}\omega_0 t} \quad . \tag{2.10}$$

It is often convenient to utilize the Cartesian components of the vector $\boldsymbol{A}(t)$

in the Van der Pol plane, i.e., in the plane rotating about the origin with an angular velocity ω_0 relative to the plane (ξ, η)

$$\boldsymbol{A}(t) = a(t) + ib(t) = A(t)\cos\varphi(t) + iA(t)\sin\varphi(t) \quad . \tag{2.11}$$

The amplitudes $a(t)$ and $b(t)$ are the projections of the vector of length $A(t)$, rotating with an instantaneous angular velocity $\dot\varphi(t)$ about the axes a and b in the Van der Pol plane. Clearly, for a modulated oscillation, i.e., slowly varying $A(t)$ and $\varphi(t)$, the functions $\boldsymbol{A}(t)$, $a(t)$, and $b(t)$ are also slowly varying. It follows from (2.10) that

$$a(t) = \xi(t)\cos\omega_0 t + \eta(t)\sin\omega_0 t \quad ,$$
$$b(t) = \eta(t)\cos\omega_0 t - \xi(t)\sin\omega_0 t \quad . \tag{2.12}$$

We now turn to *random* modulated oscillations when all the functions under consideration, both rapidly varying (ξ, η, ζ) and slowly varying $(a, b, \boldsymbol{A}, A, \varphi)$, are random. Let $\xi(t)$ be a *stationary* random process. Then, as we know, $\eta(t)$ and $\zeta(t)$ are stationary as well, and $\overline{\xi(t)} = \overline{\eta(t)} = \overline{\zeta(t)} = 0$; for the covariances we have (Sect. 1.1)

$$\overline{\xi\xi_\tau} = \overline{\eta\eta_\tau} = \psi_\xi(\tau) \quad , \quad \overline{\xi\eta_\tau} = -\overline{\eta\xi_\tau} = \psi_{\xi\eta}(\tau) \quad ,$$
$$\psi_\zeta(\tau) = \overline{\zeta_\tau\zeta^*} = 2[\psi_\xi(\tau) + i\psi_{\xi\eta}(\tau)] \quad , \tag{2.13}$$

where the subscript τ still marks the values at time $t + \tau$. In particular, at $\tau = 0$

$$\overline{\xi^2} = \overline{\eta^2} = \tfrac{1}{2}\overline{|\zeta|^2} \equiv \sigma^2 \quad , \quad \overline{\xi\eta} = 0 \quad . \tag{2.14}$$

Using (2.12, 13), we can readily express the covariances $a(t)$ and $b(t)$ in terms of $\psi_\xi(\tau)$ and $\psi_{\xi\eta}(\tau)$

$$\overline{aa_\tau} = \overline{bb_\tau} = \psi_\xi(\tau)\cos\omega_0\tau + \psi_{\xi\eta}(\tau)\sin\omega_0\tau \equiv \psi_a(\tau) \quad ,$$
$$\overline{ab_\tau} = -\overline{ba_\tau} = \psi_{\xi\eta}(\tau)\cos\omega_0\tau - \psi_\xi(\tau)\sin\omega_0\tau \equiv \psi_{ab}(\tau) \quad . \tag{2.15}$$

In particular

$$\overline{a^2} = \overline{b^2} = \psi_a(0) = \psi_\xi(0) = \sigma^2 \quad , \quad \overline{ab} = 0 \quad . \tag{2.16}$$

According to (2.15), the covariance for the complex amplitude $\boldsymbol{A}(t) = a(t) + ib(\tau)$ is given by

$$\psi_{\boldsymbol{A}}(\tau) = \overline{\boldsymbol{A}_\tau \boldsymbol{A}^*} = 2[\psi_a(\tau) + i\psi_{ab}(\tau)] \quad . \tag{2.17}$$

Solving (2.15) with respect to $\psi_\xi(\tau)$ and $\psi_{\xi\eta}(\tau)$ gives

$$\psi_\xi(\tau) = \psi_a(\tau)\cos\omega_0\tau - \psi_{ab}(\tau)\sin\omega_0\tau \quad ,$$
$$\psi_{\xi\eta}(\tau) = \psi_{ab}(\tau)\cos\omega_0\tau + \psi_a(\tau)\sin\omega_0\tau \quad . \tag{2.18}$$

Thus, the covariance of the initial high-frequency oscillations $\xi(t)$ and $\eta(t)$ themselves have − as functions of the time shift τ − the character of modulated oscillations.

Let us now look at the spectral representations of the covariances in question. It follows from (2.10) that

$$\psi_A(\tau) = \psi_\zeta(\tau)\mathrm{e}^{-\mathrm{i}\omega_0\tau} \quad , \tag{2.19}$$

and for $\psi_\zeta(\tau)$ we have the spectral expansion (1.54) (since $\overline{\zeta} = 0$, the mixed moment $\tilde{B}_\zeta$ coincides with ψ_ζ)

$$\psi_\zeta(\tau) = 4\int_0^\infty g(\omega)\mathrm{e}^{\mathrm{i}\omega\tau}d\omega \quad , \tag{2.20}$$

where $g(\omega)$ is the spectral density of oscillations $\xi(t)$. If $\xi(t)$ is a modulated (quasi-monochromatic) oscillation, then on the semiaxis $\omega > 0$ the density $g(\omega)$ is sharply peaked about $\omega = \omega_0$. Substituting the decomposition (2.10) into (2.19) gives

$$\psi_A(\tau) = 4\int_0^\infty g(\omega)\exp[\mathrm{i}(\omega - \omega_0)\tau]d\omega \quad .$$

By introducing the difference frequency $\Omega = \omega - \omega_0$ and the function $\tilde{g}(\Omega) = g(\omega_0 + \Omega)$ that is only nonzero in the narrow band near $\Omega = 0$, we can rewrite the previous relation as

$$\psi_A(\tau) = 4\int_{-\omega_0}^\infty \tilde{g}(\Omega)\mathrm{e}^{\mathrm{i}\Omega\tau}d\Omega \approx 4\int_{-\infty}^{+\infty} \tilde{g}(\Omega)\mathrm{e}^{\mathrm{i}\Omega\tau}d\Omega \quad , \tag{2.21}$$

which, together with (2.17), gives

$$\left.\begin{array}{c}\psi_a(\tau)\\\psi_{ab}(\tau)\end{array}\right\} = 2\int_{-\infty}^{+\infty} \tilde{g}(\Omega)\left\{\begin{array}{c}\cos\,\Omega\tau\\[12pt]\sin\,\Omega\tau\end{array}\right.d\Omega \quad . \tag{2.22}$$

It should be noted that for the case of $g(\omega)$ *symmetric* about ω_0, i.e., for an *even* function $\tilde{g}(\Omega)$, we have according to (2.22) $\psi_{ab}(\tau) = 0$, and the modulations $\psi_\xi(\tau)$ and $\psi_{\xi\eta}(\tau)$ appear to be purely *amplitude* ones

$$\psi_\xi(\tau) = \psi_a(\tau)\cos\omega_0\tau \quad , \quad \psi_{\xi\eta}(\tau) = \psi_a(\tau)\sin\omega_0\tau \quad .$$

For $\psi_a(\tau)$, just as for any covariance, the inequality $|\psi_a(\tau)| \le \psi_a(0) = \sigma^2$ holds. It is readily seen, however, that a stronger inequality is valid [2.7]. In fact, it follows from (2.18) that

$$\psi_\xi(\tau) = \sqrt{\psi_a^2(\tau) + \psi_{ab}^2(\tau)}\,\cos[\omega_0\tau + \Phi(\tau)] \quad .$$

But $|\psi_\xi(\tau)| \leq \psi_\xi(0) = \sigma^2$, hence the "envelope" $\sqrt{\psi_a^2 + \psi_{ab}^2}$ never exceeds σ^2, so that

$$|\psi_a(\tau)| \leq + \sqrt{\sigma^4 - \psi_{ab}^2(\tau)} \quad ,$$

i.e., we obtain a more stringent inequality than $|\psi_a(\tau)| \leq \sigma^2$. As regards the random oscillation (2.10), the questions that arise concern mostly the stochastic behavior (distribution functions and moments) of the initial oscillation $\xi(t)$ on the one side, and the Van der Pol variables a, b, A, and φ on the other. If $\xi(t)$ is a *normal* process, then all such questions can be answered completely for a general case and not only when $\xi(t)$ is stationary. To begin with, consider the simpler case of the *normal* and *stationary* process $\xi(t)$.

It follows from (1.5) that if $\xi(t)$ is Gaussian, the conjugate process $\eta(t)$ is also Gaussian. Accordingly, if we know covariances ψ_ξ and $\psi_{\xi\eta}$, we can write a correlation matrix B for any set of $2n$ variables $\xi(t_1), \ldots, \xi(t_n), \eta(t_1), \ldots, \eta(t_n)$ and, using B and the inverse matrix B^{-1}, we can write the $2n$-variate distribution of these variables. The following reduces to a trivial transformation of variables, the transition to $a(t_1), b(t_1), \ldots, a(t_n), b(t_n)$ or to $A(t_1), \varphi(t_1), \ldots, A(t_n),$ $\varphi(t_n)$. We will do this for a two-variate distribution, i.e., the joint distribution of $\xi(t)$ and $\eta(t)$.

By (2.14), the matrix B has the form

	ξ	η
ξ	σ^2	0
η	0	σ^2

so that

$$w_\zeta(\xi,\eta)d\xi\, d\eta = \frac{1}{2\pi\sigma^2} \exp\left[-(\xi^2 + \eta^2)/2\sigma^2\right]d\xi\, d\eta \tag{2.23}$$

(ξ and η are noncorrelated and therefore independent). Considering now that

$$\xi^2 + \eta^2 = a^2 + b^2 = A^2 \quad , \quad d\xi\, d\eta = da\, db = A\, dA\, d\varphi \quad ,$$

we obtain from (2.23) the joint distribution of the amplitudes a and b

$$w_A(a,b)da\, db = \frac{1}{2\pi\sigma^2} \exp\left[-(a^2 + b^2)/2\sigma^2\right]da\, db \quad , \tag{2.24}$$

and the joint distribution of the envelope A and phase φ

$$\tilde{w}_A(A,\varphi)dA\, d\varphi = \frac{A\, dA}{\sigma^2}e^{-A^2/2\sigma^2}\frac{d\varphi}{2\pi} \quad . \tag{2.25}$$

Thus, a and b are independent and normally distributed, and A and φ are independent, with the envelope A being distributed according to the Rayleigh law, and the phase uniformly distributed in the interval $(0, 2\pi)$.

Similarly, we can write the correlation matrix B

	ξ	η	ξ_τ	η_τ
ξ	σ^2	0	$\psi_\xi(\tau)$	$\psi_{\xi\eta}(\tau)$
η	0	σ^2	$-\psi_{\xi\eta}(\tau)$	$\psi_\xi(\tau)$
ξ_τ	$\psi_\xi(\tau)$	$-\psi_{\xi\eta}(\tau)$	σ^2	0
η_τ	$\psi_{\xi\eta}(\tau)$	$\psi_\xi(\tau)$	0	σ^2

and its inverse B^{-1}, to write the four-variate distribution for ξ, η, ξ_τ, and η_τ. In terms of the correlation coefficients $K(\tau) = \psi_\xi(\tau)/\sigma^2$ and $L(\tau) = \psi_{\xi\eta}(\tau)/\sigma^2$, we get

$$w_{\zeta 2}\,(\xi, \eta, \xi_\tau, \eta_\tau) d\xi\, d\eta\, d\xi_\tau\, d\eta_\tau$$
$$= \frac{d\xi\, d\eta\, d\xi_\tau\, d\eta_\tau}{4\pi^2\sigma^4(1 - K^2 - L^2)} \exp\left\{ -\frac{1}{2\sigma^2(1 - K^2 - L^2)} \right.$$
$$\left. \times \left[\xi^2 + \eta^2 + \xi_\tau^2 + \eta_\tau^2 - 2K(\xi\xi_\tau + \eta\eta_\tau) - 2L(\xi\eta_\tau - \xi_\tau\eta)\right] \right\} \quad . \; (2.26)$$

The change of variables $\xi = A \cos\varphi$, $\eta = A \sin\varphi$, $\xi_\tau = A_\tau \cos\varphi_\tau$, $\eta_\tau = A_\tau \sin\varphi_\tau$ gives

$$\tilde{w}_{A2}(A, A_\tau, \varphi, \varphi_\tau) dA\, dA_\tau\, d\varphi\, d\varphi_\tau$$
$$= \frac{AA_\tau\, dA\, dA_\tau\, d\varphi\, d\varphi_\tau}{4\pi^2\sigma^4(1 - p^2)} \exp\left\{ -\frac{1}{2\sigma^2(1 - p^2)} \right.$$
$$\left. \times \left[A^2 + A_\tau^2 - 2AA_\tau(K \cos(\varphi_\tau - \varphi) + L \sin(\varphi_\tau - \varphi))\right] \right\} \quad . \qquad (2.27)$$

where we have introduced $p^2 = K^2 + L^2$.

Integrating (2.27) with respect to redundant variables enables us to obtain various special cases. So, integrating with respect to A_τ (from 0 to ∞) and φ_τ (from 0 to 2π), we obtain the distribution (2.25) for A and φ. Integrating with respect to φ and φ_τ, we arrive at the distributions for A and A_τ

$$\tilde{w}_{A2}(A, A_\tau) dA\, dA_\tau = \frac{AA_\tau\, dA\, dA_\tau}{4\pi^2\sigma^4(1 - p^2)} \exp\left\{ -\frac{A^2 + A_\tau^2}{2\sigma^2(1 - p^2)} \right\}$$
$$\times \int_0^{2\pi} d\varphi \int_0^{2\pi} d\varphi_\tau \exp\left\{ \frac{AA_\tau[K \cos(\varphi_\tau - \varphi) + L \sin(\varphi_\tau - \varphi)]}{\sigma^2(1 - p^2)} \right\} \quad .$$

The integral with respect to φ_τ (or rather, with respect to $\varphi_\tau - \varphi$) gives the zero-order Bessel function of an imaginary argument, and the integral with respect to φ adds the factor 2π. Eventually, we obtain the Rice distribution (Sect. I.5.5)

$$\tilde{w}_{A2}(A, A_\tau)dA\,dA_\tau$$
$$= \frac{AA_\tau\,dA\,dA_\tau}{\sigma^4(1-p^2)}I_0\left(\frac{pAA_\tau}{\sigma^2(1-p^2)}\right)\exp\left\{-\frac{A^2+A_\tau^2}{2\sigma^2(1-p^2)}\right\} \quad . \tag{2.28}$$

According to Doob's theorem (see Exercise I.4.5), an exponential covariance, $p(\tau) = \exp(-\alpha|\tau|)$, implies that the envelope $A(t)$ is a normal Markov process. To verify this, we use (2.28) and the univariate (Rayleigh) distribution $\tilde{w}_A(A)$ [see (2.25)], to calculate the conditional probability

$$v(A|\tau, A_0)dA = \frac{\tilde{w}_{A2}(A, A_0)dA\,dA_0}{\tilde{w}_A(A_0)dA_0}$$
$$= \frac{A\,dA}{\sigma^2(1-p^2)}I_0\left(\frac{pAA_0}{\sigma^2(1-p^2)}\right)\exp\left\{-\frac{A^2+p^2A_0^2}{2\sigma^2(1-p^2)}\right\} \quad .$$

For $p(\tau) = \exp(-\alpha|\tau|)$ this probability appears to be a transition probability, i.e., it satisfies the Smoluchowski equation

$$v(A|t-t_0, A_0) = \int_0^\infty v(A|t-\theta, B)v(B|\theta-t_0, A_0)dB \quad (t_0<\theta<t) \quad .$$

This result is quite natural since the exponential correlation implies that $A(t)$ obeys a first-order stochastic differential equation with a delta-correlated random force, i.e., $A(t)$ behaves as, say, the velocity of a particle in unidimensional Brownian motion (Sect. I.6.3). It may also be said that $A(t)$ is the envelope of a process filtered from white noise with the help of a selective oscillatory circuit.

Integrating (2.27) with respect to A and A_τ gives the joint distribution of φ and φ_τ. Bivariate distributions allow us to compute the mixed moments A and φ, e.g., $\overline{AA_\tau}$ and $\overline{\varphi\varphi_\tau}$ (these depend on $p^2 = K^2 + L^2$), and also to obtain the distribution of the derivatives

$$\dot{A} = \lim_{\tau\to 0}\frac{A_\tau-A}{\tau} \quad , \quad \dot{\varphi} = \lim_{\tau\to 0}\frac{\varphi_\tau-\varphi}{\tau} \quad .$$

For these calculations we refer the reader to the literature, see [Ref. 2.8, Chap. 8].

Let us now turn to the more general *Gaussian* process $\xi(t) = a(t)\cos\omega_0 t + b(t)\sin\omega_0 t$, with the *stationary* amplitudes a and b *not satisfying* the conditions (2.16) under which the process $\xi(t)$ is stationary. Let $\bar{a} = \bar{b} = 0$, but $\overline{a^2}\neq\overline{b^2}$ and $\overline{ab}\neq 0$. It is easily seen that the moments of $\xi(t)$ then vary periodically with time[3]. For example

$$\overline{\xi^2(t)} = \overline{a^2}\cos^2\omega_0 t + \overline{b^2}\sin^2\omega_0 t + 2\overline{ab}\sin\omega_0 t\cos\omega_0 t$$
$$= \frac{\overline{a^2}+\overline{b^2}}{2} + \frac{\overline{a^2}-\overline{b^2}}{2}\cos 2\omega_0 t + \overline{ab}\sin 2\omega_0 t \quad ,$$

---

[3] The process under consideration is periodically non-stationary, see Sect. 4.5.

i.e., $\overline{\xi^2(t)}$ is a periodic function of t with period π/ω_0. What in this case is the joint distribution of A and φ?

Note that by choosing the constant initial phase appropriately, we can always introduce a and b such that $\overline{ab} = 0$ [2.9]. Really, from (2.12) we have

$$\xi(t) = a(t) \cos \omega_0 t - b(t) \sin \omega_0 t \quad ,$$
$$\eta(t) = b(t) \cos \omega_0 t + a(t) \sin \omega_0 t \quad .$$

Substituting $(\omega_0 t + \varphi_0) - \varphi_0$ for ω_0 gives

$$\xi(t) = a_1(t) \cos (\omega_0 t + \varphi_0) + b_1(t) \sin (\omega_0 t + \varphi_0) \quad , \tag{2.29}$$

where

$$a_1(t) = a(t) \cos \varphi_0 - b(t) \sin \varphi_0 \quad ,$$
$$b_1(t) = a(t) \sin \varphi_0 + b(t) \cos \varphi_0 \quad .$$

Hence,

$$\overline{a_1 b_1} = (\overline{a^2} - \overline{b^2}) \frac{\sin 2\varphi_0}{2} - \overline{ab} \cos 2\varphi_0 \quad ,$$

and selecting φ_0 such that $\tan (2\varphi_0) = 2\overline{ab}/(\overline{a^2} - \overline{b^2})$, we have $\overline{a_1 b_1} = 0$. We will, therefore, consider the process (2.29) for which

$$\overline{a_1^2} = \sigma_1^2 \neq \overline{b_1^2} = \sigma_2^2 \quad \text{and} \quad \overline{a_1 b_1} = 0 \quad . \tag{2.30}$$

As stated, the process $\xi(t)$ is Gaussian, so that

$$w_{\mathbf{A}}(a_1, b_1) da_1\, db_1 = \frac{da_1\, db_1}{2\pi \sigma_1 \sigma_2} \exp \left\{ -\frac{1}{2} \left(\frac{a_1^2}{\sigma_1^2} + \frac{a_2^2}{\sigma_2^2} \right) \right\} \quad .$$

Going over to polar coordinates ($a_1 = A \cos \varphi$, $b_1 = A \sin \varphi$), we obtain

$$\tilde{w}_{\mathbf{A}}(A, \varphi) dA\, d\varphi = \frac{A\, dA\, d\varphi}{2\pi \sigma_1 \sigma_2} \exp \left\{ -\frac{A^2}{2} \left(\frac{\cos^2 \varphi}{\sigma_1^2} + \frac{\sin^2 \varphi}{\sigma_2^2} \right) \right\} \quad , \tag{2.31}$$

i.e., here A and φ are not independent and the distribution of φ is not uniform.

Integrating Eq. (2.31) from 0 to ∞ with respect to A gives the distribution of φ

$$w_{\varphi}(\varphi) d\varphi = \left[2\pi \left(\frac{\sigma_2}{\sigma_1} \cos^2 \varphi + \frac{\sigma_1}{\sigma_2} \sin^2 \varphi \right) \right]^{-1} d\varphi \quad ,$$

which becomes uniform at $\sigma_1 = \sigma_2$. Integrating (2.31) from 0 to 2π with respect to φ it is convenient to introduce the variable $\psi = 2\varphi$

$$w_{\mathbf{A}}(A) dA = dA \int_0^{2\pi} \tilde{w}_{\mathbf{A}}(A, \varphi) d\varphi$$

$$= \frac{A\,dA}{\sigma_1\sigma_2}\exp\left\{-\frac{A^2}{4}\left(\frac{1}{\sigma_1^2}+\frac{1}{\sigma_2^2}\right)\right\}\frac{1}{4\pi}\int\limits_0^{4\pi}\exp\left\{-\frac{A^2}{4}\left(\frac{1}{\sigma_1^2}-\frac{1}{\sigma_2^2}\right)\cos\psi\right\}d\psi$$

$$= \frac{A\,dA}{\sigma_1\sigma_2}I_0\left[\frac{A^2}{4}\left(\frac{1}{\sigma_1^2}-\frac{1}{\sigma_2^2}\right)\right]\exp\left\{-\frac{A^2}{4}\left(\frac{1}{\sigma_1^2}+\frac{1}{\sigma_2^2}\right)\right\}\quad.$$

Let us now introduce the parameters σ and α by setting

$$\sigma_1^2 = \sigma^2(1+\alpha)\quad\text{and}\quad\sigma_2^2 = \sigma^2(1-\alpha)\quad. \tag{2.32}$$

Then, according to (2.29, 30),

$$\overline{\xi^2(t)} = \sigma^2[1+\alpha\cos 2(\omega_0 t+\varphi_0)]\quad,$$

and the distributions of φ and A take the form

$$w_\varphi(\varphi)d\varphi = \frac{\sqrt{1-\alpha^2}\,d\varphi}{2\pi(1-\alpha\cos 2\varphi)}\quad, \tag{2.33}$$

$$w_A(A)dA = \frac{A\,dA}{\sigma^2\sqrt{1-\alpha^2}}I_0\left[\frac{\alpha A^2}{2\sigma^2(1-\alpha^2)}\right]\exp\left[-\frac{A^2}{2\sigma^2(1-\alpha^2)}\right]\quad. \tag{2.34}$$

It can readily be shown that at any α (2.34) will give $\overline{A^2} = 2\sigma^2$, as is to be expected from (2.32), since $\overline{A^2} = \overline{a_1^2}+\overline{b_1^2} = \sigma_1^2+\sigma_2^2$.
If

$$\left|\frac{\alpha A^2}{2\sigma^2(1-\alpha^2)}\right|\ll 1\quad,$$

then $I_0\approx 1$, and (2.34) goes over into the Rayleigh distribution. The smaller α, the higher the values required for this to hold. At $\alpha = 0$ the Rayleigh distribution holds at any A. In the opposite limiting case, where

$$\left|\frac{\alpha A^2}{2\sigma^2(1-\alpha^2)}\right|\gg 1\quad,$$

we can make use of the asymptotic expression

$$I_0(x)\approx\frac{e^x}{\sqrt{2\pi x}}\quad(|x|\gg 1)\quad,$$

to obtain

$$w_A(A)dA\approx\frac{dA}{\sqrt{\pi\alpha}\sigma}\exp\left[-A^2/2\sigma^2(1+\alpha)\right]\quad,$$

i.e., the Gaussian law. The closer α is to unity, the smaller are the values of A

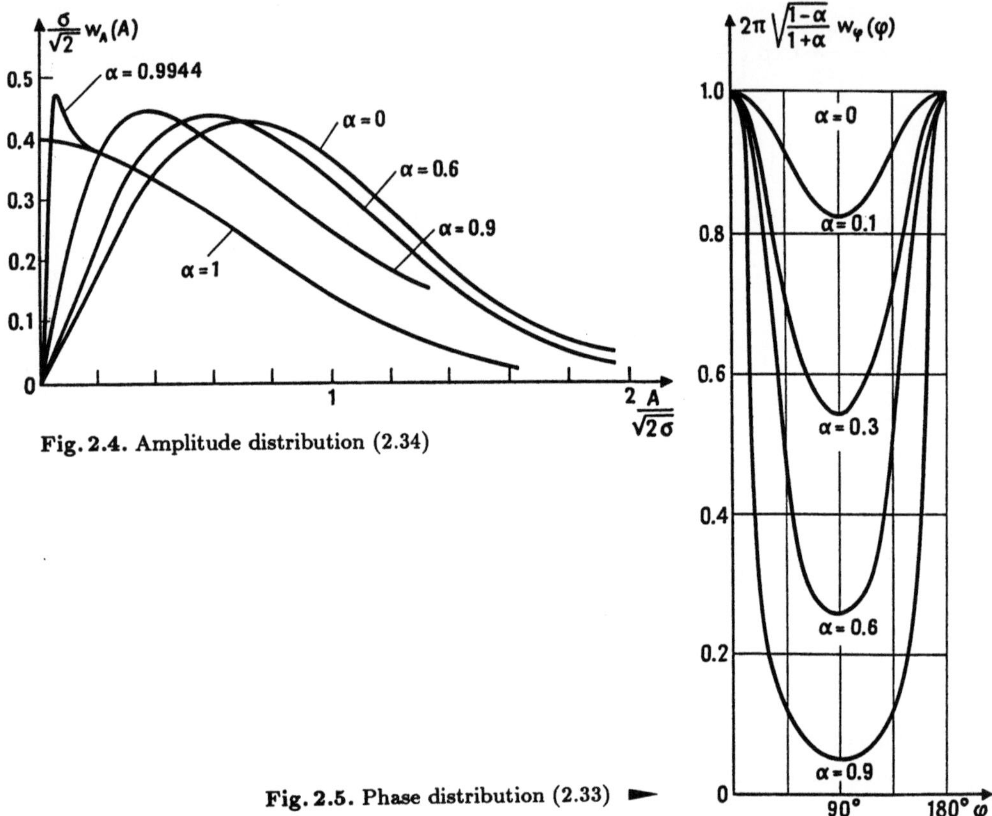

Fig. 2.4. Amplitude distribution (2.34)

Fig. 2.5. Phase distribution (2.33) ▶

required for this law to hold. At $\alpha = 1$ the Gaussian distribution holds for the entire semiaxis $A>0$.

Figure 2.4 represents the curves $w_A(A)$, and Fig. 2.5 the curves $w_\varphi(\varphi)$ [in the interval of φ from 0 to π since $w_\varphi(\varphi + \pi) = w_\varphi(\varphi)$], taken from [2.9]. In this work the moments $\overline{A^n}$ are also calculated. The even moments are shown to be the elementary functions σ_1^2 and σ_2^2, and the odd ones are expressed in terms of full elliptic integrals. Moreover, the mixed moment $\overline{A^2 A_\tau^2}$ is calculated, the four-variate distribution $\tilde{w}_{A2}(A, A_\tau, \varphi, \varphi_\tau)$ obtained (for the special form of the correlation matrix a_1, b_1, $a_{1\tau}$, $b_{1\tau}$), and the distribution found of the envelope and phase of the process (2.29) with a regular component, i.e., at $\overline{a_1}$ and $\overline{b_1}$ distinct from zero.

Now consider the relationship between the distributions of ξ and A for the case where the modulated process $\xi(t)$ is *stationary*, but *not Gaussian*.

Let us find the characteristic function of the process $\xi(t) = A(t)\cos[\omega_0 t + \varphi(t)]$

$$f_\xi(s) = \int\limits_{-\infty}^{+\infty} e^{is\xi} w_\xi(\xi)d\xi$$

$$= \int_0^\infty dA \int_0^{2\pi} \exp\left[\mathrm{i}sA \cos\left(\omega_0 t + \varphi\right)\right] \tilde{w}_{\boldsymbol{A}}(A,\varphi) d\varphi \quad .$$

Representing $\tilde{w}_{\boldsymbol{A}}$ in terms of the conditional distribution of φ

$$\tilde{w}_{\boldsymbol{A}}(A,\varphi) = w_A(A) w_\varphi(\varphi|A) \quad ,$$

and using the known Fourier expansion

$$\exp\left[\mathrm{i}sA \cos\left(\omega_0 t + \varphi\right)\right] = \sum_{-\infty}^{+\infty} \mathrm{i}^n J_n(sA) \exp\left[\mathrm{i}n(\omega_0 t + \varphi)\right]$$

gives

$$f_\xi(s) = \sum_{-\infty}^{+\infty} \mathrm{i}^n \int_0^\infty J_n(sA) w_A(A) dA \int_0^{2\pi} \exp\left[\mathrm{i}n(\omega_0 t + \varphi)\right] w_\varphi(\varphi|A) d\varphi \quad .$$

But, owing to the fact that $\xi(t)$ is stationary, the characteristic function cannot depend on t. This is only possible for

$$w_\varphi(\varphi|A) = \text{const} = \frac{1}{2\pi} \quad ,$$

i.e., with φ independent of A and uniformly distributed in the interval $(0, 2\pi)$,

$$\tilde{w}_{\boldsymbol{A}}(A,\varphi) dA \, d\varphi = w_A(A) dA \frac{d\varphi}{2\pi} \quad . \tag{2.35}$$

The characteristic function $f_\xi(s)$ will then be

$$f_\xi(s) = \int_0^\infty w_A(A) J_0(sA) dA = \int_0^\infty \frac{w_A(A)}{A} J_0(sA) A \, dA \quad , \tag{2.36}$$

i.e., it is the Hankel transform of $w_A(A)/A$. The inverse transform is

$$\frac{w_A(A)}{A} = \int_0^\infty f_\xi(s) J_0(sA) s \, ds \quad . \tag{2.37}$$

The distribution of ξ is thus the Fourier transform of the characteristic function $f_\xi(s)$

$$w_\xi(\xi) = \frac{1}{2\pi} \int_{-\infty}^{+\infty} f_\xi(s) \mathrm{e}^{-\mathrm{i}s\xi} ds \quad , \tag{2.38}$$

and the distribution of the envelope A is the Hankel transform of the same function. This elegant result was first obtained by *Blanc-Lapierre* et al. [2.10], but here we will give the later, more consistent derivation due to *Bunkin* and *Gudzenko* [2.11].

Using the relationships (2.36) and (2.38) we may, however, go a step further and deduce the *direct* relation between $w_\xi(\xi)$ and $w_A(A)$ [2.12]. For this purpose, it is sufficient to insert (2.36) into (2.38) and use the values of the discontinuous integral

$$\frac{1}{2\pi} \int_{-\infty}^{+\infty} J_0(sA)e^{-is\xi}ds = \begin{cases} \dfrac{1}{\pi\sqrt{A^2 - \xi^2}} & \text{for} \quad A \geq |\xi| \quad , \\ 0 & \text{for} \quad A < |\xi| \quad . \end{cases}$$

Thus,

$$w_\xi(\xi) = \frac{1}{2\pi} \int_{-\infty}^{+\infty} f_\xi(s)e^{-is\xi}ds$$

$$= \int_0^\infty w_A(A)dA \frac{1}{2\pi} \int_{-\infty}^{+\infty} J_0(sA)e^{-is\xi}ds = \frac{1}{\pi} \int_{|\xi|}^\infty \frac{w_A(A)dA}{\sqrt{A^2 - \xi^2}} \quad . \tag{2.39}$$

If we introduce the integration variable x by setting $A = |\xi| \cosh x$, then we can write (2.39) in a more concise form, which is also more convenient in certain specific cases

$$w_\xi(\xi) = \frac{1}{\pi} \int_0^\infty w_A(|\xi| \cosh x)dx \quad . \tag{2.40}$$

It is easily verified that for the Rayleigh distribution formula (2.40) will give the Gaussian law for ξ. For a uniform distribution of A in the interval $(0, A_0)$ we obtain, by (2.39),

$$w_\xi(\xi) = \frac{1}{2\pi A_0} \ln \left(\frac{A_0 + \sqrt{A_0^2 - \xi^2}}{A_0 - \sqrt{A_0^2 - \xi^2}} \right) \quad ,$$

and for the exponential distribution $w_A(A) = \alpha \exp(-\alpha A)$, according to (2.40), we have

$$w_\xi(\xi) = \frac{\alpha}{\pi} \int_0^\infty \exp(-\alpha|\xi| \cosh x)dx = \frac{\alpha}{\pi} K_0(\alpha\xi) \quad ,$$

where K_0 is the zero-order MacDonald function with a logarithmic singularity at zero, decreasing at infinity as $\exp(-\alpha\xi)/\sqrt{\xi}$.

2.3 Spectrum of Oscillations with Fluctuating Frequency

Consider the oscillation

$$\xi(t) = A_0 \cos\left[\omega_0 t + \varphi(t) + \varphi_0\right] \quad , \tag{2.41}$$

where the frequency deviation $\Omega(t) = \dot{\varphi}(t)$ is a stationary random process, and φ_0 a random variable uniformly distributed in the interval $(0, 2\pi)$. Let us look at the way in which the spectrum of $\xi(t)$ is related to that of the frequency deviation $\Omega(t)$. This subject has been the theme of a number of studies of which we will follow mainly those of *Malakhov* [2.13], *Elkin* and *Rodak* (see [2.14] and the references therein; and also [2.15]).

It seems only natural to put $\overline{\Omega(t)} = 0$. Then the covariance for $\Omega(t)$ will be

$$\psi_\Omega(\tau) = \overline{\Omega(t+\tau)\Omega(t)} = \int\limits_0^\infty g_\Omega(\omega)\cos\omega\tau\,d\omega \quad , \tag{2.42}$$

where $g_\Omega(\omega)$ is the spectral density of $\Omega(t)$ for positive frequencies. The random phase shift during the time period from t to $t + \tau$ is

$$\varphi_\tau - \varphi = \int\limits_t^{t+\tau} \Omega(t)dt \quad . \tag{2.43}$$

Obviously, it is this phase shift which adequately characterizes the quality of the "clock" that may be represented by process (2.41). The mean square $\varphi_\tau - \varphi$, designated by $F(\tau)$, will, by (2.42, 43), be

$$\begin{aligned}
F(\tau) = \overline{(\varphi_\tau - \varphi)^2} &= \int\limits_t^{t+\tau} dt \int\limits_t^{t+\tau} \psi_\Omega(t - t')dt' \\
&= \int\limits_0^\infty g_\Omega(\omega)d\omega \int\limits_t^{t+\tau} dt \int\limits_t^{t+\tau} \cos\omega(t - t')dt' \\
&= 2\int\limits_0^\infty g_\Omega(\omega)\frac{1 - \cos\omega\tau}{\omega^2}d\omega \quad .
\end{aligned} \tag{2.44}$$

The mean square phase-shift is thus completely defined by the spectrum of the random frequency deviation $\Omega(t)$[4].

---

[4] Relation (2.44) expresses the *spectrum of* $\varphi(\tau)$ as understood for random functions with stationary increments, see Sect. 4.2.

Suppose, further, that $\Omega(t)$, and hence the phase shift $\chi = \varphi_\tau - \varphi$ too, are normal processes. Since $\overline{\chi} = 0$ by (2.43), and the variance of χ is expressed from (2.44), the distribution of χ will be

$$w_1(\chi)d\chi = \frac{1}{\sqrt{2\pi F}}e^{-\chi^2/2F}d\chi \quad . \tag{2.45}$$

Averaging the product

$$\xi(t+\tau)\xi(t) = \frac{A_0^2}{2}\{\cos\,[\omega_0\tau - (\varphi_\tau - \varphi)]$$
$$+ \cos\,[\omega_0(2t+\tau) - (\varphi_\tau + \varphi) + 2\varphi_0]\} \quad ,$$

using the distribution for χ (2.45) and the uniform distribution for φ_0, we get

$$\begin{aligned}
\psi_\xi(\tau) &= \frac{A_0^2}{2}\,\cos\,\omega_0\tau\cdot\overline{\cos\chi} \\
&= \frac{A_0^2}{2}e^{-\overline{\chi^2}/2}\,\cos\,\omega_0\tau \\
&= \frac{A_0^2}{2}e^{-F(\tau)/2}\,\cos\,\omega_0\tau \quad .
\end{aligned} \tag{2.46}$$

This differs from (1.80) (on ignoring amplitude fluctuations), only by more general law for the phase diffusion: i.e., $F(\tau)$ instead of $2D\tau$.

According to (1.49), the spectral density of the oscillation (2.41) (for $\omega > 0$) is

$$\begin{aligned}
g_\xi(\omega) &= \frac{A_0^2}{2\pi}\int\limits_0^\infty e^{-F(\tau)/2}[\cos\,(\omega - \omega_0)\tau + \cos\,(\omega + \omega_0)\tau]d\tau \\
&\approx \frac{A_0^2}{2\pi}\int\limits_0^\infty e^{-F(\tau)/2}\,\cos\,(\omega - \omega_0)\tau d\tau \quad ,
\end{aligned} \tag{2.47}$$

where the integral with $\cos\,(\omega + \omega_0)\tau$ can be ignored on account of its being small. The relationship between the spectra $g_\xi(\omega)$ and $g_\Omega(\omega)$ therefore appears to be fairly complicated. If we wanted to obtain from the spectrum of $\xi(t)$ the spectrum of its frequency deviation $\Omega(t)$, then, by (2.44) and (2.47), we would have to solve the nonlinear integral equation

$$g_\xi(\omega) = \frac{A_0^2}{2\pi}\int\limits_0^\infty \exp\left\{-\int\limits_0^\infty g_\Omega(u)\frac{1 - \cos\,u\tau}{u^2}du\right\}\cos\,(\omega - \omega_0)\tau d\tau \quad . \tag{2.48}$$

We will not attempt to solve this difficult problem, but turn instead to those results that might be derived from (2.47) or (2.48) in certain limiting cases. Let us look first of all at expression (2.44) for the mean square of the phase shift.

We introduce the mean square of the fluctuations $\Omega(t)$

$$\overline{\Omega^2} = \int\limits_0^\infty g_\Omega(\omega)d\omega \quad , \tag{2.49}$$

and assume first that $\Omega(t)$ has a correlation time τ_Ω such that $g_\Omega(\omega)$ differs markedly from zero only up to $\omega\tau_\Omega \sim 1$. We consider next the following two extreme cases:

1) Let $\overline{\Omega^2}\tau_\Omega^2 \gg 1$, i.e., the frequency of the oscillation (2.41) undergoes *large* and *slow* (long-correlated) shifts. In other words, the spectrum $g_\Omega(\omega)$ has a "height" $\overline{\Omega^2}\tau_\Omega$, which is very much larger than its width $1/\tau_\Omega$. If we assume that in this case, for $g_\xi(\omega)$ only $\tau \ll \tau_\Omega$ (i.e., $\Omega\tau \ll 1$), are significant then in (2.44) we may set $1 - \cos\omega\tau \approx \omega^2\tau^2/2$, with the result that

$$F(\tau) \approx \tau^2 \int\limits_0^\infty g_\Omega(\omega)d\omega = \overline{\Omega^2}\tau^2 \quad . \tag{2.50}$$

According to (2.47), we have

$$g_\xi(\omega) = \frac{A_0^2}{2\pi} \int\limits_0^\infty e^{-\overline{\Omega^2}\tau^2/2} \cos\,(\omega - \omega_0)\tau d\tau \quad ,$$

from which it is seen that the exponential function limits the integrand at $\tau^2 \sim 1/\overline{\Omega^2} \ll \tau_\Omega^2$, which justifies the assumption that $\tau \ll \tau_\Omega$. Integration gives

$$g_\xi(\omega) = \frac{A_0^2}{2\sqrt{2\pi\overline{\Omega^2}}} \exp\left[-(\omega - \omega_0)^2/2\overline{\Omega^2}\right] \quad , \tag{2.51}$$

i.e., we get a *Gaussian* (bell-shaped) form of the spectral line, which is, for example, characteristic of the Doppler broadening of optical spectral lines. The half-width at $1/e$ of the maximum is $\sqrt{2}$ times higher than root-mean-square frequency fluctuations.

2) Let $\overline{\Omega^2}\tau_\Omega^2 \ll 1$, i.e., the frequency of the oscillation (2.41) undergoes *small* and *rapid* (short-correlated) fluctuations (the height $\overline{\Omega^2}\tau_\Omega$ of the spectrum $g_\Omega(\omega)$ is small compared with its width $1/\tau_\Omega)^5$. In that case, in (2.44) those values of $g_\Omega(\omega)$ are significant that lie close to the main maximum of the factor $(1 - \cos\omega\tau)/\omega^2$, i.e., in the interval from $\omega = 0$ to $\omega \sim 1/\tau$. Therefore, if $g_\Omega(0) \neq 0$ and $\tau \gg \tau_\Omega$, we get

$$F(\tau) \approx 2g_\Omega(0) \int\limits_0^\infty \frac{1 - \cos\omega\tau}{\omega^2}d\omega = \pi g_\Omega(0)\tau \quad , \tag{2.52}$$

---

[5] This case has been considered by *Wiener* [Ref. 2.16, Lecture 5].

i.e., the diffusion law with the diffusion coefficient

$$2\mathcal{D} = \pi g_\Omega(0) \quad . \tag{2.53}$$

Of course, this result also occurs in the limiting case of the delta-correlated fluctuations $\Omega(t)$.

Formula (2.47) now gives

$$g_\xi(\omega) = \frac{A_0^2}{2\pi} \int\limits_0^\infty e^{-\mathcal{D}\tau} \cos\,(\omega - \omega_0)\tau d\tau \quad . \tag{2.54}$$

The exponential function limits the integrand at $\tau \sim 1/\mathcal{D}$, which agrees with the assumption that $\tau \gg \tau_\Omega$ subject to

$$\mathcal{D}\tau_\Omega = \frac{\pi}{2} g_\Omega(0)\tau_\Omega \ll 1 \quad . \tag{2.55}$$

If the spectrum $g_\Omega(\omega)$ falls off with ω more or less monotonically, then, by (2.49), $\overline{\Omega^2} \approx g_\Omega(0)/\tau_\Omega$ and the condition (2.55) is met by virtue of the initial assumption that $\overline{\Omega^2}\tau_\Omega^2 \ll 1$. Integrating in (2.54) gives

$$g_\xi(\omega) = \frac{A_0^2}{2\pi} \frac{\mathcal{D}}{\mathcal{D}^2 + (\omega - \omega_0)^2} \quad , \tag{2.56}$$

i.e., the line has a *resonance* (Lorentzian) shape as obtained in Sect. 1.5 for the natural line width of the oscillator as a result of the action of delta-correlated impacts. Spectral lines in optics have similar shapes when their width is conditioned by atomic collisions, with free path times being distributed exponentially (Poisson process). The resonance shape (but with a smaller width) is also peculiar to the natural broadening of optical spectral lines due to the finite life time of the excited state. In the classical treatment it is due to the exponential damping of oscillations of the atomic oscillators (a pulse process with exponential pulses).

We will now proceed to give a more accurate estimate of the line shape $g_\xi(\omega)$ for small and rapid fluctuations $\Omega(t)$. For this purpose, we separate the constant part $g_\Omega(0)$ from $g_\Omega(\omega)$. Then (2.44) gives

$$F(\tau) = 2\mathcal{D}|\tau| + \tilde{F}(\tau) \quad ,$$

where $\mathcal{D}$ is known from (2.53). The function

$$\tilde{F}(\tau) = 2 \int\limits_0^\infty \frac{g_\Omega(\omega) - g_\Omega(0)}{\omega^2} (1 - \cos\,\omega\tau)d\omega \tag{2.57}$$

is uniformly bounded

$$|\tilde{F}(\tau)| \leq 4 \int\limits_0^\infty \frac{|g_\Omega(\omega) - g_\Omega(0)|}{\omega^2} d\omega = \text{const} \quad .$$

Thus, (2.47) takes the form

$$g_\xi(\omega) = \frac{A_0^2}{\pi} \int\limits_0^\infty e^{-\mathcal{D}\tau} e^{-\tilde{F}(\tau)/2} \cos(\omega - \omega_0)\tau \, d\tau \quad . \tag{2.58}$$

To this expression we now apply the converse (in the sense of the Fourier transform) of the theorem (2.49). According to (2.39), the spectral amplitude of the integral

$$I(\tau) = \int\limits_{-\infty}^{+\infty} x_1^*(t) x_2(t + \tau) dt$$

is $\tilde{I}(\omega) = 2\pi \tilde{x}_1^*(\omega)\tilde{x}_2(\omega)$, where $\tilde{x}_{1,2}(\omega)$ are spectral amplitudes of the functions $x_{1,2}(t)$. Similarly, it can readily be shown that in the Fourier expansion

$$G(t) = x_1^*(t)x_2(t) = \int\limits_{-\infty}^{+\infty} \tilde{G}(\omega)e^{i\omega t} d\omega \quad ,$$

the spectral amplitude $\tilde{G}(\omega)$ is expressed by the integral

$$\tilde{G}(\omega) = \frac{1}{2\pi} \int\limits_{-\infty}^{+\infty} x_1^*(t)x_2(t)e^{-i\omega t} dt = \int\limits_{-\infty}^{+\infty} \tilde{x}_1^*(\Omega)\tilde{x}_2(\Omega + \omega)d\Omega \quad . \tag{2.59}$$

For real and even functions $x_{1,2}(t)$ [and according to (2.44) the function $F(\tau)$ that is of interest to us is exactly of this type] the theorem is readily seen to take the form

$$\tilde{G}(\omega) = \frac{1}{\pi} \int\limits_0^\infty x_1(t)x_2(t) \cos \omega t \, dt = \int\limits_{-\infty}^{+\infty} \tilde{x}_1(\Omega)\tilde{x}_2(\Omega + \omega)d\Omega \quad , \tag{2.60}$$

where

$$\tilde{x}_{1,2}(\omega) = \frac{1}{\pi} \int\limits_0^\infty x_{1,2}(t) \cos \omega t \, dt \quad . \tag{2.61}$$

Inserting in (2.58)

$$x_1(\tau) = e^{-\tilde{F}(\tau)/2} \quad , \quad x_2(\tau) = \frac{A_0^2}{2} e^{-\mathcal{D}|\tau|} \quad ,$$

we thus have

$$g_\xi(\omega) = \frac{1}{2\pi} \int\limits_{-\infty}^{+\infty} \tilde{x}_1(\Omega)\tilde{x}_2(\Omega + \omega - \omega_0)d\Omega \quad , \tag{2.62}$$

where

$$\tilde{x}_1(\omega) = \frac{1}{\pi} \int\limits_0^\infty e^{-\tilde{F}(\tau)/2} \cos(\omega - \omega_0)\tau\, d\tau \quad ,$$

$$\tilde{x}_2(\omega) = \frac{A_0^2}{2\pi} \int\limits_0^\infty e^{-\mathcal{D}\tau} \cos(\omega - \omega_0)\tau\, d\tau = \frac{A_0^2 \mathcal{D}}{2\pi[\mathcal{D}^2 + (\omega - \omega_0)^2]} \quad ,$$

see (2.54) and (2.56).

When considering case (2) above, we completely ignored $\tilde{F}(\tau)$ in the relation for $F(\tau)$. Now assume that this term is small and include it to first order. Using (2.57) we obtain

$$\tilde{x}_1(\omega) \approx \frac{1}{\pi} \int\limits_0^\infty \left[1 - \frac{\tilde{F}(\tau)}{2}\right] \cos(\omega - \omega_0)\tau\, d\tau$$

$$= \frac{1}{\pi} \int\limits_0^\infty \left[1 - \int\limits_0^\infty \frac{g_\Omega(u) - g_\Omega(0)}{u^2}du\right] \cos(\omega - \omega_0)\tau\, d\tau$$

$$+ \frac{1}{\pi} \int\limits_0^\infty \int\limits_0^\infty \frac{g_\Omega(u) - g_\Omega(0)}{u^2} \cos u\tau \, \cos(\omega - \omega_0)\tau\, d\tau\, du$$

$$= \left[1 - \int\limits_0^\infty \frac{g_\Omega(u) - g_\Omega(0)}{u^2}du\right]\delta(\omega - \omega_0)$$

$$+ \frac{1}{2} \int\limits_0^\infty \frac{g_\Omega(u) - g_\Omega(0)}{u^2}[\delta(u + \omega - \omega_0) + \delta(u - \omega + \omega_0)]du$$

$$\approx \delta(\omega - \omega_0) + \frac{g_\Omega(\omega - \omega_0) - g_\Omega(0)}{2(\omega - \omega_0)^2} \quad .$$

Substituting the expressions derived for $\tilde{x}_{1,2}(\omega)$ into (2.62), we should take into account that owing to $\mathcal{D}$ being small, the spectrum of $\tilde{x}_2(\omega)$ is much narrower than the second term in $\tilde{x}_1(\omega)$. Therefore,

$$g_\xi(\omega) = \frac{A_0^2\mathcal{D}}{2\pi} \int\limits_{-\infty}^{+\infty} \left[\delta(\Omega - \omega_0) + \frac{g_\Omega(\Omega - \omega_0) - g_\Omega(0)}{(\Omega - \omega_0)^2}\right] \frac{d\Omega}{\mathcal{D}^2 + (\Omega + \omega - 2\omega_0)^2}$$

$$\approx \frac{A_0^2}{4}\left\{\frac{\frac{2}{\pi}\mathcal{D}}{\mathcal{D}^2 + (\omega - \omega_0)^2} + \frac{g_\Omega(\omega - \omega_0) - g_\Omega(0)}{(\omega - \omega_0)^2}\right\} \quad . \tag{2.63}$$

At frequencies close to the line maximum, i.e., at $|\omega - \omega_0| \sim \mathcal{D}$, the second term can be ignored, so that we return to the resonance spectral density (2.56). In contrast, for $|\omega - \omega_0| \gg \mathcal{D}$, discarding $\mathcal{D}^2$ in the denominator of the first term in (2.63) and taking (2.53) into account, we obtain

$$g_\xi(\omega) \approx \frac{A_0^2}{4} \frac{g_\Omega(\omega - \omega_0)}{(\omega - \omega_0)^2} \ . \tag{2.64}$$

In other words, somewhat away from the maximum, or as opticians put it, on the wings of the spectral line, the spectrum of $\xi(t)$ reproduces (to within a factor $A_0^2/4(\omega - \omega_0)^2$) the spectrum of frequency fluctuations. If the condition $|\tilde{F}(\tau)| \ll 1$ is not met, the spectrum of $\xi(t)$ becomes complex and lends itself to simple calculation only in the other limiting case of slow and large fluctuations. As we have seen, it then has the form of a Gaussian curve. It can however be shown that for sufficiently large $|\omega - \omega_0|$, equation (2.64) holds at all times [2.14]. In examining the spectrum of $\xi(t)$, i.e., $g_\xi(\omega)$, at the wings of the spectral line [i.e., at frequencies more distant from ω_0 than the width of the central part or peak of $g_\xi(\omega)$], it is thus possible to obtain the high-frequency region of the spectrum of frequency fluctuations $\Omega(t)$.

In conclusion we should point out that the above analysis of the extreme forms of the spectrum of $g_\xi(\omega)$ was based on the assumption of a relatively simple and monotonic behavior of the spectrum of $g_\Omega(\omega)$. If, for instance, $g_\Omega(\omega)$ falls off stepwise, i.e., $\Omega(t)$ features several distinct correlation times, or $g_\Omega(\omega)$ has a sufficiently narrow maximum at a high frequency $\tilde{\omega}$, so that $\psi_\Omega(\tau)$ is an oscillating function with the "carrier" frequency $\tilde{\omega}$, then the criteria of the closeness of $g_\xi(\omega)$ to a resonance, or Gaussian, form call for a refinement[6]. In particular, in the case of the narrow high-frequency spectrum $g_\Omega(\omega)$ considered above the form of $g_\xi(\omega)$ will not be Gaussian.

2.4 Spectra of Pulse Processes with Independent Intervals

The Poisson process, in which pulses are independent, is, of course, a special case of a more general type of pulse process, discussed in Sect. I.3.5. In the more general processes the occurrence of the νth pulse within $(t, t + dt)$ is determined by the times $t_\mu \le t$ of the occurrence of a certain number of previous pulses ($\mu = \nu - 1, \nu - 2, \ldots$). The probability of the νth pulse occurring within $(t, t + dt)$ is thus *conditional* here, i.e., it is dependent of $t_{\nu-1}$, $t_{\nu-2}$, $\ldots$. But for a Poisson process the probability of the same event occurring is *absolute*, since it is independent of any previous t_μ.

As in Sect. I.3.5, we will confine our discussion to the case where the probability of the νth pulse is only affected by the $(\nu-1)$th pulse, i.e., it depends solely on the time $\tau_{\nu-1}$ elapsed since the $(\nu-1)$th pulse. The intervals $\tau_\nu = t_{\nu+1} - t_\nu$ betwen pulses are thus independent of one another.

---

[6] See [2.14], and also [2.17, 18].

We now consider the following example of a *pulse process with independent intervals:*

$$\zeta(t) = \sum_\nu F_\nu\left(\frac{t - t_\nu}{\vartheta_\nu}\right) \quad . \tag{2.65}$$

Regarding the *random functions* $F_\nu(\theta)$ and random parameters ϑ_ν that enter into $\zeta(t)$, we make the following assumptions:

1) The pulse shape described by the random functions $\{F_\nu(\theta)\}$, pulse lengths $\{\vartheta_\nu\}$ and intervals between pulses $\tau_\nu = t_{\nu+1} - t_\nu$ are independent of one another. Note that for $F_\nu(\theta)$ we need not, generally speaking, introduce a separate parameter ϑ_ν, but it will shortly become clear that it is an advantage to have such parameters $\{\vartheta_\nu\}$ independent of $\{F_\nu(\theta)\}$.

2) We know the univariate und bivariate distribution functions of $F_\nu(\theta)$, pulse separations $w_\tau(\tau)\,d\tau$ and pulse durations $w_\vartheta(\vartheta)\,d\vartheta$, *all* these distributions being *independent of* ν.

3) The probability distribution for any pulse, let us say the zeroth (at time t_0) to be found within any time interval $(-T/2, T/2)$ is uniform, i.e., it has the form dt_0/T. This implies that all the relations with some fixed interval lengths τ_ν, but shifted relative to $t = 0$ by a time interval, t_0 confined within $(-T/2, T/2)$, are equiprobable. Otherwise, the process would not be stationary. Clearly,

$$\begin{aligned}
t_\nu &= t_0 + (t_1 - t_0) + (t_2 - t_1) + \ldots + (t_\nu - t_{\nu-1}) \\
&= t_0 + \tau_0 + \tau_1 + \ldots + \tau_{\nu-1} \quad ,
\end{aligned} \tag{2.66}$$

so that

$$t_\mu - t_\nu = \begin{cases} -(\tau_{\nu-1} + \tau_{\nu-2} + \ldots + \tau_\mu) & \text{for} \quad \mu < \nu \quad , \\ \tau_{\mu-1} + \tau_{\mu-2} + \ldots + \tau_\nu & \text{for} \quad \mu > \nu \quad . \end{cases} \tag{2.67}$$

This formulation and the solution of the problem are due to *Khurgin* [2.19], who in the case of the deterministic pulse shape $F(t)$ sought for the spectral density $g(\omega)$ directly, and not for the covariance (here this approach is simpler). Khurgin's general approach to the theory of random pulse processes is significant in that the examination is based not on the statistical behavior of instantaneous values of $\zeta(t)$ (i.e., the multivariate distribution functions), but on a stochastic treatment of the behavior of the process along the time axis (distributions of pulse lengths, separations and so forth are taken into account).

Processes of the type (2.65) occur, for example, when pulses are formed by fluctuation noise overshoots that exceed a certain threshold[7], and also in problems concerned with fluctuations in pulsed self-excited oscillators, which will be discussed below.

We now write the spectral amplitude density for a part of the process (2.65) in a sufficiently long interval $(-T/2, T/2)$

---

[7] Overshoots of normal stationary noise give, on raising the threshold, a Poisson process if, and only if, the covariance of the normal noise is subject to certain conditions (see [2.20]).

$$c(\omega) = \frac{1}{2\pi} \int\limits_{-\infty}^{+\infty} \xi(t) e^{-i\omega t} dt = \frac{1}{2\pi} \sum_{\nu=1}^{n} \int\limits_{-\infty}^{+\infty} F_\nu\left(\frac{t - t_\nu}{\vartheta_\nu}\right) e^{-i\omega t} dt \quad,$$

where the summation includes the (random) number of pulses, n, fully contained in the interval $(-T/2, T/2)$. In terms of the function $\tilde{F}_\nu(\alpha)$, which is a Fourier-conjugate of $F_\nu(\theta)$, we have

$$c(\omega) = \sum_{\nu=1}^{n} \vartheta_\nu \tilde{F}_\nu(\omega\vartheta_\nu) e^{-i\omega t_\nu} \quad.$$

We now find the covariance $\overline{c(\omega)c^*(\omega')} = g_B(\omega)\delta(\omega - \omega')$, taking into account the mutual independence of the random parameters ϑ_n, t_ν and the random function F_ν (and hence its spectral amplitude density $\tilde{F}_\nu$)

$$g_B(\omega)\delta(\omega - \omega')$$
$$= \sum_{\nu,\mu=1}^{n} \overline{\vartheta_\nu\vartheta_\mu \langle \tilde{F}_\nu(\omega\vartheta_\nu)\tilde{F}_\mu^*(\omega'\vartheta_\mu)\rangle \cdot \exp\left[i(\omega' t_\mu - \omega t_\nu)\right]} \quad. \tag{2.68}$$

The angle brackets signify the averaging over an ensemble of realizations $\tilde{F}_\nu$, and the bar the averaging over distributions of the random variables ϑ_ν and t_ν.

Since the time t_0 is uniformly distributed in the interval $(-T/2, T/2)$, we have by (2.66),

$$\overline{\exp\left[i(\omega - \omega')t_\nu\right]}$$
$$= \overline{\exp\left[i(\omega - \omega')(\tau_0 + \ldots + \tau_{\nu-1})\right]}\frac{1}{T}\int\limits_{-T/2}^{T/2} \exp\left[i(\omega - \omega')t_0\right]dt_0$$
$$= \overline{\exp\left[i(\omega - \omega')(\tau_0 + \ldots + \tau_{\nu-1})\right]}\frac{\sin(\omega - \omega')T/2}{(\omega - \omega')T/2} \to \frac{2\pi}{T}\delta(\omega - \omega') \quad.$$

This last change is permissible for $T \to \infty$ and will be carried out below. Further, we have

$$\overline{\exp\left[i(\omega' t_\mu - \omega t_\nu)\right]} = \overline{\exp\left[i\omega'(t_\mu - t_\nu)\right]}\;\overline{\exp\left[i(\omega' - \omega)t_\nu\right]}$$
$$= \overline{\exp\left[i\omega(t_\mu - t_\nu)\right]}\frac{2\pi}{T}\delta(\omega - \omega') \quad, \tag{2.69}$$

where in the factor at $\delta(\omega - \omega')$ we have, of course, put $\omega' = \omega$. Substituting (2.69) into (2.68) and discarding $\delta(\omega - \omega')$ on either side gives

$$g_B(\omega) = \frac{2\pi}{T} \sum_{\nu,\mu=1}^{n} \overline{\vartheta_\nu\vartheta_\mu \langle \tilde{F}_\nu(\omega\vartheta_\nu)\tilde{F}_\mu^*(\omega\vartheta_\mu)\rangle \cdot \exp\left[i\omega(t_\mu - t_\nu)\right]} \quad.$$

If we now introduce the notation

$$\overline{\vartheta^2\langle|\tilde{F}(\omega\vartheta)|^2\rangle} = \int_0^\infty \vartheta^2\langle|\tilde{F}(\omega\vartheta)|^2\rangle w_\vartheta(\vartheta)d\vartheta \equiv K(\omega) \quad,$$

$$\overline{\vartheta\langle\tilde{F}(\omega\vartheta)\rangle} = \int_0^\infty \vartheta\langle F(\omega\vartheta)\rangle w_\vartheta(\vartheta)d\vartheta \equiv H(\omega) \quad, \tag{2.70}$$

and in the relation for $g_B(\omega)$ separate the terms with $\mu \gtrless \nu$ from the sum over μ, then the expression takes the form

$$g_B(\omega) = \frac{2\pi}{T}\left\{nK(\omega) + |H(\omega)|^2 \sum_{\nu=1}^{n}\sum_{\mu\neq\nu}\overline{\exp\left[i\omega(t_\mu - t_\nu)\right]}\right\} \quad. \tag{2.71}$$

But, according to (2.67),

$$\exp\left[i\omega(t_\mu - t_\nu)\right] = \begin{cases} \exp\left[-i\omega(\tau_{\nu-1} + \tau_{\nu-2} + \ldots + \tau_\mu)\right] & \text{for } \mu<\nu \quad, \\ \exp\left[i\omega(\tau_{\mu-1} + \tau_{\mu-2} + \ldots + \tau_\nu)\right] & \text{for } \mu>\nu \quad, \end{cases}$$

where the intervals (whose sums enter into the exponents) are assumed to be mutually independent. Therefore, introducing the characteristic function of the pulse separation

$$\varphi(\omega) = \overline{e^{i\omega\tau}} = \int_0^\infty e^{i\omega\tau}w_\tau(\tau)d\tau \quad, \tag{2.72}$$

we have

$$\overline{\exp\left[i\omega(t_\mu - t_\nu)\right]} = \begin{cases} [\varphi^*(\omega)]^{\nu-\mu} & \text{for } \mu<\nu \quad, \\ [\varphi(\omega)]^{\mu-\nu} & \text{for } \mu>\nu \quad. \end{cases}$$

Accordingly, we get for $\varphi(\omega)\neq1$, i.e., clearly *save for the point* $\omega = 0$,

$$\sum_{\nu=1}^{n}\left(\sum_{\mu=1}^{\nu-1} + \sum_{\mu=\nu+1}^{n}\right)\overline{\exp\left[i\omega(t_\mu - t_\nu)\right]}$$

$$= 2n\,\mathrm{Re}\left\{\frac{\varphi(\omega)}{1-\varphi(\omega)}\right\} - 2\,\mathrm{Re}\left\{\frac{\varphi(\omega)[1-\varphi^n(\omega)]}{[1-\varphi(\omega)]^2}\right\} \quad.$$

As $T \to \infty$, the random number of pulses n in the interval T will, in the majority of cases, also grow indefinitely (on average, proportionally to T). Given this limiting process, we may only retain the first term on the right. The formula for $g_B(\omega) = g(\omega)$ (recalling that the point $\omega = 0$ is excluded) will then give

$$g(\omega) = \frac{2\pi n}{T}\left[K(\omega) + 2|H(\omega)|^2\,\mathrm{Re}\left\{\frac{\varphi(\omega)}{1-\varphi(\omega)}\right\}\right] \quad.$$

This spectral density corresponds, of course, to the *conditional* mean, given that there have been just n pulses in the interval T. A further averaging over the random parameter n is required. But $\bar{n} = T/\bar{\tau}$, where

$$\bar{\tau} = \frac{1}{n_1} = \int\limits_0^\infty \tau w_\tau(\tau) d\tau \tag{2.73}$$

is the average pulse separation, which is the reciprocal of the mean of the number of pulses n_1 per time unit ($\bar{\tau} = 1/n_1$). Therefore, we eventually have[8]

$$g(\omega) = 2\pi n_1 \left[K(\omega) + 2|H(\omega)|^2 \operatorname{Re}\left\{ \frac{\varphi(\omega)}{1 - \varphi(\omega)} \right\} \right] \quad . \tag{2.74}$$

Consider several special cases.

If the random functions $F_\nu(\theta)$ are defined as deterministic function F of θ and m-variate random variable a_ν

$$F_\nu(\theta) = F(\theta, a_\nu) \quad ,$$

then only the sense of the angle brackets in (2.70) will change. These will now be understood as an averaging over the distribution $w_a(a)da$ of the random vector a_ν

$$\langle |\tilde{F}(\omega\vartheta, a)|^2 \rangle = \int |\tilde{F}(\omega\vartheta, a)|^2 w_a(a) da \quad ,$$

$$\langle \tilde{F}(\omega\vartheta, a) \rangle = \int \tilde{F}(\omega\vartheta, a) w_a(a) da \quad , \tag{2.75}$$

and the Fourier transform $\tilde{F}$ of F will then surely be determnistic as well.

If we have a special case where a_ν is a univariate random variable that has the meaning of the amplitude of a deterministic pulse

$$F(\theta, a_\nu) = a_\nu F(\theta) \quad ,$$

then, taking for simplicity a_ν to be a real parameter and denoting

$$\bar{a}_\nu = a \quad , \quad D[a_\nu] = \overline{a_\nu^2} - a^2 = \sigma^2 \quad , \tag{2.76}$$

we get, from (2.75)

$$\langle |\tilde{F}(\omega\vartheta, a)|^2 \rangle = |\tilde{F}(\omega\vartheta)|^2 \int a^2 w_a(a) da = (a^2 + \sigma^2)|F(\omega\vartheta)|^2 \quad ,$$

$$\langle \tilde{F}(\omega\vartheta, a) \rangle = \tilde{F}(\omega\vartheta) \int a w_a(a) da = a\tilde{F}(\omega\vartheta) \quad .$$

Accordingly, equations (2.70) will take the form

$$K(\omega) = (a^2 + \sigma^2) \int\limits_0^\infty \vartheta^2 |\tilde{F}(\omega\vartheta)|^2 w_\vartheta(\vartheta) d\vartheta \equiv (a^2 + \sigma^2) K_1(\omega) \quad ,$$

---

[8] A more rigorous transition to (2.73) is to be found in [2.19].

$$H(\omega) = a \int_0^\infty \vartheta \tilde{F}(\omega\vartheta) w_\vartheta(\vartheta) d\vartheta \equiv a H_1(\omega) \quad , \tag{2.77}$$

so that the spectral density (2.74) will be

$$g(\omega) = 2\pi n_1 \left\{ \sigma^2 K_1(\omega) + a^2 \left[K_1(\omega) + 2|H_1(\omega)|^2 \operatorname{Re}\left\{ \frac{\varphi(\omega)}{1 - \varphi(\omega)} \right\} \right] \right\} \quad . \tag{2.78}$$

The various factors in this expression are conveniently separated. The parameters a^2 and σ^2 entered through the random amplitudes a_ν. The random pulse separations τ_ν are characterized by n_1 and $\varphi(\omega)$. Lastly, the random durations of the pulses, ϑ_ν, are included in $K_1(\omega)$ and $H_1(\omega)$. If, say, all the pulses have the same height ($\sigma^2 = 0$), then

$$g(\omega) = 2\pi n_1 a^2 \left[K_1(\omega) + 2|H_1(\omega)|^2 \operatorname{Re}\left\{ \frac{\varphi(\omega)}{1 - \varphi(\omega)} \right\} \right] \quad . \tag{2.79}$$

If, moreover, the pulse duration $\vartheta_\nu = \vartheta_0$ is also constant so that $w_\vartheta(\vartheta) = \delta(\vartheta - \vartheta_0)$, then, by (2.77),

$$K_1(\omega) = |H_1(\omega)|^2 = \vartheta_0^2 |\tilde{F}(\omega\vartheta_0)|^2 \quad ,$$

and (2.79) takes the form

$$g(\omega) = 2\pi n_1 a^2 K_1(\omega) \left[1 + 2\operatorname{Re}\left\{ \frac{\varphi(\omega)}{1 - \varphi(\omega)} \right\} \right] \quad . \tag{2.80}$$

In the special case of a Poisson process the pulse separations have the exponential distribution

$$\omega_\tau(\tau) = n_1 e^{-n_1 \tau} \quad ,$$

the characteristic function (2.72) is

$$\varphi(\omega) = \frac{n_1}{n_1 - i\omega} \quad ,$$

and hence

$$\operatorname{Re}\left\{ \frac{\varphi}{1 - \varphi} \right\} = 0 \quad .$$

Then, (2.78) gives

$$g(\omega) = 2\pi n_1 (\sigma^2 + a^2) K_1(\omega) \quad , \tag{2.81}$$

which coincides with the Campbell theorem (I.3.44). [It should be noted that in (2.77) the pulse durations are fixed, and, moreover, unlike the density $\tilde{F}(\omega\vartheta)$

in (2.70), $\tilde{F}(\omega)$ is the spectral amplitude density written in terms of the *dimensional* frequency ω.]

Now let τ have the gamma-distribution discussed in the first example of Sect. I.3.5

$$w_\tau(\tau) = \frac{\kappa^\alpha \tau^{\alpha-1}}{\Gamma(\alpha)} e^{-\kappa\tau} \quad .$$

Here $\kappa/\alpha = n_1$ is the thickness of the pulses and, in particular, $\kappa = n_1^0$ is their thickness at $\alpha = 1$, i.e., without any correlation between the times of their emergence (Poisson process). It will be recalled that for $\alpha < 1$ the correlation is "attractive", and for $\alpha > 1$ "repulsive", i.e., in the former case a pulse increases, and in the latter case decreases, the probability of the next pulse. The characteristic function (2.72) for $\alpha > 0$ is

$$\varphi(\omega) = \frac{\kappa^\alpha}{\Gamma(\alpha)} \int\limits_0^\infty \tau^{\kappa-1} \exp\left[-(\kappa - i\omega)\tau\right] d\tau = \frac{1}{(1 - i\omega/\kappa)^\alpha} \quad ,$$

which gives

$$\mathrm{Re}\left\{ \frac{\varphi(\omega)}{1 - \varphi(\omega)} \right\} = \frac{R^\alpha \cos \alpha\psi - 1}{R^{2\alpha} - 2R^\alpha \cos \alpha\psi + 1} \quad ,$$

$$R = \frac{1}{\cos\psi} = \sqrt{1 + \left(\frac{\omega}{\kappa}\right)^2} \quad .$$

Here ψ varies, from zero at $\omega = 0$, to $\pi/2$ at $|\omega| = \infty$. If α is so small that $\alpha\psi \ll \pi/2$, then

$$\mathrm{Re}\left\{ \frac{\varphi(\omega)}{1 - \varphi(\omega)} \right\} \approx \frac{\ln R}{(\psi^2 + \ln^2 R)\alpha} \quad ,$$

i.e., the second term in (2.78) will be large. On the contrary, as $\alpha \to \infty$ it will be very small, and hence

$$\mathrm{Re}\left\{ \frac{\varphi(\omega)}{1 - \varphi(\omega)} \right\} \approx \frac{\cos \alpha\psi}{R^\alpha} \quad .$$

We now turn our attention to the spectrum of pulse self-oscillations.

In periodic pulse generators, random actions may shift the moment of pulse generation. If the correlation time of these actions is small as compared with the pulse repetition period, then the shift of the νth pulse may, apart from the random impact at t_ν, be affected only by the time $\tau_\nu = t_\nu - t_{\nu-1}$ elapsed since the $(\nu-1)$th pulse. And the lengths of adjacent intervals seem not to be interrelated. The pulse self-oscillations thus fit nicely into the developed scheme[9].

---

[9] This was indicated by G.S. Gorelik as cited in [2.19]. In [2.21] the variance of intervals τ due to the shot current of vacuum tubes and thermal noise of grid resistances is found for a specific pulse generator.

Deviations τ_ν from the mean τ_0 are generally not large, i.e., $w_\tau(\tau)$ is a sharply peaked function concentrated near τ_0. Expanding $\exp(i\omega\tau)$ in powers of $\tau - \tau_0$ gives

$$\varphi(\omega) = \left[1 - \frac{\omega^2\beta^2}{2} + o(\omega^2\beta^2)\right]e^{i\omega\tau_0} \quad ,$$

where $\beta^2 = \overline{(\tau - \tau_0)^2}$ is the variance of pulse separations. Consequently,

$$\mathrm{Re}\left\{\frac{\varphi(\omega)}{1 - \varphi(\omega)}\right\} \approx \frac{\left(1 - \frac{\omega^2\beta^2}{2}\right)\left(\cos\omega\tau_0 - 1 + \frac{\omega^2\beta^2}{2}\right)}{\left(\frac{\omega^2\beta^2}{2}\right)^2 - 2\left(1 - \frac{\omega^2\beta^2}{2}\right)(\cos\omega\tau_0 - 1)} \quad .$$

This quantity shows sharp maxima at all harmonics $\omega = n\omega_0$ of the fundamental frequency $\omega_0 = 2\pi/\tau_0$. For $\omega = n\omega_0$ and $(n\omega_0\beta)^2 \ll 1$ we have

$$2\,\mathrm{Re}\left\{\frac{\varphi(\omega)}{1 - \varphi(\omega)}\right\} \approx \left(\frac{2}{n\omega_0\beta}\right)^2 \gg 1 \quad .$$

In between the periodically occurring values $\omega = n\omega_0$ subject to the same condition $(n\omega_0\beta)^2 \ll 1$ we obtain

$$2\,\mathrm{Re}\left\{\frac{\varphi}{1 - \varphi}\right\} \approx -1 \quad .$$

Setting $\omega = n\omega_0 + \delta$, where δ is small compared with ω_0, it is trivial to verify that in the vicinity of the nth harmonic

$$2\,\mathrm{Re}\left\{\frac{\varphi(\omega)}{1 - \varphi(\omega)}\right\} \approx \frac{(n\omega_0\beta)^2 - (\tau_0\delta)^2}{(n^2\omega_0^2\beta^2/2)^2 + (\tau_0\delta)^2} \quad .$$

Thus, the half-width of an overshoot at half-maximum (Fig. 2.6a) is

$$\delta_1 = (n\omega_0\beta)^2/2\tau_0 \quad ,$$

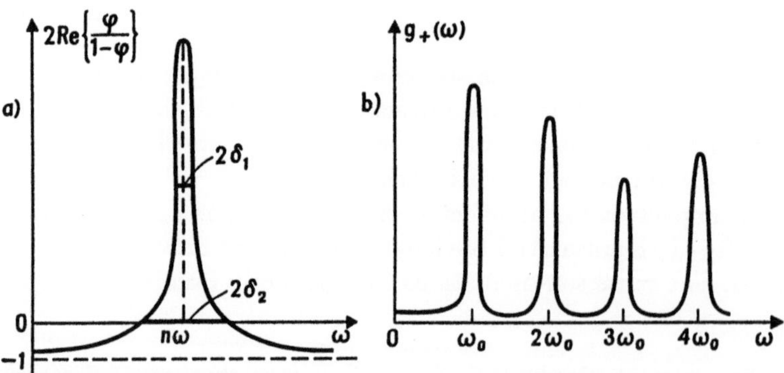

Fig. 2.6a,b. To the theory of spectrum of the pulse generator

and at zero level

$$\delta_2 = n\omega_0\beta/\tau_0 \quad .$$

In the case of short pulses the functions $H_1(\omega)$ and $K_1(\omega)$ vary slowly in the intervals ω_0 between the harmonics. Near $\omega = n\omega_0$, therefore, the behavior of the spectral density (2.78) will be determined mainly by that of the term $\mathrm{Re}\,\{\varphi/(1-\varphi)\}$, and the spectrum will be as shown in Fig. 2.6b. Of course, the "discrete" lines become increasingly lower and wider with n (note that δ_1 grows faster than δ_2), and for sufficiently high harmonics the line structure smears out.

In the process just considered the pulse durations ϑ_ν were independent of the separations τ_ν. Another problem that is often of interest is one which arises, in particular, in handling continuous self-oscillations in strongly nonlinear systems, i.e., nonsinusoidal oscillations with a fluctuating "period" [2.22]. A certain idealization allows this quasi-periodic process to be reduced to a sequence of closely spaced pulses, i.e., the random durations are *equal* to the separations

$$\vartheta_\nu = \tau_\nu = t_{\nu+1} - t_\nu \quad .$$

Thus, in this instance there is no independent parameter ϑ_ν with a distribution function $w_\vartheta(\vartheta)$. We will confine ourselves to defining the pulse shape using the *deterministic* function $F(\theta, \boldsymbol{a}_\nu)$, where $\boldsymbol{a}_\nu$ is a random vector

$$\zeta(t) = \sum_\nu F\left(\frac{t - t_\nu}{\tau_\nu}, \boldsymbol{a}_\nu\right) \quad . \tag{2.82}$$

The difference between the processes (2.65) and (2.82) is illustrated in Fig. 2.7.

The spectrum of (2.82) can be derived in the same way as in the case considered above [2.23]. The only significant difference is that in (2.82) we may no longer average the functions of ϑ_ν and of τ_ν separately, since $\vartheta_\nu = \tau_\nu$. Therefore, in addition to functions (2.70), which have now the form

$$K(\omega) = \int_0^\infty \tau^2 \langle|\tilde{F}(\omega\tau, \boldsymbol{a})|^2\rangle w_\tau(\tau)d\tau \quad ,$$

$$H(\omega) = \int_0^\infty \tau \langle\tilde{F}(\omega\tau, \boldsymbol{a})\rangle w_\tau(\tau)d\tau \quad , \tag{2.83}$$

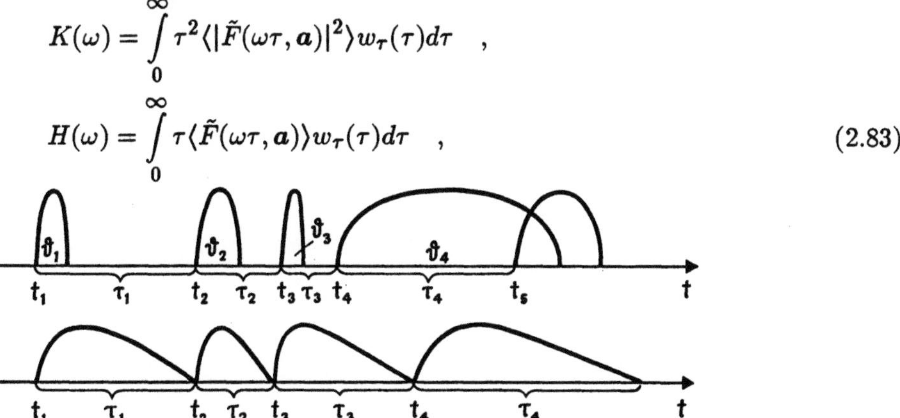

Fig. 2.7. Pulse processes with possible overlaps of pulses (*top*) and without separation between them (*bottom*)

73

we have to introduce one more function

$$I(\omega) = \int\limits_0^\infty \tau \langle \tilde{F}(\omega\tau, \boldsymbol{a})\rangle e^{i\omega\tau} w_\tau(\tau) d\tau \quad . \tag{2.84}$$

(The angle brackets denote averaging over the distribution function of $\boldsymbol{a}$). It will be noted that in the present case (where the pulse duration and separation coincide), we need to calculate the spectral amplitude $\tilde{F}(\alpha, \boldsymbol{a})$, assuming that in the dimensionless variable $\theta = t/\tau$ the function $F(\theta, \boldsymbol{a})$ is nonzero in an interval of a fixed (unit) length:

$$\tilde{F}(\alpha, \boldsymbol{a}) = \frac{1}{2\pi} \int\limits_0^1 F(\theta, \boldsymbol{a}) e^{-i\alpha\theta} d\theta \quad . \tag{2.85}$$

In terms of K, H, and I the spectral density of the process (2.82) is represented by

$$g(\omega) = 2\pi n_1 \left[K(\omega) + 2\,\mathrm{Re}\left\{ \frac{I(\omega)H^*(\omega)}{1 - \varphi(\omega)} \right\} \right] \quad . \tag{2.86}$$

In particular, if $\boldsymbol{a}\,'$is a one-dimensional pulse "amplitude"

$$\tilde{F}(\omega\tau, \boldsymbol{a}) = a\tilde{F}(\omega\tau) \quad ,$$

then, considering (2.76), we get

$$g(\omega) = 2\pi n_1 \left[(a^2 + \sigma^2) K_1(\omega) + 2a^2\,\mathrm{Re}\left\{ \frac{I_1(\omega)H_1^*(\omega)}{1 - \varphi(\omega)} \right\} \right] \quad , \tag{2.87}$$

where

$$K_1(\omega) = \overline{\tau^2 |\tilde{F}(\omega\tau)|^2} \quad ,$$

$$H_1^*(\omega) = \overline{\tau \tilde{F}(\omega\tau)} \quad ,$$

$$I_1(\omega) = \overline{\tau \tilde{F}(\omega\tau) e^{i\omega\tau}} \tag{2.88}$$

(the average is over the distribution of τ). This spectrum generally behaves as already described, specifically, all the arguments concerning sharp overshoots of $g(\omega)$ obtained at a small scattering of τ about the mean τ_0 apply completely.

Note that in both cases, as expected with closed systems, the oscillator "phase" $\Phi = \sum_{\nu=0}^N \tau_\nu$ drifts in a diffusive manner away from its dynamic value $\overline{\Phi} = N\tau_0$ which corresponds to strict periodicity of the pulses. Considering the mutual independence of τ_ν, we have

$$\overline{(\Phi - \overline{\Phi})^2} = \sum_{\nu=0}^N \overline{(\tau_\nu - \tau_0)^2} = \beta^2 N \quad ,$$

i.e., the mean square shift Φ accumulates with the number of periods.

2.5 Correlation Theory of Coherence

Though it was understood long ago (E. Verdet, 1869) that inadequacies exist in the elementary concept of perfect coherence and perfect incoherence, it was not until recently that these were eliminated from texts on optics. The general quantitative treatment of coherence is due to Wolf, Blanc-Lapierre and others[10] who have been developing it since the 1950s, using the theory of random functions and, in particular, correlation theory (the coherence theory, in its more general form, considers the moments not only of the second, but also of higher orders). This general theory is concerned with the properties of *wave fields*, i.e., functions of time and space, so that coherence theory, even if it stays within the framework of the correlation theory, should deal with *random fields* (space-time coherence) to be discussed in [2.31, 32]. Here we will confine ourselves to the question of *time* coherence, i.e., random oscillatory processes.

The concept of coherence is equally applicable to oscillations and waves of any physical nature and any frequency range, so that the optical terminology is by no means obligatory. Giving optics its due as a branch of physical sciences where the concept of coherence was conceived, we shall often draw on optical interference phenomena as examples.

Consider now one of the simplest examples of this kind — the double-slit Young-Rayleigh interferometer. The interferometer (Fig. 2.8) consists of a screen S with two parallel slits 1 and 2 and a lens L, whose principal focal plane F shows the interference pattern[11]. Note that we could just as well not install the lenses and consider the interference "at infinity", i.e., at distances $R \gg d^2/\lambda$, where d is the distance between the slits and λ the length of the light wave (the so-called Fraunhofer zone). We will take the slits to be narrow enough to enable us not to make allowances for variations of the amplitude and

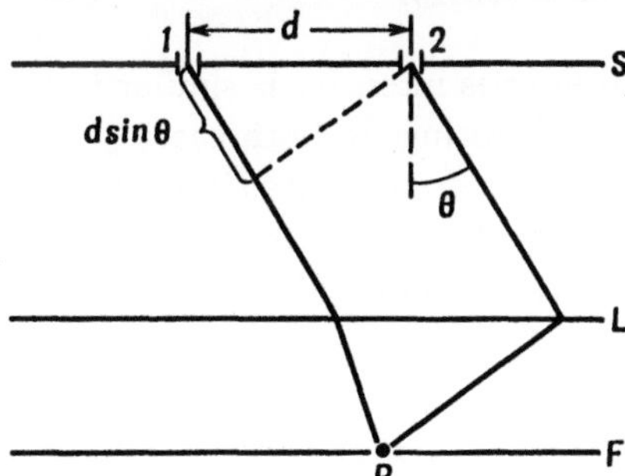

Fig. 2.8. Principle of operation of the Young-Rayleigh interferometer

---

[10] Of the many works in this field we mention [2.24–40].

[11] This interferometer "splits the wave front", unlike those devices which produce two (or more) interference waves from a single one using partial reflection and transmission ("amplitude splitting"). In the *stellar* Michelson interferometer the slits are replaced by spaced mirrors, and in radiointerferometers, by spaced antennas. The Young-Rayleigh interferometer was considered by G.S. Gorelik in connection with coherence problems as early as 1950, and later by *Thomson* and *Wolf* [2.33]. A similar discussion of the relationship between the radio signal band and the angular pattern was given by *Nodtvedt* [2.34].

phase of the incident wave over the widths of the slits. The flied in the slits will be described just by two oscillations: at the first slit by $\xi_1(t)$ and at the second by $\xi_2(t)$. The lens is assumed to be ideal, and its finite aperture is ignored.

Waves propagating from the slits may be thought of as a superposition of travelling and so-called inhomogeneous plane waves. The latter do not contribute to energy transfer and decay exponentially farther away from the screen S. They practically vanish at distances of the order of λ. The travelling waves that propagate from both slits at the same angle are focused by the lens at an appropriate point P in the plane F. The oscillations $\xi_1(t)$ and $\xi_2(t)$ thus add up at P, one of them being delayed by the time $\tau = d \sin \theta / c$, where c is the velocity of light.

We write the resultant oscillation at P

$$\xi(t,\tau) = \xi_1(t+\tau) + \xi_2(t) \quad ,$$

in terms of analytical signals

$$\xi_j(t) = \tfrac{1}{2}[\zeta_j(t) + \zeta_j^*(t)] \quad , \quad j = 1,2 \quad .$$

We are interested in the *intensity* (illuminance) distribution over F, which is described by the dependence of $\xi^2(t,\tau)$ on the coordinate $x \sim \tan \theta$ of the point in the plane, i.e., on the delay $\tau \sim \sin \theta$ to which x is uniquely related. For the instantaneous intensity we have

$$
\begin{aligned}
\xi^2(t,\tau) &= \tfrac{1}{4}[\zeta_1(t+\tau) + \zeta_2(t) + \text{complex conjugate}]^2 \\
&= \tfrac{1}{4}[2|\zeta_{1\tau}|^2 + 2|\zeta_2|^2 + 2(\zeta_{1\tau}\zeta_2^* + \zeta_{1\tau}^*\zeta_2)] \\
&\quad + [\zeta_{1\tau}^2 + \zeta_{1\tau}^{*2} + \zeta_2^2 + \zeta_2^{*2} + 2(\zeta_{1\tau}\zeta_2 + \zeta_{1\tau}^*\zeta_2^*)] \quad .
\end{aligned}
\tag{2.89}
$$

It is obvious that the transition to the instantaneous intensity is similar to the principle of the square-law detector in radio engineering. In both cases $\xi^2(t,\tau)$ contains second-order combination frequencies: sum frequencies (in particular, doubled ones) and difference frequencies (in particular, zero ones). If we regard $\zeta_j(t)$ as modulated oscillations (we will see in what follows that this is necessary for interference obsevations), i.e.,

$$\zeta_j(t) = A_j(t)e^{i\omega_j t} \quad , \quad A_j(t) = A_j(t)e^{i\varphi_j(t)} \quad , \quad j = 1,2 \quad ,$$

then

$$
\begin{aligned}
\xi^2(t,\tau) = \tfrac{1}{4}\big\{ & 2|A_{1\tau}|^2 \\
& + 2|A_2|^2 + 2[A_{1\tau}A_2^*\, e^{i\omega_1 \tau + i(\omega_1 - \omega_2)t} + \text{complex conjugate}] \\
& + [A_{1\tau}^2\, e^{2i\omega_1(t+\tau)} + A_2^2 e^{2i\omega_2 t} \\
& + 2A_{1\tau}A_2\, e^{i\omega_1 \tau + i(\omega_1 + \omega_2)t} + \text{complex conjugate}]\big\} .
\end{aligned}
\tag{2.90}
$$

In radio engineering a variety of combination frequencies produced by non-linear elements are used. Specifically, it is well known that the *intermediate frequency* $\omega_1 - \omega_2$ is widely used, which is the difference between the frequency ω_1 of the oscillation of interest to us and the heterodyne frequency ω_2. But interferometers (both optical and radio) mostly use oscillations with the same carrier frequency ($\omega_1 = \omega_2 = \omega_0$). Then $\xi^2(t, \tau)$ takes the form

$$\xi^2(t, \tau) = \tfrac{1}{2}[|A_{1\tau}|^2 + |A_2|^2 + (A_{1\tau}A_2^* e^{i\omega_0\tau} + \text{complex conjugate})]$$
$$+ \tfrac{1}{4}[(A_{1\tau}^2 + A_2^2 + 2A_{1\tau}A_2 e^{i\omega_0\tau})e^{2i\omega_0 t} + \text{complex conjugate}] \quad , \quad (2.91)$$

i.e., it contains the low-frequency range of the spectrum (the upper line) and the modulated oscillations with carrier frequency $2\omega_0$ (the lower line; see Fig. 3.9).

Admittedly, in optics too we can consider oscillations with different frequencies ω_1 and ω_2, including such similar ones that the difference frequency $\omega_1 - \omega_2$ lies, say, in the microwave range. The principle and the technical feasibility of such optical *heterodyning* was first indicated by *Gorelik* [2.35] (even before lasers came in). However, here we will be concerned with the normal conditions of interference observations, where $\omega_1 = \omega_2 = \omega_0$ and $\xi^2(t, \tau)$ is given by (2.91).

By intensity we understand the *low-frequency* component of $\xi^2(t, \tau)$, as it is this component that is separated by any real observational technique. The point is that the *observation time* T, even if very small compared to the time-scale τ_c of the modulating functions $A_j(t)$, is always far longer than the carrier frequency period $T_0 = 2\pi/\omega_0$. This is also true of the *moving averaging*, which may be modelled by the operation

$$\widetilde{\xi_T^2(t, \tau)} = \frac{1}{T} \int\limits_t^{t+T} \xi^2(t, \tau)dt \quad , \tag{2.92}$$

and of the *accumulation,* which differs from (2.92) solely in that it involves multiplying by T and that t is understood to be a certain fixed moment when the accumulation begins. In optics, for example, moving averaging is performed by the eye or an electric filter after the photomultiplier, in radio engineering by the videofilter. In optics the accumulation is accomplished by a photographic plate or bolometer, in radio engineering by a variety of accumulating and integrating circuits. Making allowance for the strong inequality $T \gg T_0$, averaging over a period T_0 may be thought to occur *a fortiori* and be included in the *definition* of the "instantaneous" intensity $I(t, \tau)$

$$I(t, \tau) = \widetilde{\xi_{T_0}^2}(t, \tau) = \tfrac{1}{2}[|\zeta_{1\tau}|^2 + |\zeta_2|^2 + 2\,\text{Re}\,\{\zeta_{1\tau}\zeta_2^*\}]$$
$$= \tfrac{1}{2}[|A_{1\tau}|^2 + |A_2|^2 + 2\,\text{Re}\,\{A_{1\tau}A_2^* e^{i\omega_0\tau}\}] \quad . \tag{2.93}$$

But the *observed* intensity distribution is the result of the subsequent averaging of $I(t, \tau)$ over the observation time T

$$\widetilde{\overline{I_T(t,\tau)}} = \frac{1}{T}\int\limits_t^{t+T} I(t,\tau)dt \quad . \tag{2.94}$$

In the general case, it is clearly completely dependent on the relationship between T and the modulation time-scale τ_c.

We will focus on the phenomena occurring in *random* modulation. Equation (2.93) then describes a certain instantaneous realization of the intensity distribution over the plane F, i.e., as a function of the delay τ.

In a majority of interference experiments and measurements we deal with *stationary* and *stationarily related* oscillations $\zeta_1(t)$ and $\zeta_2(t)$. If these stationary processes also meet the second-order ergodicity condition, then (as we saw in Sect. I.4.7), as the observation time T is increased, we arrive at the same result as in averaging over the realization ensemble (2.93), since ergodicity suggests convergence in probability

$$\widetilde{\overline{I_T(t,\tau)}} \xrightarrow{\text{in pr.}} \langle I(t,\tau)\rangle \quad , \quad \text{as} \quad T \to \infty \quad .$$

If we make use of the relationship (1.1) for second moments, then the statistical average of (2.93) has the form

$$\begin{aligned}
I(\tau) \equiv \langle I(t,\tau)\rangle &= \tfrac{1}{2}[B_1(0) + B_2(0) + 2\,\mathrm{Re}\,\{B_{12}(\tau)\}] \\
&= \tfrac{1}{2}[B_{A1}(0) + B_{A2}(0) + 2\,\mathrm{Re}\,\{B_{A12}(\tau)e^{i\omega_0\tau}\}] \quad , \tag{2.95}
\end{aligned}$$

where B are the moments of the modulated oscillations $\zeta_j(t)$, and B_A are the moments of the complex amplitudes $A_j(t)$ $(j = 1,2)$.

Note that the statistical average of the instantaneous intensity (2.89) in the general (nonstationary) case contains both the moments B and the "second" moments $\tilde{B}$, see (1.2). However, in the case of the stationary analytical signals under consideration $\tilde{B}$ are zero[12]. Thus, statistical averaging, which is equivalent to time averaging over a long enough period T, automatically incorporates averaging over the period T_0 of the carrier frequency.

Clearly, the quantities

$$I_{1,2} \equiv \tfrac{1}{2}B_{1,2}(0) = \tfrac{1}{2}B_{A1,2}(0)$$

are the intensities of uniform exposure due to each of the slits 1 and 2. Deviations from the uniform total illuminance $I_1 + I_2$ are completely dictated by the real part of the covariance

$$B_{12}(\tau) = \langle \zeta_1(t+\tau)\zeta_2^*(t)\rangle = B_{A12}(\tau)e^{i\omega_0\tau} = \langle A_1(t+\tau)A_2^*(t)\rangle e^{i\omega_0\tau} \quad .$$

---

[12] It is readily seen that not only the "second" covariance of the stationary analytical signal [see (1.11)] vanishes, but also the "second" *mutual* covariance for two stationarily related signals.

Uncorrelated oscillations at the interferometer slits (e.g., illumination of each of them by a wave from an independent source) will produce no interference pattern, i.e., there will be no coherence.

In optics the covariance $B_{12}(\tau)$ is generally called the *complex coherence function of second order* (here, the *time* coherence function) and is denoted by $\Gamma^{(2)}(1,2) \equiv B_{12}(\tau)$. But, as noted above, there are essentially no grounds for abandoning the general terminology used in the theory of random functions. Equation (2.95) shows that the intensity distribution over plane F represents, up to the constant $I_1 + I_2$, the covariance of the interfering oscillations.

By introducing the mutual correlation coefficients for the complex amplitudes $A_j(t)$ and high-frequency oscillations $\zeta_j(t)$

$$K_{A12}(\tau) = \frac{B_{A12}(\tau)}{2\sqrt{I_1 I_2}} = |K_{12}(\tau)| \exp\left[i\Theta_{12}(\tau)\right] \quad ,$$

$$K_{12}(\tau) = \frac{B_{12}(\tau)}{2\sqrt{I_1 I_2}} = |K_{12}(\tau)| \exp\left\{i[\Theta_{12}(\tau) + \omega_0\tau]\right\} \quad , \tag{2.96}$$

we may write (2.95) as

$$I(\tau) = I_1 + I_2 + 2\sqrt{I_1 I_2}|K_{12}(\tau)| \cos\left[\Theta_{12}(\tau) + \omega_0\tau\right] \quad . \tag{2.97}$$

Thus, as is to be expected for modulated oscillations (Sect. 2.2), the distribution $I(\tau)$ itself is a modulated oscillation: interference fringes recur in τ with frequency $\omega_0 + \dot\Theta_{12}(\tau) \approx \omega_0$ and are inscribed into the envelope $|K_{12}(\tau)|$.

In optics it has long been the custom to characterize the degree of contrast of a pattern by the so-called *visibility* of fringes. This easily measurable quantity is by definition

$$V = \frac{I_{\max} - I_{\min}}{I_{\max} + I_{\min}} \quad , \tag{2.98}$$

where $I_{\max}$ and $I_{\min}$ are the intensities (illuminance of screen F) at a neighboring maximum and minimum, respectively. For modulated oscillations we can consider that for these neighboring values $I_{\max}$ and $I_{\min}$, which correspond to a change of $\Theta_{12}(\tau) + \omega_0\tau$ by π, the value of $|K_{12}(\tau)|$ is the same. It then follows from (2.97, 98) that

$$V(\tau) = \frac{2\sqrt{I_1 I_2}}{I_1 + I_2}|K_{12}(\tau)| \quad , \tag{2.99}$$

i.e., up to a constant factor, the visibility coincides with the envelope $|K_{12}(\tau)|$. It is thus clear that the correlation coefficient represents an adequate and convenient quantitative measure for the degree of coherence ranging from complete coherence $|K_{12}| = 1$ to complete absence of coherence $K_{12} = 0$. In optics, $K_{12}(\tau)$ is said to be the *complex degree of coherence*. At the same time, interference serves as a pictorial illustration of the "uncertainty relation" between the width of the whole observed interference pattern and the degree of monochromaticity of the light used (since the spectral density of the light is also directly observable using for example a spectroscope).

For simplicity, consider the case of a plane wave striking normally the interferometer slits so that $\zeta_1(t) = \zeta_2(t) = \zeta(t)$. Accordingly, $I_1 = I_2 = I$ and (2.97) takes the form

$$I(\tau) = 2I\{1 + |K(\tau)| \cos\left[\Theta(\tau) + \omega_0\tau\right]\} \quad , \tag{2.100}$$

where $K(\tau) = |K(\tau)| \exp\left[i\Theta(\tau)\right]$ is the autocorrelation coefficient for the complex amplitude $A(t)$ of $\zeta(t)$. According to (1.54), $K(\tau)$ and the spectral density $g_\zeta(\omega)$ are related by the Fourier transform[13]

$$K(\tau) = \frac{B(\tau)}{B(0)} = \frac{\displaystyle\int_0^\infty g_\zeta(\omega)e^{i\omega\tau}\,d\omega}{\displaystyle\int_0^\infty g_\zeta(\omega)\,d\omega} \quad , \tag{2.101}$$

thus leading to the "uncertainty relation" (Sect. 1.5). Specifically, for a Gaussian spectral line (of width Ω at $1/e$ of the maximum at $\omega - \omega_0$) the envelope of the correlation coefficient will be Gaussian, too:

$$K(\tau) = \exp\left(-\Omega^2\tau^2/16 + i\omega_0\tau\right) \quad ,$$
$$(|K(\tau)| = \exp\left(-\Omega^2\tau^2/16\right) \quad , \quad \Theta(\tau) = 0) \quad . \tag{2.102}$$

The Lorentzian line shape (of width Ω at half height) has the exponential envelope

$$K(\tau) = \exp\left(-\Omega|\tau|/2 + i\omega_0\tau\right) \quad ,$$
$$(|K(\tau)| = \exp\left(-\Omega|\tau|/2\right) \quad , \quad \Theta(\tau) = 0) \quad . \tag{2.103}$$

For the rectangular spectrum[14] uniform within the frequency band $(\omega_0 - \Omega/2, \omega_0 + \Omega/2)$ we find

$$K(\tau) = \frac{\sin\left(\Omega\tau/2\right)}{\Omega\tau/2}e^{-i\omega_0\tau} \quad ,$$
$$\left(|K(\tau)| = \left|\frac{\sin\left(\Omega\tau/2\right)}{\Omega\tau/2}\right| \quad , \quad \Theta(\tau) = 0 \quad \text{or} \quad \pi\right) \tag{2.104}$$

[0 or π depending on the sign of $\sin\left(\Omega\tau/2\right)$].

---

[13] Recall that $g_\zeta(\omega) = 2g_+(\omega) = 4g(\omega)$ for $\omega > 0$ and $g_\zeta(\omega) = 0$ for $\omega < 0$, where $g(\omega)$ is the spectral (*even* in ω) density of the initial *real* oscillation $\xi(t)$.

[14] Generally speaking, it is impossible to prepare such a case for oscillations in time (Sect. 4.4).

The visibility $V = |K(\tau)|$ depends at all times on $\Omega\tau$, so that the interference pattern as a function of τ broadens with decreasing spectrum width Ω, and vice versa. As $\Omega \to 0$ (ideal monochromaticity) $V = |K| = 1$, i.e., in our idealized device the interference fringes fill the entire sector $|\theta| \leq \pi/2$ with a constant maximum contrast (complete coherence).

If a plane wave strikes screen S normally, the visibility is only related to the monochromaticity of the oscillation $\zeta(t) = \zeta_1(t) = \zeta_2(t)$. In the general case, oscillations at spaced slits may differ not only in that one of them is delayed by τ, but also in that they represent temporal behavior of the wave *field* $\zeta(t, \mathbf{r})$ at *different* points: $\zeta_1(t) = \zeta(t, \mathbf{r}_1)$, $\zeta_2(t) = \zeta(t, \mathbf{r}_2)$. Thereby, these oscillations (and hence visibility) depend on the *space structure* of the field. For a point source the structure is only determined by its location, but for an extended source it is in addition governed by the statistical behavior of the entire ensemble of oscillations produced by the "point" elements of the source. Moreover, the space-time structure of the field may be influenced both by deterministic and random irregularities of the medium along the path from the source to the interferometer, and also by fluctuations in the device itself, e.g., fluctuations of "slit" positions. Of course, the last factor is essentially excluded in the double-slit Young-Rayleigh interferometer, but in the stellar Michelson interferometer it is always present since here the mirrors are separated by several meters so that vibrations of the apparatus can no longer be ignored. Consider several illustrations of this last statement.

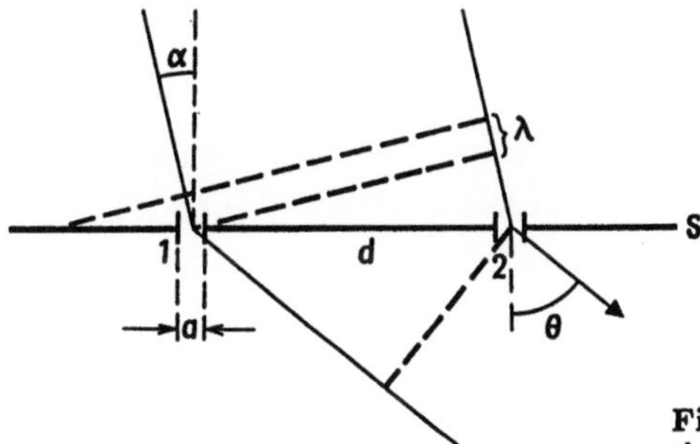

Fig. 2.9. Oblique incidence of a plane wave on the slits of the interferometer

Let a plane wave be incident obliquely upon an interferometer screen S, angle α to the normal being sufficiently small (Fig. 2.9). A wave from a point source may also be regarded as plane, if the source "seen" at an angle α lies within the Fraunhofer zone $(R \gg d^2/\lambda)^2$. For oblique incidence the oscillation at slit 1 is now delayed with respect to the oscillation at slit 2 by $\tau - \tau_\alpha$, where

---

[15] For the nearest star $R \approx 4$ light years and the radius of the zeroth Fresnel zone $\sqrt{\lambda R}$ at $\lambda = 5.000\,\text{Å}$ is $140\,\text{km}$. When we say that α is small we mean that the oscillations of the field $\zeta(t, \mathbf{r})$ are in-phase within the width a of each slit. Figure 2.9 shows that the condition is met if $a \ll \lambda / \sin \alpha$. For the same wavelength and $\alpha = 1''$ we get $a \ll \lambda/\alpha \approx 10\,\text{m}$

$$\tau = d \sin \theta / c \quad , \quad \tau_\alpha = d \sin \alpha / c \quad . \tag{2.105}$$

Clearly, this brings about a shift of the entire interference pattern, since in the preceding equations the difference $\tau - \tau_\alpha$ will be substituted for τ.

If α is a random variable independent of $\zeta_1(t)$ and $\zeta_2(t)$ (say, α are fluctuations, due to medium irregularities, of the direction of wave arrival), then averaging over the ensemble of realizations of ζ_1 and ζ_2 in (2.95) will now give the *conditional* mean, i.e., the mean at a fixed α, for example, $B_{12}(\tau - \tau_\alpha)$. The unconditional covariance (indicated by a bar) results from subsequent averaging over the distribution $w_\alpha(\alpha)$ of the random variable α

$$\overline{B}_{12}(\tau) = \int B_{12}(\tau - \tau_\alpha) w_\alpha(\alpha) d\alpha \quad , \tag{2.106}$$

where τ_α is given by (2.105)[16]. If, apart from the additional shift τ_α, the oscillations $\zeta_1(t)$ and $\zeta_2(t)$ show no distinctions, we should substitute the covariance B for B_{12} in (2.106).

We rewrite the equation in other terms. Using (2.101) (of course, with τ replaced by $\tau - \tau_\alpha$) and introducing the characteristic function $\varphi_\alpha(u)$ for the random angle α

$$w_\alpha(\alpha) = \frac{1}{2\pi} \int\limits_{-\infty}^{+\infty} \varphi_\alpha(u) e^{-iu\alpha} du \quad ,$$

we obtain, by (2.106),

$$\begin{aligned}
\overline{K}(\tau) &= \frac{\overline{B}(\tau)}{\overline{B}(0)} \\
&= \frac{1}{2\pi\overline{B}(0)} \int\limits_0^\infty g_\zeta(\omega) e^{i\omega\tau} d\omega \int\limits_{-\infty}^{+\infty} \varphi_\alpha(u) du \int\limits_{-\infty}^{+\infty} \exp\left[-i(\omega\tau_\alpha + u\alpha)\right] d\alpha \quad .
\end{aligned}$$

If the distribution $w_\alpha(\alpha)$ is sharp enough (and hence only of importance for small α for which we may use $\sin \alpha \approx \alpha$ and $\tau_\alpha \approx \alpha d/c$), the integral with respect to α is equal to $2\pi\delta(u + \omega d/c)$. Integrating with respect to u gives

$$\overline{K}(\tau) = \frac{1}{\overline{B}(0)} \int\limits_0^\infty g_\zeta(\omega) \varphi_\alpha\left(-\frac{\omega d}{c}\right) e^{i\omega\tau} d\omega \quad . \tag{2.107}$$

Now consider an *extended* source consisting of incoherent (i.e., uncorrelated) "point" elements. Each such element seen at an angle α generates a

---

[16] If α fluctuates not in the wave field itself, but owing to vibration of the interferometer, then the function $\overline{B}_{12}$ given by (2.106) is naturally viewed as a *distorted* covariance for the field.

plane wave arriving at an angle α. The corresponding elementary oscillation $\zeta(t,\alpha)d\alpha$ may then, generally speaking, depend explicitly on α, if it is only due to different luminosities of different elements of the extended source. The total oscillation at the jth slit ($j = 1, 2$) produced by the extended source as a whole is $\zeta(t) = \int \zeta(t,\alpha)d\alpha$, and as the oscillations $\zeta_j(t,\alpha)$ are assumed to be delta-correlated in α, the intensity distribution over the interference pattern will be

$$\begin{aligned}
I(\tau) &= \tfrac{1}{2}\langle|\int[\zeta_1(t+\tau-\tau_\alpha,\alpha)+\zeta_2(t,\alpha)]^2 d\alpha|^2\rangle \\
&= \tfrac{1}{2}\int\langle|\zeta_1(t+\tau-\tau_\alpha,\alpha)+\zeta_2(t,\alpha)|^2\rangle d\alpha \\
&= \int[I_1(\alpha)+I_2(\alpha)+\operatorname{Re}B_{12}(\tau-\tau_\alpha,\alpha)]d\alpha \quad .
\end{aligned} \tag{2.108}$$

In other words, the intensities of the interference patterns due to the uncorrelated point elements will add up. The integration limits are determined by the interval of α where $I_1(\alpha)$ and $I_2(\alpha)$ are still nonzero, so that formally these limits can be extended to $\pm\infty$.

Since the spectral density, too, varies with α,

$$B_{12}(\tau,\alpha) = \int\limits_0^\infty g_{12}(\omega,\alpha)e^{i\omega\tau}d\omega \quad ,$$

we can write the variable (dependent on τ) part of the intensity (2.108) in the form

$$\begin{aligned}
I_{12}(\tau) &= \int\limits_{-\infty}^{+\infty}\operatorname{Re}\{B_{12}(\tau-\tau_\alpha,\alpha)\}d\alpha \\
&= \operatorname{Re}\left\{\int\limits_0^\infty d\omega\int\limits_{-\infty}^{+\infty}g_{12}(\omega,\alpha)\exp[i\omega(\tau-\tau_\alpha)]d\alpha\right\} \quad .
\end{aligned} \tag{2.109}$$

Of course, α is now not a random variable, but simply a parameter on which the elementary oscillations at the slits depend. Nonetheless, in the special case where all the elementary sources have the same spectrum, and at each frequency ω we have the same distribution $w_\alpha(\alpha)$ of the luminosity in angle, i.e., the spectral density factorizes

$$g_{12}(\omega,\alpha) = g_{12}(\omega)w_\alpha(\alpha) \quad .$$

Equation (2.109) is formally reduced to (2.107). In fact, in that case

$$I_{12}(\tau) = \operatorname{Re}\left\{\int\limits_0^\infty g_{12}(\omega)e^{i\omega\tau}d\omega\int\limits_{-\infty}^{+\infty}w_\alpha(\alpha)e^{-i\omega\tau_\alpha}d\alpha\right\} \quad .$$

Since $w_\alpha(\alpha)\geq 0$ and we can always normalize $w_\alpha(\alpha)$ to unity along the entire length of the source, the integral with respect to α (for source dimensions so small that $\tau_\alpha = d\sin\alpha/c \approx d\alpha/c$) coincides with the "characteristic function" of $w_\alpha(\alpha)$

$$\int\limits_{-\infty}^{+\infty} w_\alpha(\alpha)e^{-i\omega\tau_\alpha}\,d\alpha = \int\limits_{-\infty}^{+\infty} w_\alpha(\alpha)e^{-ik\,d\alpha}\,d\alpha = \varphi_\alpha(-kd) \quad,$$

$$k = \frac{\omega}{c} = \frac{2\pi}{\lambda} \quad,$$

and we thus arrive at an equation of the form (2.107).

It is clear without calculation that in the case of two point sources separated by an angular distance α_0,

$$w_\alpha(\alpha) = \frac{1}{2}\left[\delta\left(\alpha - \frac{\alpha_0}{2}\right) + \delta\left(\alpha + \frac{\alpha_0}{2}\right)\right] \quad.$$

The visibility is poorest when the interference fringes in both patterns are shifted by an odd number of angular half-widths of fringe. If the source fills the angular interval $(-\alpha_0/2, \alpha_0/2)$ with a constant brightness, so that the "characteristic function" is

$$\varphi_\alpha(u) = \frac{1}{\alpha_0} \int\limits_{-\alpha_0/2}^{\alpha_0/2} e^{-iu\alpha}\,d\alpha = \frac{\sin\,(u\alpha_0/2)}{u\alpha_0/2} \quad,$$

then (2.107) gives

$$K(\tau) = \frac{1}{B(0)} \int\limits_{0}^{\infty} g_\zeta(\omega)\frac{\sin\,(kd\alpha_0/2)}{kd\alpha_0/2}e^{i\omega\tau}\,d\omega \quad.$$

With ideal monochromaticity $[g_\zeta(\omega) \sim \delta(\omega - \omega_0)]$ we get

$$K(\tau) = \frac{\sin\,(\pi d\alpha_0/\lambda_0)}{\pi d\alpha_0/\lambda_0}e^{i\omega_0\tau} \quad, \qquad k_0 = \frac{2\pi}{\lambda_0} \quad. \tag{2.110}$$

Thus, the visibility of the entire interference pattern decreases with increasing α_0 or d, obeying the same law (2.104) as for a point source with increasing width Ω of its rectangular spectral band.

A similar situation occurs in the *two-dimensional* case where the source is seen at a certain *solid* angle, and the interferometer contains not slits but rather small holes (e.g., mirrors in the stellar Michelson interferometer, or reflector antennas in the astronomical radiointerferometer). For a circular monochromatic disk of uniform brightness and angular diameter α_0, the visibility in terms of the two-dimensional Fourier transform of the intensity distribution is

$$V = 2J_1\left(\frac{\pi d\alpha_0}{\lambda_0}\right)\Big/\frac{\pi d\alpha_0}{\lambda_0} \quad , \tag{2.111}$$

where J_1 is the first-order Bessel function. Increasing the base length d improves the resolving power in angle α_0. The first (and most distinct) zero of visibility in (2.110, 111) occurs at $d = \lambda_0/\alpha_0$ and $d = 1.22\lambda_0/\alpha_0$, respectively. In the first stellar interferometer installed at Mt. Wilson Observatory in 1920, the base d could be increased up to $6\,\mathrm{m}$, with the result that at $\lambda_0 = 5,400\,\mathrm{\mathring{A}}$ the interference pattern vanished at $\alpha_0 \approx 10^{-7} \approx 0.02''$. This enabled the angular diameters of six stars in the range from $0.02''$ to $0.05''$ to be measured.

Although *radiointerferometers* operate at much longer wavelengths, they allow higher angular resolutions to be achieved than in optical interferometers, due to the use of extremely long bases (hundreds or thousands of kilometers). For example, at $d = 1,000\,\mathrm{km}$ and $\lambda_0 = 1\,\mathrm{cm}$ we obtain $\alpha_0 \approx 2.10^{-8} \approx 2.10^{-3''}$. If the base is increased up to $10,000\,\mathrm{km}$, which is technically feasible, α_0 decreases by another order of magnitude. This high resolution will enable radiointerferometers to solve astronometrical problems such as the measurement of the parallax of quasars and pulsars, the angular diameters of quasars and extragalactic nebulae, etc.

The sensitivity to atmospheric fluctuations of the phase and angle of the incident light wave, the need for exceptional precision in manufacturing the instrument, as well as the accuracy of its training and tracking, all these are major limitations of the stellar Michelson interferometer, which measures the covariance of high-frequency (light) oscillations. These difficulties are circumvented in the so-called *intensity interferometer* suggested by *Brown* and *Twiss* in 1954 for radio astronomy [2.36] and for light [2.37], see also [2.38]. Let us consider its principle.

We have until now been concerned with the mean intensity $I(\tau)$ of the interference pattern that was dependent on the covariance of the oscillations $\zeta_1(t)$ and $\zeta_2(t)$. Is it possible to derive any information about the degree of coherence of $\zeta_1(t)$ and $\zeta_2(t)$ from the correlation of the *instantaneous intensities* $I_1(t) = |\zeta_1(t)|^2$ and $I_2(t) = |\zeta_2(t)|^2$? It might appear that the answer is negative, because these squares of moduli no longer contain the phases of $\zeta_1(t)$ and $\zeta_2(t)$, but only their amplitudes. Nevertheless, the correlation between $I_1(t)$ and $I_2(t)$ allows us, as we shall see, to make a judgement about the correlation of $\zeta_1(t)$ and $\zeta_2(t)$. Note that from now on we shall be dealing with a higher (fourth) moment, i.e., we will go beyond the framework of correlation theory.

The covariance of intensities is

$$\begin{aligned}
\psi_I(\tau) &= \overline{I_1(t+\tau)I_2(t)} - \overline{I_1(t+\tau)}\cdot\overline{I_2(t)}\\
&= \langle \zeta_1(t+\tau)\zeta_1^*(t+\tau)\zeta_2(t)\zeta_2^*(t)\rangle - I_1\cdot I_2 \quad ,
\end{aligned}$$

where I_1 and I_2 are mean intensities, and τ is still the relative time delay due to the path difference for both slits. If, as is often the case, $\zeta_1(t)$ and $\zeta_2(t)$ may be treated as *normal* random functions, then the fourth moment in the expression for $\psi_I(\tau)$ is equal to the sum of pairwise products of second moments (see Exercise 1.6.2).

$$\langle \zeta_1(t+\tau)\zeta_1^*(t+\tau)\zeta_2(t)\zeta_2^*(t)\rangle$$
$$= \langle \zeta_1(t+\tau)\zeta_1^*(t+\tau)\rangle\langle \zeta_2(t)\zeta_2^*(t)\rangle$$
$$+ \langle \zeta_1(t+\tau)\zeta_2(t)\rangle\langle \zeta_1^*(t+\tau)\zeta_2^*(t)\rangle$$
$$+ \langle \zeta_1(t+\tau)\zeta_2^*(t)\rangle\langle \zeta_1^*(t+\tau)\zeta_2(t)\rangle = I_1\cdot I_2 + |\tilde{B}_{12}(\tau)|^2 + |B_{12}(\tau)|^2 \quad,$$

where $B_{12}(\tau) = \langle \zeta_1(t+\tau)\zeta_2^*(t)\rangle$ and $\tilde{B}_{12}(\tau) = \langle \zeta_1(t+\tau)\zeta_2(t)\rangle$ are the "first" and "second" mutual covariances (Sect. 1.1). But with stationarily related analytical signals $\tilde{B}_{12}(\tau) = 0$, and thus the covariance of the intensities will be simply

$$\psi_I(\tau) = |B_{12}(\tau)|^2 = I_1 I_2 |K_{12}(\tau)|^2 \quad. \tag{2.112}$$

According to (2.99), measuring $\psi_I(\tau)$ gives directly the visibility $V(\tau)$.

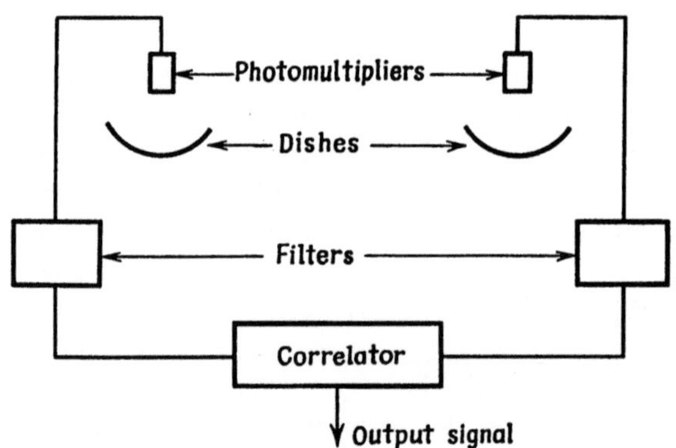

Fig. 2.10. Arrangement of the intensity interferometer of Brown and Twiss

The configuration of the intensity interferometer is shown in Fig. 2.10. Photomultipliers measure the intensity of light beams, or rather amplitude fluctuations. The transmission band is now controlled by electric filters, i.e., $\Omega/2\pi \sim 10^8$ Hz (while an optical filter with a 100 Å band is rated at $\Omega/2\pi \sim 10^{13}$ Hz). For definiteness, let us take a rectangular band, i.e., the relationship

$$V(\tau) = \left| \frac{\sin\,(\Omega\tau/2)}{\Omega\tau/2} \right|^2 \quad.$$

It is easily seen that a path difference of 30 cm ($\tau = 10^{-9}$ s) impairs the visibility by 2 % only. Hence high angular resolution can be achieved without the need for exceptional dimensional precision of the device. A second advantage is the insensitivity to the phases of light oscillations: only the phases of relatively low-frequency intensity fluctuations are important. Therefore, the performance of

the interferometer is essentially unaffected by atmospheric fluctuations, with
the fluctuations of angles of arrival not exceeding $1''$, and of arrival times,
10^{-13} s.

The disadvantages of the device include, first, its low sensitivity: the highest stellar magnitude still measurable with short exposures, is determined by the noise at the correlator input. Second, fairly high spectral densities (the number of photons per unit spectral range) which are only provided by hot stars, are required.

In conclusion, we point to another application of radio wave interference for measurements. The extension of interference to include radio range was prompted by the high monochromaticity of radio generators, the possibility of achieving good stability of their frequencies (reducing the technical drifts) and of mutually synchronizing widely separated transmitters. In the 1930s Mandelshtam and Papaleksi developed various radio interference methods for measuring the velocity of radio wave propagation under real conditions (and, given the velocity, for measuring distances). The methods use the additional potentialities of the radio-frequency range: the ease of transforming the radio-wave frequencies into simple fractional ratios, in relaying these waves, and the ease of directly measuring the *phase differences* of interfering oscillations, not of the modulus of the correlation coefficient (visibility) but its argument $\omega_0\tau + \Theta_{12}(\tau)$, see (2.96). One version of this method is in the location of a receiver using a hyperbolic graticule created on the ground by three synchronized transmitters. These techniques have found their application in geodesy, cartography, navigation and other fields (see [2.39]).

2.6 Nonstationary Interference. Source Correlation

With stationary and stationarily coupled oscillations $\zeta_{1,2}(t)$ the interference pattern is absolutely stable. It persists as the averaging (or accumulation) time T is arbitrarily increased and is described — if the processes $\zeta_{1,2}(t)$ are ergodic — by the statistical mean of the instantaneous intensity distribution (2.95). If the oscillations are nonstationary, or at least nonstationarily coupled, averaging over the realization ensemble no longer coincides with averaging over a sufficiently long time interval.

By (2.93, 94), the pattern observed for an averaging time T is described by

$$\widetilde{I_T(t,\tau)} = \frac{1}{2T} \int\limits_{t}^{t+T} [|\zeta_1(t+\tau)|^2 + |\zeta_2(t)|^2$$
$$+ 2\,\mathrm{Re}\,\{\zeta_1(t+\tau)\zeta_2^*(t)\}]dt \quad . \tag{2.113}$$

The intensity distribution over the screen (as a function of τ) now varies with t and the averaging time T.

Take, for example, the simple case of two harmonic oscillations of *different frequencies* $\zeta_1(t) = \exp(i\omega_0 t)$, $\zeta_2(t) = \exp[i(\omega_0 + \Omega)t]$. In this case, it follows from (2.113) that

$$\widetilde{I_T(t,\tau)} = \frac{1}{T} \int\limits_{t}^{t+T} [1 + \cos(\Omega t - \omega_0 \tau)]dt$$

$$= 1 + \frac{\sin(\Omega T/2)}{\Omega T/2} \cos\left(\Omega t - \omega_0 \tau + \frac{\Omega T}{2}\right) \quad . \tag{2.114}$$

It is seen that fringes recurrent in τ with period $2\pi/\omega_0$ are not stationary; they travel in τ with a velocity equal to

$$d\tau/dt = \Omega/\omega_0 \quad .$$

The visibility of these fringes is $|\sin(\Omega T/2)/(\Omega T/2)|$, i.e., it is determined by the product of the averaging time T and the beat frequency Ω. For $\Omega T \ll 2\pi$, the visibility is close to unity. The underlying idea of this averaging is that the fringe shift during T, equal (in τ) to $T(d\tau/dt) = \Omega T/\omega_0$, must be considerably smaller than the period $2\pi/\omega_0$. Otherwise, the pattern will be smeared out with the result that the intensity distribution will become uniform. Thus, (2.114) describes the continuous transition from complete coherence as $\Omega T \to 0$ to complete incoherence as $\Omega T \to \infty$.

It should be stressed that when coherence is partial ($\Omega T \neq 0$), we observe *moving* fringes. Hence, for total exposure to be unifrom, the pattern need not necessarily be fixed. The key point is the relationship between device response (time T) and nonmonochromaticity (here, the difference Ω between the two frequencies). The eye ($T \sim 0.1\,\text{s}$) may fail to detect fringes but a camera, say, with an exposure $T = 0.01\,\text{s}$ can register them.

Clearly, this applies to any modulated oscillations, both deterministic and random. If

$$\zeta_j(t) = A_j(t) \exp\{i[\omega_0 t + \varphi_j(t)]\} \quad , \quad j = 1,2 \quad , \tag{2.115}$$

where A_j and φ_j differ only slightly during a high-frequency period $T_0 = 2\pi/\omega_0$, then the difference $\varphi_{1\tau} - \varphi_2 \equiv \varphi_1(t + \tau) - \varphi_2(t)$ will also be a slowly varying function of t. Taking for simplicity $A_1 = A_2 = 1$ (the interference pattern is especially sensitive to phases) and discarding the terms varying in time t with frequency $2\omega_0$, we obtain for instantaneous intensity

$$I(t,\tau) = 1 + \cos(\omega_0 \tau + \varphi_{1\tau} - \varphi_2) \quad .$$

Synchronously with the variation of $\varphi_{1\tau} - \varphi_2$, the fringes will shift to and fro and/or travel in one direction with the instantaneous velocity

$$\frac{d\tau}{dt} = \frac{\dot{\varphi}_2 - \dot{\varphi}_{1\tau}}{\omega_0 + \dot{\varphi}_{1\tau}} \approx \frac{\dot{\varphi}_2 - \dot{\varphi}_{1\tau}}{\omega_0} \quad .$$

The condition that the shift be small during time T will clearly reduce to the condition that the change in $\varphi_2 - \varphi_{1\tau}$ during T be far smaller than 2π.

To sum up: in the general case it is advisable to relate the *very definition* of the notion of coherence to the time constant of a device. Such a definition is both more general and more flexible, as it does not assume that the interference pattern is absolutely unchangeable. If, however, we deal with a chaotic (random) modulation of interfering oscillations, then, as always in such cases, the *statistical* behavior of the phenomenon observed is of interest. Its calculation certainly requires an ensemble of realizations obtained either in the same time interval $(t, t + T)$ using a large number of identical interferometers, or, which may be more realistic, using one interferometer but in successive time intervals $(t_i, t_i + T)$ with the same conditions reproduced at each initial time t_i.

We may either first find the ensemble average of (2.89) and then perform the time averaging, or proceed the other way round. In the first case we will get

$$\langle \xi^2(t, \tau) \rangle = \tfrac{1}{4}(2B_1(t + \tau, t + \tau) + 2B_2(t, t) + 2\,\mathrm{Re}\,\{B_{12}(t + \tau, t)\}$$
$$+ [\tilde{B}_1(t + \tau, t + \tau) + \tilde{B}_2(t, t)$$
$$+ 2\,\mathrm{Re}\,\{\tilde{B}_{12}(t + \tau, t)\} + \text{complex conjugate}]) \quad . \qquad (2.116)$$

The "second" moments $\tilde{B}$ that enter into the expression are now distinct from zero, but, unlike the "first" moments B, they contain oscillations with frequency $2\omega_0$, for instance,

$$\tilde{B}_1(t + \tau, t + \tau) = \overline{A_{1\tau}^2} \exp\,(2i\omega_0 t) \quad ,$$
$$\tilde{B}_{12}(t + \tau, t) = \overline{A_{1\tau} A_2} \exp\,[i\omega_0(2t + \tau)] \quad .$$

Therefore, averaging (2.116) over the interval $T \gg T_0 = 2\pi/\omega_0$ practically eliminates the terms with $\tilde{B}$ to yield

$$\widetilde{\langle I(t, \tau) \rangle_T} = \frac{1}{2T} \int\limits_{t}^{t+T} [B_1(t + \tau, t + \tau) + B_2(t, t)$$
$$+ 2\,\mathrm{Re}\,\{B_{12}(t + \tau, t)\}]dt \quad . \qquad (2.117)$$

The same result is also obtained with the reverse sequence of operations, i.e., by ensemble-averaging (2.113). Whether or not there will be an interference pattern depends on the last term in (2.117), i.e., on averaging over T of a mixed moment. The contrast (visibility) of fringes may be determined as for stationary oscillations in Sect. 2.5. In the latter case all the moments B are independent of t so that the wavy line is no longer necessary (*moving* averaging changes nothing at all, and accumulation comes down to multiplication by T)[17]. In other words, (2.117) becomes (2.95).

---

[17] Recall that the "second" moments $\tilde{B}$ for stationary analytical signals become zero in their own right, i.e., their absence in (2.117) has nothing to do with the condition that $T \gg T_0$.

Finally, consider the "mechanism" of those statistical phenomena in oscillation *sources* that are responsible for the incoherence of these sources. By a source we mean a dynamic system in which – apart from possible deterministic actions and interactions – random actions and interactions also occur. The latter may be inherently unremovable (one example is natural fluctuations in the Thomson vacuum tube oscillator, see Sect. I.5.8). The forces and interactions vary in nature. It may be that constituent systems of the source are connected in a deterministic way, but the random forces acting on them are uncorrelated. Nevertheless, the existence of a deterministic connection may introduce a mutual correlation into oscillations of the constituent systems. An example is coupled oscillators, each of which is acted upon by a force uncorrelated with that applied to the other, see Exercise 2.8.7. Also, the very excitation of oscillations in the constituent systems may be due to random interactions between them, as exemplified by collisions in a gas of oscillator-"atoms", see Exercise 2.8.3. There is such a rich variety of conditions which may be encountered that it is hardly possible to give an exhaustive classification.

We will only consider one illustration of how the coherence of the oscillations of two coupled sources deteriorates. Our treatment will for the most part be qualitative, the problem formulation being simplified as much as possible. We will consider two coupled *Thomson self-oscillatory systems* of the type discussed in Sect. I.5.8. Unlike the two coupled oscillators of Exercise 2.8.7, i.e., *linear dissipative* systems in which oscillations are *supported* by fluctuating forces, we will here concern ourselves with *nonlinear self-oscillatory* systems. Oscillations also occur in the absence of fluctuating forces (the former being ideally monochromatic), whose role is only to produce *nonmonochromaticity*, i.e., to predetermine the *finite* duration of the coherent train.

Instead of (I.5.94) for a single oscillator

$$\ddot{x} + x = \mu[\dot{x}(p - \tfrac{3}{4}\dot{x}^2) + F(t)] \quad , \tag{2.118}$$

we now have the system of two equations for two identical, weakly (of the order of the small parameter μ) coupled oscillators

$$\ddot{x} + x = \mu[\kappa y + \dot{x}(p - \tfrac{4}{3}\dot{x}^2) + F_1(t)] \quad ,$$
$$\ddot{y} + y = \mu[\kappa x + \dot{y}(p - \tfrac{4}{3}\dot{y}^2) + F_2(t)] \quad . \tag{2.119}$$

The stationary white noises $F_1(t)$ and $F_2(t)$ are mutually uncorrelated

$$\langle F_1(t)F_1(t')\rangle = C_1\delta(t - t') \quad ,$$

$$\langle F_2(t)F_2(t')\rangle = C_2\delta(t - t') \quad ,$$

$$\langle F_1(t)F_2(t')\rangle = 0 \quad .$$

Since a complete presentation of the theory of self-oscillatory system (2.119) lies beyond the scope of the book, we will confine ourselves to those results that are

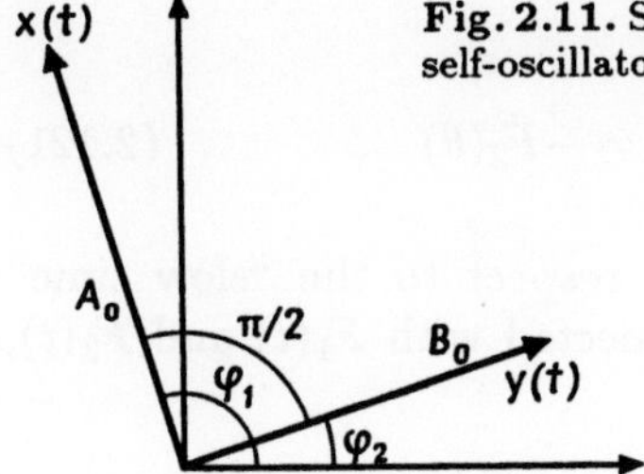

Fig. 2.11. Stationary dynamic mode of two identical coupled Thomson self-oscillatory systems

directly related to the discussion of the issue of interest to us. Note, however, that for small fluctuations about established self-oscillations $x(t)$ and $y(t)$, the *correlation* theory of self-oscillations can be utilized. This was first done in [2.40] as applied to a single self-oscillatory system, i.e., equation (2.118) (see Sect. 3.4). The generalization of a similar theory to more involved Thomson-type self-oscillatory systems, in particular to coupled self-oscillators (2.119), poses no major difficulties. A number of such problems have been worked in detail in [2.41].

Figure 2.11 illustrates stationary (noiseless) dynamic modes. In the Van der Pol plane the vectors of both oscillations are shown

$$x(t) = A_0 \cos (t + \varphi_1) \quad ,$$
$$y(t) = B_0 \cos (t + \varphi_2) \quad .$$

Two equally justified stable modes are possible such that the phase difference is $\Phi = \varphi_1 - \varphi_2 = \pm\pi/2$ and the ratio $\lambda \equiv A_0/B_0 \gtrless 1$, respectively. The equality $\lambda = 1$ only occurs in the absence of any coupling ($\kappa = 0$), and then the angle Φ is arbitrary, (i.e., it is solely determined by the initial conditions). To be more specific, we select the mode with $\Phi = +\pi/2$ ($\lambda \geq 1$), in which A_0 and B_0 are given by

$$A_0(p - A_0^2) + \kappa B_0 = 0 \quad ,$$
$$B_0(p - B_0^2) - \kappa A_0 = 0 \quad . \tag{2.120}$$

The oscillators are assumed to be excited ($p>0$). If there is no coupling ($\kappa = 0$), we will then have $A_0 = B_0 = \sqrt{p}$ (Sect. I.5.8).

Given the presence of fluctuating forces $F_1(t)$ and $F_2(t)$, we will for simplicity completely ignore *amplitude* fluctuations, assuming A_0 and B_0 to be constant and equal to their dynamic values. As to the phases, we will introduce their deviations α_1 and α_2 in reference to the initial phases φ_{10} and φ_{20} at $t = 0$

$$\varphi_1 = \varphi_{10} + \alpha_1 \quad , \quad \varphi_2 = \varphi_{20} + \alpha_2 \quad .$$

In so doing, we will impose on the initial values the condition of the stationary dynamic relation: $\Phi_0 = \varphi_{10} - \varphi_{20} = \pi/2$. Then α_1 and α_2 ($\alpha = \alpha_1 - \alpha_2 =$

$\Phi - \pi/2$) obey the equations

$$\alpha_1' - \frac{\kappa}{2\lambda}\sin\alpha = -\tilde{F}_1(\theta) \quad , \quad \alpha_2' - \frac{\kappa\lambda}{2}\sin\alpha = -\tilde{F}_2(\theta) \quad , \tag{2.121}$$

where the prime stands for differentiation with respect to the "slow time" $\theta = \mu t$. The forces $\tilde{F}_1(\theta)$ and $\tilde{F}_2(\theta)$, linearly connected with $F_1(t)$ and $F_2(t)$, are also mutually uncorrelated white noises

$$\langle \tilde{F}_1(\theta)\tilde{F}_1(\theta')\rangle = \frac{\mu C_1}{2A_0^2}\delta(\theta - \theta') \quad ,$$

$$\langle \tilde{F}_2(\theta)\tilde{F}_2(\theta')\rangle = \frac{\mu C_2}{2B_0^2}\delta(\theta - \theta') \quad ,$$

$$\langle \tilde{F}_1(\theta)\tilde{F}_2(\theta')\rangle = 0.$$

Subtracting the second of equations (2.121) from the first, we obtain for the fluctuations of the angle between the vectors $\boldsymbol{x}$ and $\boldsymbol{y}$ the following equation:

$$\alpha' + \gamma\sin\alpha = \tilde{F}_2(\theta) - \tilde{F}_1(\theta) \quad , \quad \gamma \equiv \frac{\kappa}{2}\left(\lambda - \frac{1}{\lambda}\right) \geq 0 \quad . \tag{2.122}$$

With strong coupling (large κ) the fluctuations of α are small, so that (2.122) can be linearized ($\sin\alpha \approx \alpha$)

$$\alpha' + \gamma\alpha = \tilde{F}_2(\theta) - \tilde{F}_1(\theta) \quad .$$

In that case, $\boldsymbol{x}$ and $\boldsymbol{y}$ behave as follows (Fig. 2.11). The angle $\Phi = \pi/2 + \alpha$ between them fluctuates as if the vectors were linked by a spring (with an equilibrium state at $\alpha = 0$) and moved in a viscous medium with a friction coefficient γ. At the same time, the entire couple of the vectors $\boldsymbol{x}$ and $\boldsymbol{y}$ rotates in a random (diffusive) way and may drift indefinitely from its initial position. As the coupling weakens (κ reduces), so does the "spring" linking the vectors $\boldsymbol{x}$ and $\boldsymbol{y}$ and the fluctuations of the angle α increase, i.e., the diffusion rotation of one vector will depend less and less on the position of the other. In the long run, at $\kappa = 0$, each vector undergoes a diffusion rotational motion of its own, as must be the case with uncoupled self-oscillatory systems.

We are interested in the interference pattern produced by the superposition of oscillations $x(t)$ and $y(t)$, i.e., in the intensity distribution[18]

$$\widetilde{I_{xy}(t,\tau)_T} = \tfrac{1}{2}\overline{[x(t+\tau) + y(t)]^2} = \tfrac{1}{2}\{A_0^2 + B_0^2 + 2A_0 B_0 \widetilde{\cos\,(\tau + S)}\} \quad ,$$

where

---

[18] We assume that $\tau \geq 0$. Otherwise, in the variance equations below we would need to write $|\tau|$ instead of τ.

$$S = \varphi_{1\tau} - \varphi_2 = \varphi_{1\tau} - \varphi_1 + (\varphi_1 - \varphi_2) = \varphi_{1\tau} - \varphi_1 + \tfrac{\pi}{2} + \alpha \quad .$$

The subscript τ means, as usual, that the quantity is taken at time $t + \tau$, and the wavy line (and the subscript T) indicate that the averaging is over the time interval T. With equations (2.121, 122) linearized in α the fluctuations α_1, α_2 and $\alpha = \alpha_1 - \alpha_2$ are normal processes, so that S, too, is a Gaussian variable (with mean $\overline{S} = \tfrac{\pi}{2}$). Therefore,

$$\overline{\cos\,(\tau + S)} = \overline{\cos\,[\tau + \overline{S} + (S - \overline{S})]}$$
$$= \cos\,(\tau + \overline{S})e^{-D[S]/2} = -e^{-D[S]/2}\,\sin\,\tau \quad ,$$

and the intensity is

$$\widetilde{\overline{I_{xy}(t,\tau)}}_T = \tfrac{1}{2}\{A_0^2 + B_0^2 - 2A_0 B_0 \widetilde{e^{-D[S]/2}}\,\sin\,\tau\} \quad . \tag{2.123}$$

It is interesting to compare (2.123) with the pattern produced by either of the two coupled oscillators, say by $x(t)$:

$$\widetilde{\overline{I_x(t,\tau)}}_T = \tfrac{1}{2}\widetilde{\overline{[x(t+\tau) + x(t)]^2}} = A_0^2[1 + \widetilde{\overline{\cos\,(\tau + S)}}] \quad ,$$

where now $S = \varphi_{1\tau} - \varphi_1$ and $\overline{S} = 0$. Linearizing the equations for phases, we arrive at

$$\widetilde{\overline{I_x(t,\tau)}}_T = A_0^2[1 + \widetilde{e^{-D[S]/2}}\,\cos\,\tau] \quad . \tag{2.124}$$

In both cases the problem reduces to calculating $D[S]$. We will provide the result of the calculation using the linearized phase equations and not too large delays τ of one of the two oscillations (namely, we will assume that $\gamma\vartheta \ll 1$, where $\vartheta = \mu\tau$ is the delay in the slow time θ). This suggests that we should consider the fringes of the interference pattern that are not very distant from its center ($\tau = 0$). We thus obtain for I_{xy}

$$D_{xy}[S] = \frac{p(\lambda^2 - 1)}{(\lambda^2 + 1)\mu\gamma}$$
$$\times \left\{(\lambda^2 - 1 + \gamma\vartheta)\left(\frac{\mathcal{D}_1}{\lambda^2} + \mathcal{D}_2\right)\frac{1 - e^{-2\gamma\theta}}{2\gamma} + \frac{\lambda^2 - 1}{\lambda^2}\mathcal{D}_1\vartheta\right\} \quad ,$$
$$\tag{2.125}$$

and for I_x

$$D_x[S] = \frac{p}{(\lambda^2 + 1)\mu\gamma}\left\{\gamma^2\vartheta^2\left(\frac{\mathcal{D}_1}{\lambda^2} + \mathcal{D}_2\right)\frac{1 - e^{-2\gamma\theta}}{2\gamma} + \frac{(\lambda^2 - 1)^2}{\lambda^2}\mathcal{D}_1\vartheta\right\} \quad . \tag{2.126}$$

Here $\mathcal{D}_j = \mu^2 C_j/4p$ $(j = 1, 2)$ are phase diffusion coefficients for each oscillator in the absence of coupling.

But given the presence of coupling ($\kappa\neq0$) we have $\lambda\neq1$ and $\gamma\neq0$. Thus, with increasing time $\theta = \mu t$ the exponential function $\exp(-2\gamma\theta)$ decays and stationary variances $D^{(\infty)}[S]$ appear. Moreover, in (2.126) we may ignore the first term in the curly brackets as it is small[19]

$$D_{xy}^{(\infty)}[S] = \frac{p(\lambda^2 - 1)}{(\lambda^2 + 1)\gamma^2}\left\{\frac{\lambda^2 - 1}{\mu}\left(\frac{\mathcal{D}_1}{\lambda^2} + \mathcal{D}_2\right) + \left[(2\lambda^2 - 1)\frac{\mathcal{D}_1}{\lambda^2} + \mathcal{D}_2\right]\gamma\tau\right\} \quad , \tag{2.127}$$

$$D_{x}^{(\infty)}[S] = \frac{\lambda^4 + 1}{\lambda^2(\lambda^2 + 1)}\cdot 2\mathcal{D}_1\tau \quad . \tag{2.128}$$

Strictly speaking, in (2.125–128) we are not allowed to go to the decoupled case ($\kappa \to 0$), as this deprives us of the possibility of linearizing the phase equations in α. Qualitatively, however, the results remain meaningful.

At small κ we have $\lambda^2 - 1 \approx 2\kappa/p$ and $\gamma = \kappa^2/p$, so that (2.127, 128) take the form

$$D_{xy}^{(\infty)}[S] \approx \frac{p}{2\kappa^2}(\mathcal{D}_1 + \mathcal{D}_2)\left(\frac{2}{\mu} + \kappa\tau\right) \quad , \quad D_{x}^{(\infty)}[S] \approx 2\mathcal{D}_1\tau \quad .$$

Owing to the factor $p/2\kappa^2$ and the summand $2/\mu$, the stationary variance $D_{xy}^{(\infty)}[S]$ is extremely large at any shifts τ, including the center of the interference pattern ($\tau = 0$). This means, according to (2.123), that there are practically no fringes. However, in the self-interference pattern of a single oscillator (2.124), the fringes ($\cos\tau$) are inscribed into an exponentially decaying envelope $\exp(-2\mathcal{D}_1/\tau)$ [with an arbitrary sign for τ implied, we write $\exp(-2\mathcal{D}_1/|\tau|)$]. In that case a very weak coupling with the other oscillator only introduces negligible corrections, which vanish at $\kappa = 0$.

At $\kappa = 0$ we have $D_{xy}^{(\infty)} = \infty$, but this "stationary value" is only achieved during an infinitely long relaxation time $\theta \sim 1/2\gamma \to \infty$. For any finite t (and $\theta = \mu t$) the transition to uncoupled oscillators, i.e., to $\kappa = 0$ in (2.125, 126), gives

$$D_{xy}[S] = 2(\mathcal{D}_1 + \mathcal{D}_2)t + 2\mathcal{D}_1\tau = 2\mathcal{D}_1(t + \tau) + 2\mathcal{D}_2t \quad , \tag{2.129}$$

$$D_{x}[S] = 2\mathcal{D}_1\tau \quad . \tag{2.130}$$

The *self*-interference is thus stationary and the same as above, but the interference *between* the oscillators is *nonstationary* and is easily seen to be due to the independent phase diffusion of both oscillators: of the first during time $t + \tau$, and of the second during time t.

---

[19] In the calculations the relationships between $\lambda = A_0/B_0$ and γ, κ, and p that follow from (2.120) and (2.122) were used.

If the wavy line in (2.123) is taken to mean accumulative operation during the time from 0 to T, then, by (2.129) and (2.123), we obtain (since at $\kappa = 0$ the amplitudes are the same, $B_0 = A_0$)

$$\widetilde{I_{xy}(t,\tau)}_T = TA_0^2 \left\{ 1 - \frac{1 - \exp\left[-(\mathcal{D}_1 + \mathcal{D}_2)T\right]}{(\mathcal{D}_1 + \mathcal{D}_2)T} e^{-\mathcal{D}_1\tau} \sin \tau \right\} \ .$$

The pattern is the same as with a single oscillator (save for the shift by a half-fringe). It vanishes once the accumulation time T has appreciably exceeded the total diffusion time $1/(\mathcal{D}_1 + \mathcal{D}_2)$.

Lastly, it is quite trivial to point out that, when the random forces are "switched off", the phase diffusion stops and the oscillations become strictly monochromatic ($\mathcal{D}_1 = \mathcal{D}_2 = 0$), both interference patterns (2.123, 124) are stationary and feature the constant visibilities $V = 2A_0B_0/(A_0^2 + B_0^2)$ and $V = 1$, respectively. The strictly monochromatic oscillations know nothing about whether or not their sources are independent.

2.7 Statistical Properties of Polarization of Modulated Oscillations

In handling the interference phenomena it was tacitly assumed that in the case of transverse (in particular, electromagnetic) waves, polarization of the oscillations being added was the same. This allowed us to use the *scalar* form of the interference theory of modulated oscillations. In the general case of transverse waves it is necessary to consider the *vector* oscillation with the vector $\boldsymbol{\xi}(t)$ lying in the plane perpendicular to the wave normal. Introducing the Cartesian coordinates $x = x_1, y = x_2$ into that plane, we can write the projections of $\boldsymbol{\xi}(t)$, i.e., two mutually orthogonal modulated oscillations

$$\xi_j(t) = A_j(t) \cos\left[\omega_0 t + \varphi_j(t)\right] \ , \quad j = 1,2 \ . \tag{2.131}$$

With ideal monochromaticity (at constant A_j and φ_j), the polarization is perfect: the vector $\boldsymbol{\xi}$, rotating with angular frequency ω_0, circumscribes an ellipse with the semiaxis ratio

$$\pm\frac{b}{a} = \tan\chi \ , \quad \sin 2\chi = \frac{2A_1A_2 \sin(\varphi_1 - \varphi_2)}{A_1^2 + A_2^2} \ , \tag{2.132}$$

the angle γ between the semimajor axis a and the axis x_1 being obtainable from the equation

$$\tan 2\gamma = \frac{2A_1A_2 \cos(\varphi_1 - \varphi_2)}{A_1^2 - A_2^2} \ . \tag{2.133}$$

If $\varphi_1 - \varphi_2 = n\pi$, where n is an integer, then the polarization is linear. At $\varphi_1 - \varphi_2 = (2n+1)\pi/2$ the principal axes of the ellipse coincide with the x_1 and x_2 axes. If, moreover, $A_1 = A_2$, the polarization is circular.

In the case of modulated oscillations, when A_j, φ_j vary with time in a deteministic or random manner, the various kinds of polarization alternate in a deterministic or random manner, respectively.

In the vector case it is also convenient to go over from real processes $\xi_j(t)$ to analytical signals,

$$\zeta_j(t) = A_j(t)\exp\{i[\omega_0 t + \varphi_j(t)]\} = \boldsymbol{A}_j(t)e^{i\omega_0 t} \quad , \quad j = 1,2 \quad , \qquad (2.134)$$

where $\boldsymbol{A}_j(t) = A_j(t)\exp[i\varphi_j(t)]$ are complex amplitudes. From the very beginning we assume that $\xi_j(t)$, and hence $\zeta_j(t)$ too, are *stationary and stationarily connected* random processes. Their joint description within the framework of correlation theory is given by the correlation matrix

$$\psi_{\zeta jk}(\tau) = \langle\zeta_j(t+\tau)\zeta_k^*(t)\rangle = \langle\boldsymbol{A}_{j\tau}\boldsymbol{A}_k^*\rangle e^{i\omega_0\tau} = \psi_{Ajk}(\tau)e^{i\omega_0\tau}$$

which characterizes both the average *polarization* properties of the vector $\zeta(t)$, and its *spectral* behavior, since in the stationary case the spectral density matrix $g_{\xi jk}(\omega)$ exists

$$\psi_{\zeta jk}(\tau) = 4\int_0^\infty g_{\xi jk}(\omega)e^{i\omega\tau}d\omega \quad , \quad \psi_{Ajk}(\tau) \approx 4\int_{-\infty}^{+\infty} \tilde{g}_{jk}(\Omega)e^{i\Omega\tau}d\Omega \quad .$$

Here the function $\tilde{g}_{jk}(\Omega) = g_{\xi jk}(\omega_0 + \Omega)$ is only nonzero in the neighborhood of $\Omega = 0$, which is narrow as compared with ω_0.

But if we are interested in the polarization properties of $\zeta(t)$ alone, then we can in many cases confine ourselves to a quasi-stationary (or rather quasi-static) approximation in which time delays τ are far shorter than the length of the coherent train $2\pi/\Omega_m$ (Ω_m is the band width where $\tilde{g}_{jk}(\Omega)$ is markedly distinct from zero). Thus, $\omega_0\tau \ll 2\pi\omega_0/\Omega_m$, which does not, of course, exclude the possibility of having $\omega_0\tau \gg 2\pi$, i.e., the shift τ may include many periods of the high frequency ω_0. In such short times τ the complex amplitudes $\boldsymbol{A}_j(t)$, and thereby polarization as well, remain almost unchanged. Therefore, to a first approximation we can totally ignore these variations and describe the polarization by mean values of pairwise products of $\boldsymbol{A}_j(t)$ taken at the *same* instant of time

$$\begin{aligned} J_{jk} &= \langle\boldsymbol{A}_j(t)\boldsymbol{A}_k^*(t)\rangle \\ &= \langle A_j(t)A_k(t)\exp\{i[\varphi_j(t) - \varphi_k(t)]\}\rangle \quad , \quad j,k = 1,2 \quad . \end{aligned} \qquad (2.135)$$

Variables J_{jk} form the so-called *polarization matrix (tensor)*

$$\hat{J} = \begin{pmatrix} \langle A_1^2\rangle & \langle A_1 A_2 e^{i(\varphi_1 - \varphi_2)}\rangle \\ \langle A_1 A_2 e^{-i(\varphi_1 - \varphi_2)}\rangle & \langle A_2^2\rangle \end{pmatrix} \quad . \qquad (2.136)$$

The matrix is Hermitian ($J_{21} = J_{12}^*$) and its trace and determinant are readily seen to be real, nonnegative

$$\mathrm{tr}\{\hat{J}\} = J_{11} + J_{22} = \langle A_1^2 \rangle + \langle A_2^2 \rangle \equiv 2J_0 \geq 0 \quad ,$$

$$|\hat{J}| = J_{11}J_{22} - J_{12}J_{21} \geq 0 \quad , \tag{2.137}$$

and invariant under any orthogonal rotations of the axes in the plane (x_1, x_2). Here I_0 denotes the total intensity of the vector oscillation $\xi(t)$:

$$I_0 = \overline{\xi_1^2} + \overline{\xi_2^2} = \tfrac{1}{2}(\overline{|\zeta_1|^2} + \overline{|\zeta_2|^2}) = \tfrac{1}{2}(\overline{A_1^2} + \overline{A_2^2})$$
$$= \tfrac{1}{2}(J_{11} + J_{22}) \equiv \tfrac{1}{2}\,\mathrm{tr}\{\hat{J}\} \quad . \tag{2.138}$$

Now we will find the average intensity due to both oscillations if we projected them on a certain direction $N = (\cos\theta,\ \sin\theta)$ in the plane (x_1, x_2)[20]. Assume that one of the oscillations, say $\zeta_1(t)$, has been delayed by a time $\tau \ll 2\pi/\Omega_m$, i.e., it lags behind in phase by $\varepsilon = \omega_0 \tau$. As noted above, we do not take account of this small delay in the amplitude $A_1(t)$. Then

$$I(\theta, \varepsilon) = \tfrac{1}{2}\langle |A_1(t)e^{i\varepsilon}\cos\theta + A_2(t)\sin\theta|^2 \rangle$$
$$= \tfrac{1}{2}\{J_{11}\cos^2\theta + J_{22}\sin^2\theta + (J_{12}e^{i\varepsilon} + J_{21}e^{-i\varepsilon})\sin\theta\cos\theta\} \quad . \tag{2.139}$$

If we introduce the coefficient of mutual correlation

$$\mu_{12} = \frac{J_{12}}{\sqrt{J_{11}J_{22}}} = |\mu_{12}|\exp(i\beta_{12}) \quad , \tag{2.140}$$

then

$$J_{12}e^{i\varepsilon} + J_{21}e^{-i\varepsilon} = J_{12}e^{i\varepsilon} + J_{12}^*e^{-i\varepsilon}$$
$$= 2\sqrt{J_{11}J_{22}}|\mu_{12}|\cos(\beta_{12} + \varepsilon), \tag{2.141}$$

hence

$$I(\theta, \varepsilon) = \tfrac{1}{2}\{J_{11}\cos^2\theta + J_{22}\sin^2\theta$$
$$+ 2\sqrt{J_{11}J_{22}}|\mu_{12}|\cos(\beta_{12} + \varepsilon)\sin\theta\cos\theta\} \quad . \tag{2.142}$$

This expression can be presented in the form (2.92) if we introduce the intensities $I_1 = J_{11}\cos^2\theta$ and $I_2 = J_{22}\sin^2\theta$, which correspond to the amplitudes $A_1\cos\theta$ and $A_2\sin\theta$ of the projections of each of these oscillations on the direction N. A significant distinction from (2.92) is that there the correlation coefficient K_{12} was a function of τ (we were interested in the influence on the interference of the *spectrum* of the oscillations added), but in (2.141) μ_{12} is a complex *number* which characterizes the correlation of simultaneous values of amplitudes and phases.

---

[20] In optics the operation is carried out by a Nicol's prism or polaroid, in radio by a linear antenna, e.g., a linear vibrator.

Completely *unpolarized* oscillation (in optics, natural light) manifests axial symmetry, i.e., the intensity $I(\theta, \varepsilon)$ is independent of the angle θ and remains unchanged as the delay in phase ε varies (subject, of course, to the above condition that $\varepsilon \ll 2\pi\omega_0/\Omega_m$). It is clear from (2.141) that this is only possible at $J_{11} = J_{22}$ and $\mu_{12} = 0$ ($J_{12} = J_{21} = 0$). Since $J_{11} + J_{22} = 2I_0$, where I_0 is the total intensity, the polarization matrix takes the form

$$\hat{J} = I_0 \begin{pmatrix} 1 & 0 \\ 0 & 1 \end{pmatrix} \; . \tag{2.142}$$

In the opposite case of a *monochromatic* (and hence *completely polarized*) oscillation we can remove the averaging sign $\langle \ldots \rangle$ in (2.136) (since all the realizations are identical). Then we obviously have

$$|\hat{J}| = J_{11}J_{22} - J_{12}J_{21} = 0 \quad , \quad \mu_{12} = e^{i(\varphi_1 - \varphi_2)} \quad , \tag{2.143}$$

so that $|\mu_{12}| = 1$. These relationships also hold in the special case of incomplete monochromaticity where

$$A_2(t)/A_1(t) = \text{const} \quad , \quad \varphi_1(t) - \varphi_2(t) = \text{const} \quad .$$

Under these conditions it is only the size of the polarization ellipse (i.e., the total intensity) that changes; its excentricity and slope remain unchanged.

With linear polarization ($\varphi_1 - \varphi_2 = n\pi$) all the elements of $\hat{J}$ are real, and with circular polarization

$$\hat{J} = I_0 \begin{pmatrix} 1 & \pm i \\ \pm i & 1 \end{pmatrix} \; , \tag{2.144}$$

where the upper sign corresponds to the right polarization, and the lower to the left.

Clearly, with superimposed *uncorrelated* oscillations $[\zeta_j(t) = \sum_n \zeta_j^{(n)}(t),\ j = 1, 2]$ the polarization matrices add up

$$J_{jk} = \sum_n J_{jk}^{(n)} \; .$$

Generally speaking, the inverse expansion of $\zeta_j(t)$ into a sum of uncorrelated oscillations is not unique. For example, the polarization matrix for a completely unpolarized oscillation can be represented as the sum

$$\begin{pmatrix} 1 & 0 \\ 0 & 1 \end{pmatrix} = \begin{pmatrix} 1 & 0 \\ 0 & 0 \end{pmatrix} + \begin{pmatrix} 0 & 0 \\ 0 & 1 \end{pmatrix} \; ,$$

i.e., $\zeta_j(t)$ can be expanded into uncorrelated oscillations linearly polarized along the x_1 and x_2 axes. Or it can be written

$$\begin{pmatrix} 1 & 0 \\ 0 & 1 \end{pmatrix} = \frac{1}{2} \begin{pmatrix} 1 & i \\ -i & 1 \end{pmatrix} + \frac{1}{2} \begin{pmatrix} 1 & -i \\ i & 1 \end{pmatrix} ,$$

which corresponds to the sum of two uncorrelated, right and left, circularly polarized oscillations. But the expansion can be made unique with some additional constraints.

An important expansion of this kind is the representation of a given vector oscillation as a sum of two uncorrelated oscillations: completely unpolarized and completely polarized, so that

$$\hat{J} = \hat{J}^{(u)} + \hat{J}^{(p)} \quad , \quad \hat{J}^{(u)} = \begin{pmatrix} A & 0 \\ 0 & A \end{pmatrix} \quad , \quad \hat{J}^{(p)} = \begin{pmatrix} B & D \\ D^* & C \end{pmatrix} \quad (2.145)$$

where A, B, and C are nonnegative and, by (2.143), we have $BC - DD^* = 0$. According to (2.145) the components are

$$J_{11} = A + B \quad , \quad J_{12} = D \quad , \quad J_{21} = D^* \quad , \quad J_{22} = A + C \quad ,$$

hence

$$\begin{aligned} BC - DD^* &= (J_{11} - A)(J_{22} - A) + J_{12}J_{21} \\ &= A^2 - (J_{11} + J_{22})A + |\hat{J}| = 0 \quad . \end{aligned}$$

Of the two roots of this equation for A (both of them are easily seen to be real and nonnegative) we select the root

$$\begin{aligned} A &= \frac{1}{2}\left\{ J_{11} + J_{22} - \sqrt{(J_{11} + J_{22})^2 - 4|\hat{J}|} \right\} \\ &= \frac{1}{2}\left\{ J_{11} + J_{22} - \sqrt{(J_{11} - J_{22})^2 + 4J_{12}J_{21}} \right\}, \end{aligned} \qquad (2.146)$$

as it is the only one for which B and C are nonnegative

$$\begin{aligned} B &= \frac{1}{2}\left\{ J_{11} - J_{22} + \sqrt{(J_{11} - J_{22})^2 + 4J_{12}J_{21}} \right\} \quad , \\ C &= \frac{1}{2}\left\{ J_{22} - J_{11} + \sqrt{(J_{11} - J_{22})^2 + 4J_{12}J_{21}} \right\}. \end{aligned} \qquad (2.147)$$

As a result, the expansion (2.145) appears to be unique. In addition, it is invariant under orthogonal rotations of the axes in the (x_1, x_2) plane.

The *degree of polarization P* is said to be the ratio of the intensity of the polarized part

$$I^{(p)} = \tfrac{1}{2}\operatorname{tr}\{\hat{J}^{(p)}\} = \tfrac{1}{2}(B + C) = \tfrac{1}{2}\sqrt{(J_{11} + J_{22})^2 - 4|\hat{J}|}$$

to the total intensity

$$I_0 = \tfrac{1}{2}\operatorname{tr}\{\hat{J}\} = \tfrac{1}{2}(J_{11} + J_{22}) \quad .$$

Thus

$$P = \frac{I^{(p)}}{I_0} = \sqrt{1 - \frac{4|\hat{J}|}{(J_{11} + J_{22})^2}} \quad . \tag{2.148}$$

Clearly, the degree of polarization is an invariant with respect to the orthogonal rotations of the axes, as it is expressed in terms of the invariants $\mathrm{tr}\{\hat{J}\}$ and $|\hat{J}|$. It ranges from $P = 0$ (no polarization, $J_{11} = J_{22}$, $J_{12} = J_{21} = 0$) to $P = 1$ (complete polarization, $|\hat{J}| = 0$).

If in the expression for the intensity (2.141) we go over to the sine and cosine of the angle 2θ, we get

$$I(\theta, \varepsilon) = \tfrac{1}{4}\{ J_{11} + J_{22} + \sqrt{(J_{11} - J_{22})^2 + 4 J_{12} J_{21} \cos^2(\beta_{12} + \varepsilon)} \cos(2\theta + \alpha)\} \quad ,$$

whence it is clear that $I(\theta, \varepsilon)$ as a function of θ changes the most at a phase delay ε, such that $\cos^2(\beta_{12} + \varepsilon) = 1$. In that case

$$I(\theta, \varepsilon) = \tfrac{1}{4}(J_{11} + J_{22})\{1 + P \cos(2\theta + \alpha)\} \quad .$$

Hence

$$P = \frac{I(\theta, \varepsilon)_{\max} - I(\theta, \varepsilon)_{\min}}{I(\theta, \varepsilon)_{\max} + I(\theta, \varepsilon)_{\min}} \quad ,$$

which enables the degree of polarization P to be found by appropriate measurements.

At times, besides the complete degree of polarization P, opticians use the degree of circular polarization

$$P_c = 2\frac{\langle A_1 A_2 \sin(\varphi_1 - \varphi_2)\rangle}{\langle A_1^2 \rangle + \langle A_2^2 \rangle} = \frac{\mathrm{i}(J_{21} - J_{12})}{J_{11} + J_{22}} \quad , \tag{2.149}$$

and the degree of linear polarization

$$P_l = \frac{1}{\langle A_1^2 \rangle + \langle A_2^2 \rangle} \sqrt{[\langle A_1^2 \rangle - \langle A_2^2 \rangle]^2 + 4\langle A_1 A_2 \cos(\varphi_1 - \varphi_2)\rangle^2}$$

$$= \frac{1}{J_{11} + J_{22}} \sqrt{(J_{11} - J_{22})^2 + (J_{12} + J_{21})^2} \quad . \tag{2.150}$$

It is easily seen that P is

$$P = \sqrt{P_l^2 + P_c^2} \quad . \tag{2.151}$$

Other measures of the degree of polarization have also been suggested. They are suitable given certain additional conditions (e.g., in the case of normally distributed $\zeta_1(t)$ and $\zeta_2(t)$, the *polarization number* $R_{12} = \langle r_{12} \rangle = \langle \zeta_2/\zeta_1 \rangle$, see [2.42]). But we shall not discuss them here. We only indicate the association of the quantities considered above with the parameters introduced by Stokes as early as 1852.

The polarization matrix contains four real quantities: J_{11}, J_{22}, $\mathrm{Re}\{J_{12}\}$, and $\mathrm{Im}\{J_{12}\}$. The Stokes parameters widely used in optics are related to the elements of the polarization matrix[21] in the following way:

$$
\begin{aligned}
I &= \langle A_1^2 \rangle + \langle A_2^2 \rangle = J_{11} + J_{22} \quad , \\
U &= 2\langle A_1 A_2 \cos(\varphi_1 - \varphi_2)\rangle = J_{12} + J_{21} \quad , \\
Q &= \langle A_1^2 \rangle - \langle A_2^2 \rangle = J_{11} - J_{22} \quad , \\
V &= 2\langle A_1 A_2 \sin(\varphi_1 - \varphi_2)\rangle = \mathrm{i}(J_{21} - J_{12}),
\end{aligned}
\tag{2.152}
$$

hence

$$
\begin{aligned}
J_{11} &= \tfrac{1}{2}(I + Q) \quad , \quad J_{12} = \tfrac{1}{2}(U + \mathrm{i}V) \quad , \\
J_{22} &= \tfrac{1}{2}(I - Q) \quad , \quad J_{21} = \tfrac{1}{2}(U - \mathrm{i}V) \quad .
\end{aligned}
\tag{2.153}
$$

All the dimensionless characteristics of *polarization* include, of course, the *ratios* of parameters, since these characteristics must be independent of the total intensity. In other words, both the polarization matrix elements and Stokes parameters can always be normalized to the quantity $I = J_{11} + J_{22}$ (or we can simply assume $I = 1$), and then only three independent parameters remain. Still the use of four parameters offers a certain advantage, as no renormalization is required in expanding or adding uncorrelated oscillations. The Stokes parameters, being linearly related to polarization matrix elements, also appear to be additive when summing up the uncorrelated oscillations.

Using (2.153), we can readily express the polarization characteristics (2.148–150) in terms of the Stokes parameters

$$
P = \frac{1}{I}\sqrt{Q^2 + U^2 + V^2} \quad , \quad P_c = \frac{V}{I} \quad , \quad P_l = \frac{1}{I}\sqrt{Q^2 + U^2} \quad .
$$

The fact that $|\hat{J}|$ is nonnegative, i.e., that $P \leq 1$, is expressed by the inequality

$$
I^2 \geq Q^2 + U^2 + V^2 \quad .
$$

For complete polarization this becomes an equality and for the no-polarization conditions $Q = U = V = 0$. The polarization ellipse characterstics (2.132, 133) for the completely polarized part of the oscillations are given in terms of the Stokes parameters as follows:

$$
\sin 2\chi = V/I \quad , \quad \tan 2\gamma = U/Q \quad .
$$

We see that in the approximation considered ($\tau \ll 2\pi/\Omega_m$) all the statistical

---

[21] In our notation the Stokes parameter I is twice the total intensity, $I = 2I_0$. The definition of the intensity (2.138) is often given without the factor 1/2, i.e., it is considered that $I_0 = \langle A_1^2 \rangle + \langle A_2^2 \rangle$, and thus $I = I_0$.

characteristics of polarization of modulated oscillations are completely described in terms of second moments of simultaneous values of the complex amplitudes $A_1(t)$ and $A_2(t)$.

It is shown in [2.43] that for the *normal* oscillations $\zeta_1(t)$ and $\zeta_2(t)$, the degree of polarization P may also be expressed in terms of the second moments of fluctuations of the *instantaneous intensities* $I_j(t) = \zeta_j(t)\,\zeta_j^*(t)$, $(j = 1, 2)$. Let us introduce for the intensities the correlation matrix $\hat{\sigma}$ with elements σ_{jk} obeying

$$\sigma_{jk}^2 = \langle \Delta I_j(t) \Delta I_k(t) \rangle = \langle (I_j - \overline{I_j})(I_k - \overline{I_k}) \rangle$$
$$= \langle \zeta_j \zeta_j^* \zeta_k \zeta_k^* \rangle - \langle \zeta_j \zeta_j^* \rangle \langle \zeta_k \zeta_k^* \rangle \quad .$$

Since for normal variables we have

$$\langle \zeta_j \zeta_j^* \zeta_k \zeta_k^* \rangle = \langle \zeta_j \zeta_j^* \rangle \langle \zeta_k \zeta_k^* \rangle + \langle \zeta_j \zeta_k \rangle \langle \zeta_j^* \zeta_k^* \rangle + \langle \zeta_j \zeta_k^* \rangle \langle \zeta_j^* \zeta_k \rangle \quad ,$$

and for the analytical signal $\langle \zeta_j \zeta_k \rangle = 0$, then

$$\sigma_{jk}^2 = \langle \zeta_j \zeta_k^* \rangle \langle \zeta_j^* \zeta_k \rangle = J_{jk} J_{kj}.$$

This immediately leads to the equalities $|\hat{\sigma}| = |\hat{J}|$ and $\operatorname{tr}\{\hat{\sigma}\} = \operatorname{tr}\{\hat{J}\}$. Accordingly, for the normal processes ζ_1 and ζ_2 the degree of polarization (2.148) may also be written as

$$P = \sqrt{1 - \frac{4|\hat{\sigma}|}{(\operatorname{tr}\{\hat{\sigma}\})^2}} \quad ,$$

i.e., we may find P by measuring the moments of intensity fluctuation.

We note in conclusion that in the case of *nonstationary* oscillations $\zeta_1(t)$ and $\zeta_2(t)$, the polarization characteristics (polarization matrix or Stokes parameters) will be functions of time t (in the same quasi-stationary approximation, i.e., ignoring the shift τ in the complex amplitudes). Just as in the case of nonstationary interference, the question of the degree of polarization of the oscillation $\zeta(t)$ can be answered differently depending on the relationship between the characteristic time scales of variation of the statistical means $J_{jk}(t) = \langle A_j(t) A_k^*(t) \rangle$, and the duration T of the observation (averaging or accumulative). If for sufficiently small T we successfully fix the polarization ellipse having a definite size and configuration, it may appear that, as T increases,

$$\widetilde{J_{11}(t)}_T = \widetilde{J_{22}(t)}_T, \quad \widetilde{J_{12}(t)}_T = \widetilde{J_{21}(t)}_T \quad ,$$

i.e., for large T the oscillation will be completely unpolarized. That it is advisable to average over the observation time just the polarization matrix elements and not the degree of polarization P calculated from (2.148) via $J_{jk}(t)$, is seen from (2.139). In fact, the mean *observed* intensity $\widetilde{I(t, \theta, \varepsilon)}_T$ is *linearly* related to $\widetilde{J_{jk}(t)}_T$.

2.8 Exercises

2.8.1   Give the explicit form of (2.86) for the case of a Poisson process where $\varphi(\omega) = n_1/(n_1 - i\omega)$.

Solution.

$$g(\omega) = 2\pi n_1 \left[K(\omega) + 2\operatorname{Re}\{I(\omega)H^*(\omega)\} - 2\frac{n_1}{\omega}\operatorname{Im}\{I(\omega)H^*(\omega)\} \right] \quad .$$

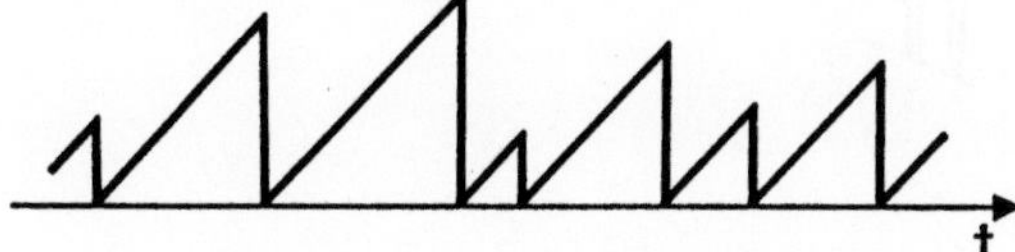

Fig. 2.12. Scheme of a process with independent intervals and triangular pulse shapes

2.8.2   Find the spectral density and covariance of a Poisson process with independent intervals for pulses that follow the law (Fig. 2.12)

$$F\left(\frac{t - t_\nu}{\tau_\nu}, a\right) = a \cdot (t - t_\nu) \quad ,$$

i.e., that start from zero at $t = t_\nu$ to grow linearly with time at fixed rate a within each time interval $(t_\nu, t_{\nu+1})$. The velocity of a charged particle in a medium behaves in this way, if during the mean free path the acceleration a is constant (motion in a constant eletric field), and owing to atomic collisions the velocity of the particle drops to zero.

Solution. Introducing $\theta = (t - t_\nu)/\tau$ we obtain, by (2.85),

$$\tilde{F}(\alpha) = \frac{a\tau}{2\pi}\int_0^1 \theta e^{-i\alpha\theta}\,d\theta = \frac{a\tau}{2\pi\alpha^2}[(1 + i\alpha)e^{-i\alpha} - 1] \quad .$$

At $w_T(\tau) = n_1\exp(-n_1\tau)$ calculation, using (2.83,84), gives

$$K(\omega) = \frac{a^2(3n_1^2 + \omega^2)}{2\pi^2 n_1^2(n_1^2 + \omega^2)^2} \quad ,$$

$$H(\omega) = \frac{a}{2\pi(n_1 + i\omega)^2} \quad ,$$

$$I(\omega) = \frac{a}{2\pi n_1(n_1 - i\omega)} \quad .$$

Considering that $\varphi(\omega) = n_1/(n_1 - i\omega)$, we find from (2.86)

$$g(\omega) = \frac{a^2}{\pi n_1(n_1^2 + \omega^2)} \quad .$$

To this spectral density corresponds the exponential covariance

$$\psi(\tau) = \frac{a^2}{n_1^2} e^{-n_1|\tau|} \quad .$$

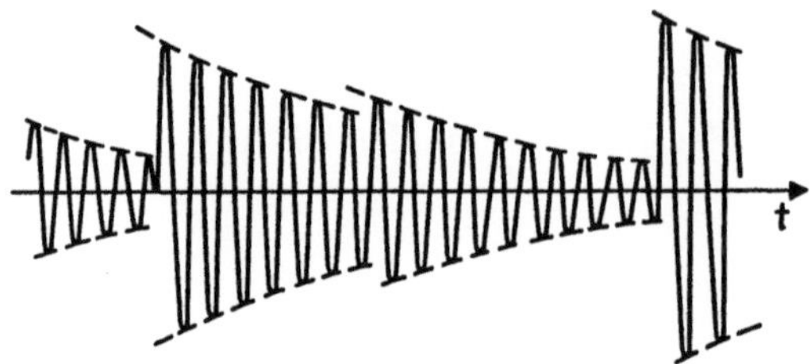

Fig. 2.13. Process with independent intervals and pulses in the form of decaying wave trains

2.8.3 Find the spectral density of the process (2.82) for Poisson pulses that have the form of exponentially decaying sinusoidal trains (Fig. 2.13):

$$F\left(\frac{t - t_\nu}{\tau_\nu} \, , a_\nu\right) = a_\nu \exp\left[-(\gamma - i\omega_{0\nu}(t - t_\nu) + \beta_\nu\right] \quad .$$

The random variables $a_\nu = \{a_\nu, \omega_{0\nu}, \beta_\nu\}$ are assumed to be mutually independent. This process models the radiation of the oscillator ("atom") excited by collisions at times t_ν. The oscillator is damped until the next collision occurs. The quantitiy $1/\gamma$ is the damping time (in quantum mechanics the life-time of the excited state), which determines the minimal ("radiation") width of the spectral line. The frequency $\omega_{0\nu}$ is different in different pulses due to the Doppler effect, since in collisions the translational velocity of the "atom" along the direction of observation v varies in a random way. The shift $\omega_{0\nu}$ with respect to the frequency radiated by an atom at rest is proportional to v.

Solution. As in the preceding problem, F is a deterministic function, and the angle brackets in (2.83, 84) denote averaging over the parameters a, ω_0, and β. Introducing $\theta = (t - t_\nu)/\tau$, we obtain from (2.85)

$$\tilde{F}(\alpha, a) = \frac{a e^{i\beta}}{2\pi} \frac{1 - \exp\left[-(\gamma - i\omega_0)\tau - i\alpha\right]}{(\gamma - i\omega_0)\tau + i\alpha} \quad .$$

Putting $\alpha = \omega\tau$ and $w_\tau(\tau) = n_1 \exp(-n_1\tau)$, we find from (2.83, 84)

$$\begin{aligned}
K(\omega) &= \frac{n_1 \overline{a^2}}{4\pi^2} \int_0^{+\infty} \left\langle \frac{1 - 2e^{-\gamma\tau} \cos(\omega_0 - \omega)\tau + e^{-2\gamma\tau}}{(\omega_0 - \omega)^2 + \gamma^2} \right\rangle e^{-n_1\tau} d\tau \\
&= \frac{(\gamma + n_1)\overline{a^2}}{2\pi^2(2\gamma + n_1)} \left\langle \frac{1}{(\omega_0 - \omega)^2 + (\gamma + n_1)^2} \right\rangle \quad ,
\end{aligned}$$

$$
H(\omega) = \frac{n_1 \overline{a e^{i\beta}}}{2\pi} \int\limits_0^\infty \left\langle \frac{\exp\left[-\gamma + i(\omega_0 - \omega)\right]\tau - 1}{i(\omega_0 - \omega) - \gamma} \right\rangle e^{-n_1 \tau} d\tau
$$

$$
= \frac{\overline{a e^{i\beta}}}{2\pi} \left\langle \frac{1}{\gamma + n_1 - i(\omega_0 - \omega)} \right\rangle \;,
$$

$$
I(\omega) = \frac{n_1 \overline{a e^{i\beta}}}{2\pi} \int\limits_0^\infty \left\langle \frac{e^{-(\gamma - i\omega_0)\tau} - e^{i\omega\tau}}{i(\omega_0 - \omega) - \gamma} \right\rangle e^{-n_1 \tau} d\tau
$$

$$
= \frac{n_1 \overline{a e^{i\beta}}}{2\pi(n_1 - i\omega)} \left\langle \frac{1}{\gamma + n_1 - i(\omega_0 - \omega)} \right\rangle \;,
$$

where the angle brackets now imply averaging over the distribution of ω_0. Taking $\varphi(\omega) = n_1/(n_1 - i\omega)$ into account, we have

$$
\text{Re}\left\{ \frac{IH^*}{1 - \varphi} \right\} = \text{Re}\left\{ \frac{n_1 \overline{a^2} |\overline{e^{i\beta}}|^2}{4\pi^2 i \omega} \left| \left\langle \frac{1}{\gamma + n_1 - i(\omega_0 - \omega)} \right\rangle \right|^2 \right\} = 0 \;,
$$

i.e., in (2.86) remains only the first term (for any distributions of a, β, and ω_0

$$
g(\omega) = \frac{n_1(\gamma + n_1)\overline{a^2}}{\pi(2\gamma + n_1)} \left\langle \frac{1}{(\omega_0 - \omega)^2 + (\gamma + n_1)^2} \right\rangle \;. \tag{2.154}
$$

Note that for $\overline{a} = 0$ and/or $\overline{\exp(i\beta)} = 0$ [i.e., for a uniform distribution of phases β in the interval $(0, 2\pi)$] the variables $H(\omega)$ and $I(\omega)$ vanish, and so in (2.86) we only have the first term if we do not assume that the flow of moments t_ν is Poissonian.

In the case of oscillator-"atoms" the ray velocity v is normally distributed and so the Doppler shift $\omega_0 - \overline{\omega}_0 \sim v$ also has a Gaussian distribution. However, the integral

$$
\left\langle \frac{1}{(\omega_0 - \omega)^2 + (\gamma + n_1)^2} \right\rangle = \frac{1}{\sqrt{2\pi}\sigma_\omega} \int\limits_{-\infty}^{+\infty} \frac{\exp\left[-(\omega_0 - \overline{\omega}_0)^2/2\sigma_\omega^2\right]}{(\omega_0 - \omega)^2 + (\gamma + n_1)^2} d\omega_0 \tag{2.155}
$$

for arbitrary ω can not be taken in closed form. Introducing into (2.155) the notation

$$
x = \frac{\omega - \overline{\omega}_0}{\sqrt{2}\sigma_\omega} \;, \quad y = \frac{\omega_0 - \overline{\omega}_0}{\sqrt{2}\sigma_\omega} \;, \quad b = \frac{\gamma + n_1}{\sqrt{2}\sigma_\omega} \;,
$$

we can write the expression (2.154) for the spectral density in the following way:

$$g(\omega) = \frac{n_1 \overline{a^2} U(b,x)}{(2\gamma + n_1)\sqrt{2}\sigma_\omega} \quad , \quad U(b,x) = \frac{b}{\pi^{3/2}} \int\limits_{-\infty}^{+\infty} \frac{e^{-y^2}\,dy}{(y-x)^2 + b^2} \quad .$$

The function $U(b,x)$ is called the (normalized) Voigt function [2.44]. If the parameter b is small, i.e., the Doppler broadening comes to the fore, then we can obtain a crude estimation by taking the value of the exponential function at $y = x$ outside the integral sign, to arrive at $U(b,x) \approx \exp(-x^2)/\sqrt{\pi}$. Thus, the line shape appears to be close to *Gaussian* (bell-shaped). On the contrary, for $b \gg 1$ we can set $y = 0$ in the denominator and take it out of the integral sign to get $U(b,x) \approx b/\pi(x^2 + b^2)$. We obtain for $g(\omega)$ the expression (2.154) with $\omega = \overline{\omega}_0$, but without the angle brackets [as $\sigma_\omega \to 0$ the Gaussian distribution ω_0 goes over into $\delta(\omega_0 - \overline{\omega}_0)d\omega_0$]. With the Doppler effect making only a small contribution, we thus obtain the *Lorentzian* line shape with the maximum at $\omega = \overline{\omega}_0$ and half-width $n_1 + \gamma$. The radiation damping γ and the mean number of collisions per unit time n_1 enter into the half-width in an additive manner. In that case, the total intensity is

$$I = \int\limits_{-\infty}^{+\infty} g(\omega)d\omega = \frac{n_1 \overline{a^2}}{2\gamma + n_1} \quad ,$$

so that the spectral density can be written in the form

$$g(\omega) = \frac{I}{\pi} \frac{n_1 + \gamma}{(\omega - \overline{\omega}_0)^2 + (n_1 + \gamma)^2} \quad .$$

2.8.4   If we add together N *independent* processes $\xi_j(t)$ of the form considered at the end of the previous problem, i.e. if we are interested in the total radiation of N independent identical oscillators (ignoring the Doppler effect), then for $\xi(t) = \sum_1^N \xi_j(t)$ the spectral density will simply be $g_\Sigma(\omega) = Ng(\omega)$, and the line shape will remain the same. But it would seem possible to reason along another line. All the independent instants of collisions for all the oscillators, t_{ν_j} $(j = 1, 2, \ldots, N)$, may be arranged in ascending order and numbered using a general subscript μ. The interval $t_{\mu+1} - t_\mu$ will on average be N times shorter than $1/n_1$ — the interval for each individual process $\xi_j(t)$, and so the mean number of collisions per second for $\xi(t)$ will be Nn_1. The process $\xi(t)$ has the same pulse shape as $\xi_j(t)$, since the sum of trains with certain amplitudes and phases varying in time by the law $\exp(-\gamma t + i\omega_0 t)$ will again give the same damped train. But then the direct application of (2.86) to the total process $\xi(t)$ must give a Lorentzian line with a half-width $\gamma + Nn_1$, and not $\gamma + n_1$, which disagrees with the expression $g_\Sigma(\omega) = Ng(\omega)$. Discuss the reasons for this discrepancy.

Solution. In deriving (2.86) it was supposed that the parameters a_ν and β_ν were *mutually independent* (the frequency $\omega = \overline{\omega}_0$ is now considered to be fixed), with the result that all the pulses were *uncorrelated.* But for the total process $\xi(t)$ the pulses are correlated, since at each t_μ only one out of the N oscillations that are to be combined ($j = 1, 2, \ldots, N$) has an amplitude and phase varying in a random fashion. The μth pulse will only become uncorrelated with adjacent ones after the occurrence of jumps for each of the oscillations that are added. This will require, on average, N intervals $\overline{t_{\mu+1} - t_\mu}$, i.e., the time $1/n_1$. Equation (2.86) is thus inapplicable to the resultant process $\xi(t)$, because the pulses are correlated.

2.8.5 Discuss the joint influence on the visibility of an interference pattern of the radiation nonmonochromaticity and small fluctuations of the wave incidence angle α, assuming a Gaussian shape of the spectral line and a normal distribution for α with $\overline{\alpha} = 0$ and $\overline{\alpha^2} = \sigma_\alpha^2$. It can also be assumed, for simplicity, that $I_1 = I_2 = I = \psi(0)/2$.

Solution. According to (2.102),

$$B(\tau - \tau_\alpha) = B(0) \exp\left\{-\frac{\Omega^2(\tau - \tau_\alpha)^2}{16} + i\omega_0(\tau - \tau_\alpha)\right\} \quad .$$

If we substitute α for $\sin\alpha$ (owing to σ_α^2 being small), then the difference

$$\chi = \tau - \tau_\alpha = \tau - \frac{d}{c}\alpha$$

will also be a normal variable with $\overline{\chi} = \tau$ and $D[\chi] \equiv \sigma^2 = d^2\sigma_\alpha^2/c^2$, i.e., it will have the distribution

$$w_\chi(\chi)d\chi = \frac{1}{\sqrt{2\pi}\sigma} \exp\left[-(\chi - \tau)^2/2\sigma^2\right]d\chi \quad .$$

From (2.106), the unconditional covariance $\overline{B}(\tau)$ will be

$$\overline{B}(\tau) = \int\limits_{-\infty}^{\infty} B(\chi)w_\chi(\chi)d\chi$$

$$= \frac{2I}{\sqrt{\beta}} \exp\left\{-\left(\frac{\Omega^2\tau^2}{16\beta} + \frac{\omega_0^2\sigma^2}{2\beta}\right) + i\frac{\omega_0\tau}{\beta}\right\} \quad , \tag{2.156}$$

where $\beta = 1 + \Omega^2\sigma^2/8$. Consequently, the intensity (2.95) averaged over α is

$$\overline{I}(\tau) = I_1 + I_2 + \mathrm{Re}\left\{\overline{B}(\tau)\right\}$$

$$= 2I\left\{1 + \frac{1}{\sqrt{\beta}} \exp\left[-\left(\frac{\Omega^2\tau^2}{16\beta} + \frac{\omega_0^2\sigma^2}{2\beta}\right)\right] \cos\frac{\omega_0\tau}{\beta}\right\}$$

and, accordingly, the visibility is

$$\overline{V}(\tau) = \frac{1}{\sqrt{\beta}} \exp\left[-\left(\frac{\Omega^2\tau^2}{16\beta} + \frac{\omega_0^2\sigma^2}{2\beta}\right)\right] \quad .$$

In the absence of fluctuations of α ($\sigma = 0, \beta = 1$), we again obtain (2.102). For $\sigma \neq 0$, but with ideal monochromaticity ($\Omega = 0, \beta = 1$), we arrive at

$$\overline{V}(\tau) = \exp\left(-\omega_0^2 \sigma^2/2\right) \quad ,$$

i.e., throughout the entire interference pattern the visibility drops uniformly and quickly with increasing $\sigma^2 = d^2 \sigma_\alpha^2/c^2$. Accordig to (2.156)

$$\overline{K}(\tau) = \frac{\overline{B}(\tau)}{\overline{B}(0)} = \exp\left\{-\frac{\Omega^2 \tau^2}{16\beta} + i\frac{\omega_0 \tau}{\beta}\right\} \quad ,$$

so that for $\sigma \neq 0$, $\overline{V}(\tau) \neq |\overline{K}(\tau)|$.

2.8.6 Calculate the visibility of the interference pattern for stationary quasi-monochromatic oscillations at interferometer slits having the form

$$\zeta_j(t) = \exp\{i[\omega_0 t + \varphi_j(t)]\} \quad , \quad j = 1, 2 \quad ,$$

and subject to the conditions that $\overline{\varphi_j(t)} = 0$, $\overline{\varphi_j^2(t)} = \sigma^2$, $\overline{\varphi_1(t+\tau)\varphi_2(t)} = \sigma^2 K(\tau)$, the difference $\varphi_1(t+\tau) - \varphi_2(t)$ being normally distributed.

Solution. Under the above conditions

$$\overline{\cos(\varphi_{1\tau} - \varphi_2)} = \exp\left[-\overline{(\varphi_{1\tau} - \varphi_2)^2}/2\right] \quad , \quad \overline{\sin(\varphi_{1\tau} - \varphi_2)} = 0 \quad ,$$
$$\overline{(\varphi_{1\tau} - \varphi_2)^2} = 2\sigma^2[1 - K(\tau)] \quad .$$

Consequently, the intensity in the interference pattern varies accordig to the law

$$I(\tau) = 2[1 + \overline{\cos(\omega_0 \tau + \varphi_{1\tau} - \varphi_2)}] = 2[1 + e^{-\sigma^2[1-K(\tau)]} \cos \omega_0 \tau] \quad ,$$

i.e., the visibility of the fringes is

$$V(\tau) = e^{-\sigma^2[1-K(\tau)]} \quad .$$

For $K(\tau) \neq 1$ the visibility falls off with increasing phase variance σ^2. The visibility is unity either at complete correlation ($K = 1$) or in the absence of phase fluctuations ($\sigma^2 = 0$), i.e., for ideal monochromaticity of oscillations.

2.8.7 Find the spectral density matrix for the oscillations of two coupled oscillators subject to mutually uncorrelated stationary white noises

$$\ddot{x}_1 + 2h_1\dot{x}_1 + n_1^2 x_1 - \kappa x_2 = f_1(t) \quad ,$$
$$\ddot{x}_2 + 2h_2\dot{x}_2 + n_2^2 x_2 - \kappa x_1 = f_2(t) \quad , \tag{2.157}$$

$$\langle \tilde{f}_j(\omega)\tilde{f}_k^*(\omega')\rangle = N_j\delta_{jk}\delta(\omega - \omega') \quad , \quad j,k = 1,2 \quad . \tag{2.158}$$

Solution. From (2.157), the equations for spectral amplitude densities $\tilde{x}_{1,2}(\omega)$ are

$$\tilde{x}_1(\omega)Z_1(\mathrm{i}\omega) - \kappa\tilde{x}_2(\omega) = \tilde{f}_1(\omega) \quad , \quad \tilde{x}_2(\omega)Z_2(\mathrm{i}\omega) - \kappa\tilde{x}_1(\omega) = \tilde{f}_2(\omega) \quad ,$$

where $Z_j(\mathrm{i}\omega) = n_j^2 - \omega^2 + 2\mathrm{i}\omega h_j$ $(j = 1,2)$. Hence

$$\tilde{x}_1(\omega) = \frac{Z_2\tilde{f}_1(\omega) + \kappa\tilde{f}_2(\omega)}{Z_1Z_2 - \kappa^2} \quad , \quad \tilde{x}_2(\omega) = \frac{\kappa\tilde{f}_1(\omega) + Z_1\tilde{f}_2(\omega)}{Z_1Z_2 - \kappa^2} \quad . \tag{2.159}$$

We next set up the correlation matrix for the amplitude densities $\tilde{x}_1(\omega)$ and $\tilde{x}_2(\omega)$

$$\langle \tilde{x}_1(\omega)\tilde{x}_k^*(\omega')\rangle = g_{jk}(\omega)\delta(\omega - \omega') \quad ,$$

to obtain, using (2.158, 159),

$$g_{11}(\omega) = \frac{|Z_2|^2 N_1 + \kappa^2 N_2}{|Z_1Z_2 - \kappa^2|^2} \quad ,$$

$$g_{22}(\omega) = \frac{\kappa^2 N_1 + |Z_1|^2 N_2}{|Z_1Z_2 - \kappa^2|^2} \quad ,$$

$$g_{12}(\omega) = g_{21}^*(\omega) = \frac{\kappa(Z_2 N_1 + Z_1^* N_2)}{|Z_1Z_2 - \kappa^2|^2} \quad .$$

Of course, the mutual correlation of the responses $x_1(t)$ and $x_2(t)$ is due to the coupling κ. It vanishes at $\kappa = 0$. If the oscillators are identical and the spectral densities of the noises are the same ($N_1 = N_2 = N$), then

$$g_{11}(\omega) = g_{22}(\omega) = \frac{N(|Z|^2 + \kappa^2)}{|Z^2 - \kappa^2|^2} \quad , \quad g_{12}(\omega) = g_{21}(\omega) = \frac{2\kappa N\,\mathrm{Re}\,\{Z\}}{|Z^2 - \kappa^2|^2} \quad .$$

The correlation matrix for $x_1(t)$ and $x_2(t)$ may be derived from $g_{jk}(\omega)$ by the Khintchine theorem:

$$B_{jk}(\tau) = \langle x_j(t + \tau)x_k^*(t)\rangle = \int\limits_{-\infty}^{+\infty} g_{jk}(\omega)\mathrm{e}^{\mathrm{i}\omega\tau}\,d\omega \quad .$$

3. Spectral Theory of Random Actions on Dynamic Systems

In Sects. I.5.8, 9 and Chap. I.6 the theory of random actions on dynamic systems was given in a *temporal* representation. It is well known, however, that in many cases it is advisable to use the *spectral* approach to problems of this sort. This is true of both deterministic and random processes. For the latter, correlation theory, which makes extensive use of spectral expansions of random processes and their covariances, is quite natural and adequate.

The subject matter of this chapter has been arranged largely in order of increasing complexity of the dynamic systems with constant parameters, the so-called harmonic systems (Sect. 3.1). We then go on to consider nonlinear, but memoryless, systems, i.e., ones that are described simply in terms of the nonlinear functions "output" vs "input" behavior (Sect. 3.2). If systems of the two kinds mentioned above form links of a chain, such that there is no feedback between a link and the previous one, then this chain can be studied using the results of Sects. 3.1, 2. Specifically, the three-link chain of the radiometer, a device used to measure the level of noise in a given frequency band (Sect. 3.3), is of this type.

In Sect. 3.4 the spectral form of correlation theory is applied to the Thomson self-oscillating system, i.e., to the problem formulated and handled in temporal language within the framework of the theory of Markov processes in Sects. I.5.8, 9.

Most of Sect. 3.5 is devoted to a special kind of dynamic systems (quasi-stationary electric networks) and a special kind of random action, the so-called equilibrium thermal noise, i.e., electrical noise of thermal origin under the conditions of thermal equilibrium of the noise source. Apart from the simplest case – the two-terminal network, for which the Nyquist formula was first derived – we also consider the general case of the branched quasi-stationary network, which enables a more general fluctuation-dissipation theorem to be formulated.

3.1 Random Actions on Harmonic Systems

In a temporal treatment of the external action on a linear system with constant parameters, the system is characterized by its response $H(t - \tau)$ to the "force" $\delta(\theta)$ (Sect. I.6.5). The respons to an arbitrary "force" $f(t)$ is expressed in terms of the Duhamel integral which under zero initial conditions has the form

$$\xi(t) = \int_0^t H(t - \theta) f(\theta)\, d\theta = \int_0^t H(\theta) f(t - \theta)\, d\theta \quad . \tag{3.1}$$

It is clear from this that to calculate any moment of the nth order for the output process $\xi(t)$ requires a knowldege of the mixed moment of the same order for the input process $f(t)$. So, the mean of $\xi(t)$ and its second mixed moment are given by (I.6.64, 65)

$$\overline{\xi(t)} = \int\limits_0^t H(\theta)\overline{f(t-\theta)}\,d\theta \quad , \tag{3.2}$$

$$B_\xi(t+\tau,t) = \overline{\xi(t+\tau)\xi(t)}$$

$$= \int\limits_0^t H(\theta)\,d\theta \int\limits_0^{t+\tau} H(\theta')B_f(t-\theta,t+\tau-\theta')\,d\theta' \quad , \tag{3.3}$$

where $B_f(t_1,t_2) = \overline{f(t_1)f(t_2)}$. If, in particular, the "force" $f(t)$ is stationary and $\overline{f(t)} = 0$, then

$$\overline{\xi(t)} = 0 \quad ,$$

$$\psi_\xi(t+\tau,t) = \int\limits_0^t H(\theta)\,d\theta \int\limits_0^{t+\tau} H(\theta')\psi_f(\tau+\theta-\theta')\,d\theta' \quad , \tag{3.4}$$

since $\psi_f(t_1,t_2) = \psi_f(t_1-t_2)$. Finally, if we are interested in the *steady-state* response to the stationary action $f(t)$, we should let the upper limit increase to $+\infty$ (in 3.1.4)[1]. For the real dissipative systems $H(t) = 0$ for $t < 0$ and we can substitute $-\infty$ for the lower limit. Then (3.4) will take the form

$$\psi_\xi(\tau) = \int\limits_{-\infty}^{+\infty} H(\theta)\,d\theta \int\limits_{-\infty}^{+\infty} H(\theta')\psi_f(\tau+\theta-\theta')\,d\theta' \quad ,$$

or

$$\psi_\xi(\tau) = \int\limits_{-\infty}^{+\infty} H_2(\chi)\psi_f(\tau-\chi)\,d\chi \quad , \tag{3.5}$$

where we have introduced the "covariance" of the pulse response (see I.6.67, 68).

$$H_2(\chi) = \int\limits_{-\infty}^{+\infty} H(\theta)H(\theta+\chi)\,d\theta \quad . \tag{3.6}$$

Let the dynamic equation of a harmonic system be represented in the form

$$\hat{L}[\xi(t)] = \hat{M}[f(t)] \quad ,$$

---

[1] This corresponds to placing the initial time t_0 at $-\infty$. In (3.1) and later equations we put $t_0 = 0$.

where $\hat{L}$ and $\hat{M}$ are linear operators independent of t. In a spectral treatment this system is described by its transfer function $k(i\omega)$ which, as we know, represents the complex amplitude of the *steady-state* forced oscillation $\xi(t) = k(i\omega)\exp(i\omega t)$ under the action of by the harmonic oscillation $f(t) = \exp(i\omega t)$. If $\hat{L}$ and $\hat{M}$ are polynomials in d/dt, then $k(i\omega)$ will be a fractional rational function of $i\omega$, but such a specialization of the form of $k(i\omega)$ is by no means necessary for what is to follow.

Knowing $k(i\omega)$ and representing $f(t)$ and $\xi(t)$ as Fourier-Stieltjes integrals

$$\xi(t) = \int\limits_{-\infty}^{+\infty} e^{i\omega t}dC_\xi(\omega), \quad f(t) = \int\limits_{-\infty}^{+\infty} e^{i\omega t}dC_f(\omega) \quad ,$$

we can immediately write the relationship between the spectral amplitudes of both processes

$$dC_\xi(\omega) = k(i\omega)\, dC_f(\omega) \quad , \tag{3.7}$$

hence

$$\overline{dC_\xi(\omega)dC_\xi^*(\omega')} = k(i\omega)k^*(i\omega')\overline{dC_f(\omega)dC_f^*(\omega')} \quad . \tag{3.8}$$

If the external action $f(t)$ is stationary, so that

$$\overline{dC_f(\omega)} = 0 \quad , \quad \overline{dC_f(\omega)dC_f^*(\omega')} = \delta(\omega - \omega')\, d\omega'\, dG_f(\omega) \quad ,$$

then according to $(3.7, 8)$ we have

$$\overline{dC_\xi(\omega)} = 0 \quad , \quad \overline{dC_\xi(\omega)\, dC_\xi^*(\omega')} = \delta(\omega - \omega')\, d\omega' dG_\xi(\omega) \quad ,$$

where

$$dG_\xi(\omega) = |k(i\omega)|^2\, dG_f(\omega) \quad . \tag{3.9}$$

Thus, with the spectral approach, if $f(t)$ is stationary, the steady-state response $\xi(t)$ will also be stationary. For a continuous spectrum we can represent (3.9) in terms of the spectral densities

$$g_\xi(\omega) = |k(i\omega)|^2 g_f(\omega) \quad . \tag{3.10}$$

Here the output spectral density is derived from the input spectral density by simply multiplying the latter by the squared modulus of the transfer function. According to the Wiener-Khintchine theorem, the covariance at the output will be

$$\psi_\xi(\tau) = \int\limits_{-\infty}^{+\infty} e^{i\omega\tau}dG_\xi(\omega) = \int\limits_{-\infty}^{+\infty} e^{i\omega\tau}|k(i\omega)|^2\, dG_f(\omega) \quad . \tag{3.11}$$

In many cases it is convenient to take the integral using the residues at the poles of $|k(i\omega)|^2$, which for the transfer function $k(i\omega)$ of a dissipative system lie in the upper semiplane of the complex frequency ω.

As noted in Sect. 1.3, localization of the intensity in frequency occurring in stationary processes becomes unambiguous only due to harmonic filtration. In fact, a filter with an adequate pass-band, i.e., with a suitable transfer function $k(i\omega)$, can in principle make it physically possible to *single out* a required interval (ω_1, ω_2) and to *check* that this interval contains the part $\int_{\omega_1}^{\omega_2} g_f(\omega)\, d\omega$ of the total "power" $|f|^2 = \int_{-\infty}^{+\infty} g_f(\omega)\, d\omega$. Manipulating with total intensity alone, we could add to $g_f(\omega)$ the functions $g_1(\omega)$, which are arbitrary in many respects save for the conditions

$$\int\limits_{-\infty}^{+\infty} g_1(\omega)\, d\omega = 0$$

(i.e., $\overline{|f|^2}$ remains unchanged) and $g_f(\omega) + g_1(\omega) \geq 0$ for all ω.

As is seen from (3.11), harmonic filtration retains the discrete-continuous character of the spectrum. If we apply to the filter input a superposition of discrete harmonic oscillations and a noise with a continuous spectrum, then we will receive at the output a similar superposition, although the continuous spectrum shape and the intensity relationships of discrete lines will vary in accordance with the weight factor $|k(i\omega)|^2$.

We will consider the relationship between the temporal and spectral approaches somewhat later. At this point we simply note that the spectral treatment is the simplest and most advisable one for stationary actions. But for nonstationary actions (to which the question of the *settling* of oscillations in the system itself comes down, assuming that the "force" $f(t)$ *began* to act at $t = 0$), the local relation between the covariances of the spectral amplitudes is lost. On the contrary, the integrals (3.2, 3) are applicable to both the nonstationary actions $f(t)$, and the nonsteady states [via initial conditions that enter explicitly, even with the stationary process $f(t)$].

Consider the simple example of an RC-cell connected into the anode circuit of a tube (Fig. 3.1). If I is the shot current through the tube, then the noise component v of the voltage across the cell is related to I by

$$\dot{v} + \frac{v}{\vartheta} = \frac{I}{C} \quad (\vartheta = \mathrm{RC}) \quad . \tag{3.12}$$

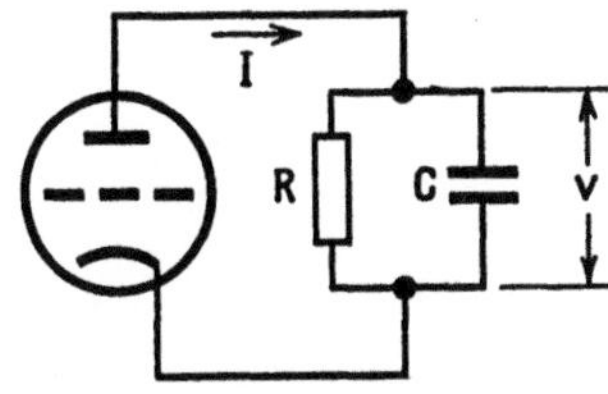

Fig. 3.1. RC-cell that filters the shot current of a tube

The transfer function $(\xi \equiv v, \ f \equiv I)$ is thus

$$k(\mathrm{i}\omega) = \frac{\vartheta}{C(1 + \mathrm{i}\omega\vartheta)}$$

and the spectral density of the voltage $v(t)$, considering that by (2.2) $g_I(\omega) = \overline{I}e/2\pi$, will be

$$g_v(\omega) = |k(\mathrm{i}\omega)|^2 g_I(\omega) = \frac{\vartheta^2 \overline{I}e}{2\pi C^2} \frac{1}{1 + \omega^2\vartheta^2} \ . \tag{3.13}$$

The covariance of $v(t)$ is

$$\psi_v(\tau) = \frac{\vartheta^2 \overline{I}e}{2\pi C^2} \int\limits_{-\infty}^{+\infty} \frac{e^{\mathrm{i}\omega\tau} d\omega}{1 + \omega^2\vartheta^2} = \overline{v^2} e^{-|\tau|/\vartheta} \ , \tag{3.14}$$

where

$$\overline{v^2} = \frac{\overline{I}eR}{2C} \ .$$

For $\tau > 0$ the integral is determined by the residue at the pole $\omega = \mathrm{i}/\vartheta$, and for $\tau < 0$, by the residue at the pole $\omega = -\mathrm{i}/\vartheta$.

We know that equation (3.12) for the delta-correlated action $I(t)$ is only symbolic, since $\dot{v}$ does not exist. What does this mean from the spectral point of view? If

$$v(t) = \int\limits_{-\infty}^{+\infty} e^{\mathrm{i}\omega t} dC_v(\omega) \ ,$$

then, formally

$$\dot{v}(t) = \int\limits_{-\infty}^{+\infty} e^{\mathrm{i}\omega t} \mathrm{i}\omega \, dC_v(\omega) \ .$$

The existence condition for $\dot{v}(t)$ is

$$\overline{|\dot{v}|^2} = \psi_{\dot{v}}(0) = -\psi_v''(0) < \infty \ .$$

But, according to (3.13),

$$\psi_v''(\tau) = -\frac{\vartheta^2 \overline{I}e}{2\pi C^2} \int\limits_{-\infty}^{+\infty} \frac{e^{\mathrm{i}\omega\tau} \omega^2 \, d\omega}{1 + \omega^2\vartheta^2} \ ,$$

whence it is clear that at $\tau = 0$ the integral diverges. If the density $g_I(\omega)$ were not constant, i.e., the shot current were not taken to be delta-correlated, then we would have

$$\psi_v''(0) = -\frac{\vartheta^2}{C^2} \int\limits_{-\infty}^{+\infty} \frac{g_I(\omega)\omega^2\,d\omega}{1 + \omega^2\vartheta^2} < \infty \quad ,$$

i.e., $\dot{v}(t)$ would exist and the stochastic equation (3.11) could not be considered symbolic.

It is a straightforward exercise to generalize this reasoning. Successive differentiation of

$$\xi(t) = \int\limits_{-\infty}^{+\infty} e^{i\omega t} k(i\omega)\,dC_f(\omega) \tag{3.15}$$

with respect to t yields that the existence condition for the derivative $\xi^{(m)}(t)$ is the requirement that

$$(-1)^m \psi_\xi^{(2m)}(0) = \int\limits_{-\infty}^{+\infty} \omega^{2m} g_\xi(\omega)\,d\omega = \int\limits_{-\infty}^{+\infty} |k(i\omega)|^2 \omega^{2m} g_f(\omega)\,d\omega \tag{3.16}$$

be finite. If $\xi(t)$ satisfies an equation of the form

$$\xi^{(n)} + a_1\xi^{(n-1)} + \ldots + a_n\xi = f(t) \quad , \tag{3.17}$$

then $|k(i\omega)|^2$ is unity divided by a polynomial of degree $2n$ in ω. For $m < n$ the integral (3.16) always converges even if the "force" $f(t)$ is a white noise , i.e., $g_f(\omega) = \text{const}$. But for $\xi^{(n)}(t)$ to exist requires now that $g_f(\omega)$ be integrable. Accordingly, if $f(t)$ is white noise, then (3.17) is a symobilic equation, since $\xi^{(n)}$ does not exist.

We will now compare the spectral and temporal approaches to the problem of the harmonic filtration of stationary processes. For the covariance of the *steady-state* response $\xi(t)$ to the *stationary* "force" $f(t)$ we have, on the one hand, the spectral expression (3.11), and on the other, equation (3.5) based on the representation of $\xi(t)$ in terms of the Duhamel integral. It is easily seen that these expressions coincide. It should be remembered that the pulse response $H(t)$ and the transfer function $k(i\omega)$ are related by the Fourier transform

$$H(\theta) = \frac{1}{2\pi} \int\limits_{-\infty}^{+\infty} k(i\omega)e^{i\omega\theta}\,d\omega \quad , \quad k(i\omega) = \int\limits_{-\infty}^{+\infty} H(\theta)e^{-i\omega\theta}\,d\theta \quad . \tag{3.18}$$

By the theorem (1.66) on the spectrum of the integral of the form (3.6), the spectral amplitude $H_2(\chi)$ is

$$2\pi \frac{k(i\omega)}{2\pi} \frac{k^*(i\omega)}{2\pi} = \frac{1}{2\pi}|k(i\omega)|^2 \quad , \tag{3.19}$$

so that

$$H_2(\chi) = \frac{1}{2\pi} \int\limits_{-\infty}^{+\infty} |k(i\omega)|^2 e^{i\omega\chi}\, d\omega \quad . \tag{3.20}$$

But, according to (3.5), the covariance $\psi_\xi(\tau)$ is the convolution of the functions $H_2(\tau)$ and $\psi_f(\tau)$, whose spectral amplitudes are (3.19) and the spectral density $g_f(\omega)$ of the process $f(t)$, respectively. Consequently, the spectral amplitude $\psi_\xi(\tau)$ is $2\pi|k(i\omega)|^2 g_f(\omega)/2\pi$, which coincides with $(3.10)^2$.

Consider the following two examples.

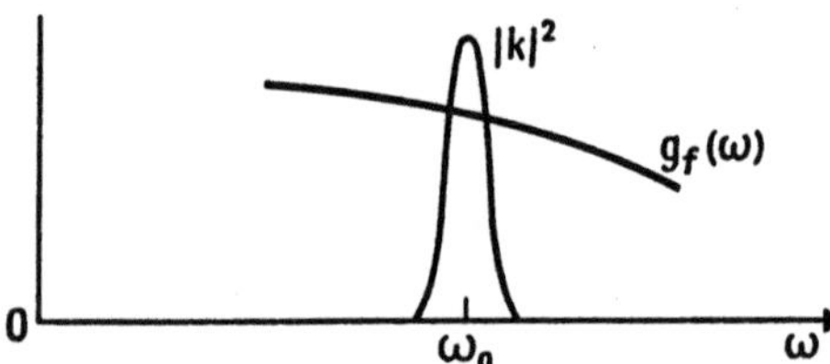

Fig. 3.2. Within the filter band $|k|^2$ the spectrum of $g_f(\omega)$ varies slightly

1) Let the spectrum $g_f(\omega)$ and the squared modulus of the transfer function, $|k(i\omega)|^2$, be sufficiently well-behaved, smooth functions such that, apart from at their major maximum frequency, they have no significant contributions far from this frequency. Let the spectrum $g_f(\omega)$ be much wider than $|k(i\omega)|^2$ (Fig. 3.2). Then within the filter band $g_f(\omega) \approx g_f(\omega_0)$ we have by (3.11) and (3.20)

$$\psi_\xi(\tau) \approx g_f(\omega_0) \int\limits_{-\infty}^{+\infty} |k(i\omega)|^2 e^{i\omega\tau}\, d\omega = 2\pi g_f(\psi_0) H_2(\tau) \quad ,$$

i.e., $\psi_\xi(\tau)$ is goverened by the filter properties. The same result follows, of course, from (5.5), if we take into account that $\psi_f(\tau)$ falls off much faster than $H_2(\tau)$:

$$\psi_\xi(\tau) = \int\limits_{-\infty}^{+\infty} H_2(\tau - \chi)\psi_f(\chi)\, d\chi \approx H_2(\tau) \int\limits_{-\infty}^{+\infty} \psi_f(\chi)\, d\chi = 2\pi g_f(0) H_2(\tau).$$

This estimate is rougher than the preceding one $[g_f(0)$ instead of $g_f(\omega_0)]$, but both yield the same variation with τ.

2) If, on the contrary, $g_f(\omega)$ is much narrower than $|k(i\omega)|^2$ (Fig. 3.3), i.e., $\psi_f(\tau)$ falls off significantly slower than $H_2(\tau)$, then $|k(i\omega)|^2 \approx |k(i\omega_0)|^2$, and

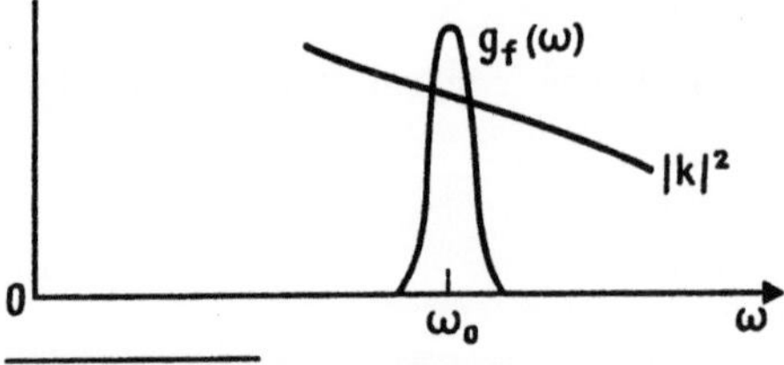

Fig. 3.3. The filter band $|k|^2$ is much wider than the spectrum of $g_f(\omega)$

---

^{2} See footnote 11, Chap. 1.

hence the spectrum and covariance at the output, are the same as at the input up to coefficient. By (3.11),

$$\psi_\xi(\tau) \approx |k(\mathrm{i}\omega_0)|^2 \int\limits_{-\infty}^{+\infty} g_f(\omega)\mathrm{e}^{\mathrm{i}\omega\tau}\,d\omega = |k(\mathrm{i}\omega_0)|^2 \psi_f(\tau) \quad ,$$

and by (3.5)

$$\psi_\xi(\tau) = \int\limits_{-\infty}^{+\infty} H_2(\chi)\psi_f(\tau - \chi)\,d\chi \approx \psi_f(\tau) \int\limits_{-\infty}^{+\infty} H_2(\chi)\,d\chi = |k(0)|^2 \psi_f(\tau) \quad .$$

Thus, the filter output spectrum is formed by the narrower of the bands of the functions $g_f(\omega)$ and $|k(\mathrm{i}\omega)|^2$ and, accordingly, the output covariance is formed by the slower of the functions $f(t)$ and $H_2(\tau)$. The correlation time at the output, ϑ_ξ, is of the order of the largest of the quantities ϑ_f (the input correlation time) and ϑ (the filter settling time). Of course, these qualitative and concise formulations are all the more correct, the larger the difference between the bands or between the times in question.

Already at first sight harmonic filtration is seen to influence the covariance of a process at the filter output. Transfer functions of real dissipative systems narrow down the spectrum of the input action to a certain degree, thereby "lengthening" the output as compared with the input correlation. In particular, with white noise at the input (i.e., $g_f = $ const and the correlation time $\vartheta_f = 0$) the behavior of the spectral density at the output is, by (3.10), completely determined by the system $[g_\xi(\omega) = g_f|k(\mathrm{i}\omega)|^2]$ itself. Therefore,

$$\psi_f(\tau) = g_f \int\limits_{-\infty}^{+\infty} |k(\mathrm{i}\omega)|^2 \mathrm{e}^{\mathrm{i}\omega\tau}\,d\omega \quad ,$$

and the nonzero correlation time ϑ_ξ is determined simply by the filter settling time. A similar situation occurred in the example of the RC-cell just considered [(3.14) where $\vartheta_\xi = \vartheta$].

This is also clear from the temporal point of view. The filter *expands* input pulses, so that at all times $\vartheta_\xi > \vartheta_f$ and the inequality is the stronger, the narrower the filter band. This suggests the following conclusion regarding the distribution function of the output process, which at first sight might also appear to be apparent.

As we know, if the process $f(t)$ is Gaussian, the output process will be Gaussian as well. This follows also from (3.1), if we remember that the sum (or in the limit, the integral) of a normal process is a normal process, too. If the input process is not Gaussian, then in the general case the output process will not be Gaussian either. But for a very narrow filter band, i.e., $\vartheta_\xi \gg \vartheta_f$, the response $\xi(t)$ which, roughly speaking, consists of responses to $f(t)$ from $t - \vartheta_\xi$ to t, embraces many intervals ϑ_f, and hence a number of mutually uncorrelated actions. If, for instance, the input process is an impulse noise whose distribution is not normal (see Sect. I.3.3),

$$n_1 \vartheta_f \lesssim 1 \quad ,$$

then, since $\vartheta_\xi \gg \vartheta_f$, it may be that at the output we get

$$n_1 \vartheta_\xi \gg 1 \quad ,$$

and the distribution of the process $\xi(t)$ at the filter output, according to the central limit theorem (Sect. I.3.3), must be closer to normal. This is really so for a wide variety of cases, namely those where the pulse process $f(t)$ at the filter input consists of pulses with a *simple* shape $F(t)$. The temporal width of the pulse $(\Delta\tau_F)$ estimated over the width of its envelope (ϑ_F) will then be of the same order of magnitude as the estimate obtained using the "temporal correlation" of the pulse (τ_{F_c}). In other words, $\vartheta_F \sim \tau_{F_c}$ and, since the spectrum width $\Delta\omega_F \sim 1/\tau_{F_c}$, we have $\Delta\omega_F \cdot \Delta\tau_F \sim \Delta\omega_F \cdot \vartheta_F \sim 1$. Under these conditions this reasoning is valid, and narrow-band filtration actually "normalizes" $\xi(t)$ at the filter output[3]. But for *complicated pulse shapes, where* $\Delta\omega_F \cdot \vartheta_F \sim \vartheta_F/\tau_{F_c} \gg 1$ (Sect. 1.5), we can contrive a filtration such that it will "denormalize" the output as compared with the input process [3.2–4].

Thus, in this case of a complicated input pulse $F(t)$ $(\tau_{F_c} \ll \vartheta_F)$ a specially selected filter *reduces* the effective duration of the output pulse to the "correlation time" τ_{F_c}. Of course, the *total* duration of the output pulse always increases $(\vartheta_\xi > \vartheta_F)$. It cannot be shorter than ϑ_F if only because, after the input pulse has terminated, output oscillations carry on during the settling time of the filter. But the bulk of the energy of the output pulse appears to be concentrated within the time interval of the order of τ_{F_c}.

Such an effect is given by a filter *matched* to the input pulse $F(t)$, i.e., a filter whose pulse response $H(t)$ is a "mirror image" of $F(t)$

$$H(t) = F(t_0 - t) \quad . \tag{3.21}$$

Here t_0 may be arbitrary but not smaller than $\vartheta_F (t_0 \geq \vartheta_F)$, since the response must not preceed the input pulse $F(t)$ (realizability condition for the filter).

---

3 In [3.1] a simple quantitative estimate of the upper boundary B of the deviation of the distribution of $\xi(t)$ from the Gaussian probability density was obtained

$$\left| w_\xi(x) - w_G(x) \right| < B \quad .$$

It was also shown that $B \to 0$ as the filter band narrows down $(\Delta\omega_H \to 0)$. It was assumed that the process at the filter input had a threshold for the time for its statistical dependence, i.e., $f(t)$ and $f(t+\tau)$ are independent for $\tau > a$. In particular, for the response $\xi(t) = \sum_\nu H(t-t_\nu)$ to the shot noise $f(t) = \sum_\nu \delta(t - t_\nu)$ it was found that $B = \Delta\omega_H/2n_1$, where n_1 is the mean temporal density of pulses.

Let the input of such a matched filter be a pulse Poisson process

$$f(t) = \sum_\nu a_\nu F(t - t_\nu) \quad , \tag{3.22}$$

such that the covariance is proportional to the "covariance" of $F(t)$ [see (1.69) and (1.79)]

$$\psi_f(\tau) = n_1 \overline{a^2} \int\limits_{-\infty}^{+\infty} F(\theta) F(\theta + \tau)\, d\theta = n_1 \overline{a^2} \Psi_F(\tau) \quad . \tag{3.23}$$

According to (3.21, 22) at the filter output we will have the process

$$\begin{aligned}
\xi(t) &= \int\limits_{-\infty}^{+\infty} f(\theta) H(t - \theta)\, d\theta = \sum_\nu a_\nu \int\limits_{-\infty}^{+\infty} F(\theta - t_\nu) F(t_0 - t + \theta)\, d\theta \\
&= \sum_\nu a_\nu \int\limits_{-\infty}^{+\infty} F(\chi) F(\chi + t - t_0 - t_\nu)\, d\chi \\
&= \sum_\nu a_\nu \Psi_F(t - t_0 - t_\nu) \quad .
\end{aligned} \tag{3.24}$$

The output pulses are thus described by the "covariance" of the input pulse

$$F_\xi(t) = \Psi_F(t - t_0) \quad .$$

Accordingly, the pulse $F_\xi(t)$ is concentrated within the correlation interval τ_{F_c}, which for complicated shapes of $F(t)$ is much smaller than ϑ_F ($\tau_{F_c} \ll \vartheta_F < \vartheta_\xi$). This "compression" of complicated pulse shapes by the matched filter is widely used in present-day devices for optimal signal detection, in particular in radar [3.5].

Figure 3.4a shows a complicated pulse $F(t)$ with a rectangular envelope filled with frequency-modulated oscillations with linearly varying frequency, and the pulse response $H(t)$ of the matched filter. Figure 3.4b represents the envelopes for $F(t)$ and the response $F_\xi(t)$ of the matched filter. The envelope $|F_\xi(t)|$ includes the main maximum containing the bulk of the energy, its width τ_{F_c} being of the order of $1/\alpha\vartheta_F$, where $\alpha\vartheta_F$ is the deviation of the instantaneous frequency $\omega(t) = \omega_0 + \alpha t$ over the width ϑ_F of the input pulse (see Exercise 1.6.6).

Why then is the distribution of the Poisson process (3.24) at the output of the matched filter denormalized? The point is that basic to the central limit theorem is the fact that each of the independent random summands makes *about the same* small contribution to the total covariance (Sect. I.3.5). But in this case, although $\xi(t)$ contains more pulses emerging over the entire width of the pulse $F_\xi(t)$, then $f(t)$, i.e.,

$$n_1 \vartheta_\xi > n_1 \vartheta_F \quad ,$$

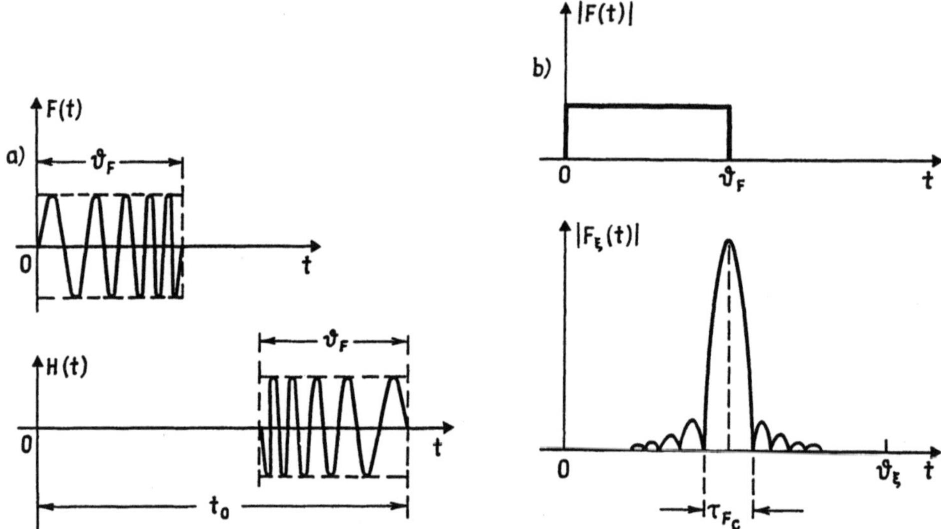

Fig. 3.4. (a) Pulse $F(t)$ with linearly varying frequency and a rectangular envelope, and the pulse response $H(t)$ of the matched filter. (b) Envelopes for $|F(t)|$ and response of the matched filter $|F_\xi(t)|$

the summands are far from being equivalent. In $\xi(t)$ small lateral lobes of pulses $F_\xi(t)$ overlap frequently and it is only rarely that there are large overshoot contributions due to overlaps of the main maxima. The mean number of random summands of this second type is of order $n_1 \tau_{F_c} \ll n_1 \vartheta_F < n_1 \vartheta_\xi$ and, if these are singled out, their distribution is substantially farther from the normal than that of $f(t)$. But these large contributions contain about 90 % of the total power of the process $\xi(t)$ [3.3]. Clearly, precisely these latter dominate the distribution of $\xi(t)$. Consequently, the denormalization phenomenon by no means contravenes the central limit theorem. For the distribution of $\xi(t)$ to be normalized as compared with $f(t)$, it is simply required that the filter increase the *effective* duration of the output pulse $F_\xi(t)$. This duration is about $\vartheta_\xi > \vartheta_f$ for simple pulses $F(t)$, but it drops to $\tau_{F_c} \ll \vartheta_F < \vartheta_\xi$ for complicated pulse shapes and their matched filters[4].

3.2 Random Actions on Memoryless Nonlinear Systems

Of interest in many cases is the action on a nonlinear system of a deterministic and/or random process (deterministic signal plus noise; random signal plus noise; noise alone). The system itself consists not only of nonlinear elements

---

[4] In [3.3] a quantitative treatment is given for the denormalization of a Poisson process consisting of pulses such as shown in Fig. 3.4a. The sensitivity of the effect to the filter matching degree is discussed, and it is shown that with accurate matching the denormalization may be very strong. The ratio of excess coefficients $\gamma = \lambda_4/\lambda_2^2$ (Sect. I.3.3) for input and output processes (γ_ξ/γ_f) may, under real conditions, be as high as 10^2–10^3.

(detectors, limiters), but also of linear circuits (filters, resonant circuits, etc.). Such a system is described by nonlinear differential equations that may contain, besides the argument t, the "delay argument" $t - \vartheta$ (delay lines are present). One of the possible problems with a fully defined input process, (i.e., one with known finite-variate distributions) is to obtain an output process that is equally fully described. The problem presents a challenge even for linear systems, and even more so for nonlinear ones.

No universal techniques exist (just as in the case of the determnistic action) and for each concrete problem we have to resort to some tricks related either to a special kind of equation describing the system or to the special nature of the action (Markov process, white noise, etc.), or to a restricted formulation of the problem.

For example, problems frequently occur in which the nonlinear terms of respective equations can be regarded as small. An approximate solution is then sought using some form of perturbation theory. We used this technique in Sect. I.5.8 when we studied natural fluctuations in the Thomson self-oscillatory system.

A problem may have another constraint: for a random process $\eta(t)$ at the output of a nonlinear system we may not need distribution funcitons, but only moments of $\eta(t)$, specifically only $\langle\eta(t)\rangle$ and $\psi_\eta(t_1, t_2)$. But in that case too, because the stochastic differential equation for $\eta(t)$ is nonlinear, the result is often an infinite system of linked equations for the moments of $\eta(t)$ of various orders, and this system has to be truncated. Interestingly, the very stochasticity of a nonlinear differential equation for $\eta(t)$ is sometimes of help in finding the moments for the process, since it allows one not to solve explicitly the original equation.

This illustrates why, although there has been some progress in the theory of nonlinear stochastic euqations, it is not as spectacular as one would wish (see, e.g., [3.6, 7]).

Memoryless nonlinear systems are easier to handle, since they deal simply with a nonlinear functional relationship between the input process $\xi(t)$ and the output process $\eta(t)$

$$\eta(t) = F[\xi(t)] \quad , \tag{3.25}$$

where the function F describes the behavior of the nonlinear element (detector, limiter, etc.). The circuit can also contain a linear but purely resistive load: any reactance would immediately lead to a nonlinear differential equation. For the case of memoryless nonlinear systems the finite-variate distributions of $\eta(t)$ are found easily, if they are known for $\xi(t)$, but we will first discuss how moments of $\eta(t)$ are derived.

To find any nth other moment of the form

$$\overline{\eta^\alpha(t_1)\eta^\beta(t_2)\ldots n^\kappa(t_k)} \quad , \quad \alpha + \beta + \ldots + \kappa = n \tag{3.26}$$

at the output of a linear system [since $\eta(t)$ is represented as a linear operator of $\xi(t)$] only requires a knowledge of the mixed moment $\xi(t)$ of the same order n of the form

$$\overline{\xi(t_1)\xi(t_2)\ldots\xi(t_n)}\quad.$$

This is no longer sufficient, even in the case of a nonlinear system as simple as (3.25). The moment of the form (3.26) can only be calculated given the k-variate distribution function for the input process $w_\eta^{(k)}(x_1, x_2, \ldots, x_k)$ [for simplicity, we will omit the parameters $(t_1, t_2, \ldots, t_k)$]. In fact,

$$\overline{\eta^\alpha(t_1)} = \int F^\alpha(x) w_\xi^{(1)}(x)\, dx \quad,$$

$$\overline{\eta^\alpha(t_1)\eta^\beta(t_2)} = \int F^\alpha(x_1) F^\beta(x_2) w_\xi^{(2)}(x_1, x_2)\, dx_1\, dx_2 \quad,$$

$$\begin{aligned}
&\overline{\eta^\alpha(t_1)\eta^\beta(t_2)\ldots\eta^\kappa(t_k)} \\
&\quad = \int F^\alpha(x_1) F^\beta(x_2) \ldots F^\kappa(x_k) w_\xi^{(k)}(x_1, \ldots, x_k)\, dx_1 \ldots dx_k \quad.
\end{aligned} \tag{3.27}$$

But once we know moments of any order for the collection of k random variables $\eta(t_1), \ldots, \eta(t_k)$, we can (barring special cases of no practical interest) also set up the k-variate distribution function $w_\eta^{(k)}(y_1, y_2, \ldots, y_k)$. Thus, $w_\eta^{(k)}$ must be determined by $w_\xi^{(k)}(x_1, x_2, \ldots, x_k)$, which is known. This is true because the dependence (3.25), which is valid at any t, reduces the determination of $w_\eta^{(k)}$ to the transition in $w_\xi^{(k)}$ from the variables $x_1, x_2, \ldots, x_k$ to $y_1, y_2, \ldots, y_k$.

In Exercise I.2.8.7 this has been done for a one-to-one transformation of the general form $\boldsymbol{\eta} = \boldsymbol{F}(\boldsymbol{\xi})$, i.e., for

$$y_1 = F_1(x_1, \ldots, x_k), \ldots, \quad y_k = F_k(x_1, \ldots, x_k) \quad. \tag{3.28}$$

The distribution function $w_\eta^{(k)}$ is expressed here in terms of $w_\xi^{(k)}$ in the following way:

$$w_\eta^{(k)}(y_1, \ldots, y_k) = w_\xi^{(k)}(x_1, \ldots, x_k) \left| \frac{\partial(x_1, \ldots, x_k)}{\partial(y_1, \ldots, y_k)} \right| \quad. \tag{3.29}$$

We inserted $x_i = F_i^{-1}(y_1, \ldots, y_k)$ into the right-hand side according to the inverse transformation $\boldsymbol{\xi} = \boldsymbol{F}^{-1}(\boldsymbol{\eta})$, which is assumed to be one-to-one. We are interested here in the special case of the transformation (3.28), where

$$y_1 = F(x_1), \ldots \quad, \quad y_k = F(x_k)$$

and, inversely,

$$x_1 = F^{-1}(y_1), \ldots \quad, \quad x_k = F^{-1}(y_k) \quad.$$

It is easily seen that (3.29) takes the form

$$w_\eta^{(k)}(y_1, \ldots, y_k) = w_\xi^{(k)}[F^{-1}(y_y), \ldots, F^{-1}(y_k)]|F^{-1'}(y_1)\ldots F^{-1'}(y_k)| \quad .$$

$$(3.30)$$

This expression would give the complete answer were it not for one complication arising from the fact that not infrequently real nonlinear apparatus are described by the *multivalued* inverse function F^{-1}. We will therefore consider this case.

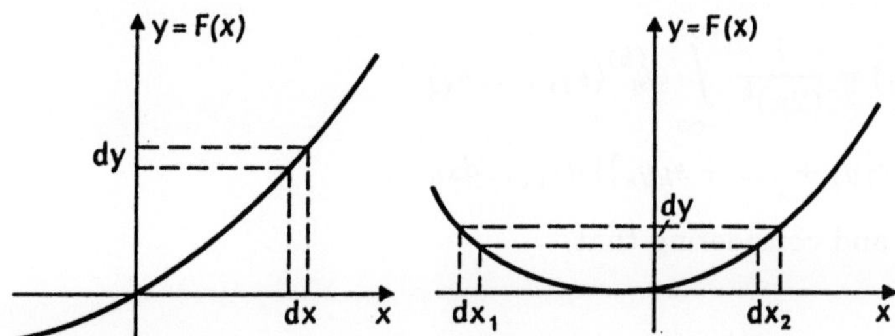

Fig. 3.5. Nonlinear behavior $y = F(x)$ for one-valued (*left*) and two-valued (*right*) inverse functions $x = F^{-1}(y)$

Figure 3.5 gives two examples of nonlinear behavior $y = F(x)$: one has a single-valued inverse function $x = F^{-1}(y)$ and the other a two-valued one. in the first case the probability for y to be found in the interval dy is the same as for x to be found in $dx = |F^{-1}(y)|dy$, so that

$$w_\eta^{(1)}(y)dy = w_\xi^{(1)}[F^{-1}(y)]|F^{-1'}(y)| \, dy \quad .$$

$$(3.31)$$

In the second case the probability that y will be found in dy equals the sum of the probabilities of two incompatible events: for x to be found either in $dx_1 = |F_1^{-1'}(y)| \, dy$, or in $dx_2 = |F_2^{-1'}(y)| \, dy$. Here F_1^{-1} and F_2^{-1} stand for two branches of the inverse function $x = F^{-1}(y)$. Accordingly, in the second case

$$w_\eta^{(1)}(y) \, dy = w_\xi^{(1)}[F_1^{-1}(y)]|F_1^{-1'}(y)| \, dy + w_\xi^{(1)}[F_2^{-1}(y)]|F_2^{-1'}(y)| \, dy \quad .$$

$$(3.32)$$

In general, if $x = F^{-1}(y)$ has m branches, then

$$w_\eta^{(1)}(y) \, dy = \sum_{\nu=1}^{m} w_\xi^{(1)}[F_\nu^{-1}(y)]|F^{-1'}(y)| \, dy \quad .$$

$$(3.33)$$

This result might also be arrived at in a somewhat different, more formal, but instructive way. Knowing $w_\xi^{(k)}$ we may derive the k-variate characteristic function for the output process

$$\varphi_\eta^{(k)}(s_1, \ldots, s_k)$$

$$= \overline{\exp\{i[s_1\eta_1 + \ldots + s_k\eta_k]\}}$$

$$= \overline{\exp\{i[s_1 F(\xi_1) + \ldots + s_k F(\xi_k)]\}}$$

$$= \int \exp\{i[s_1 F(x_1) + \ldots + s_k F(x_k)]\} w_\xi^{(k)}(x_1, \ldots, x_k)\, dx_1 \ldots dx_k \quad.$$

$$(3.34)$$

Using $\varphi_\eta^{(k)}$ we can easily find any moments of the form (3.26) (see Sect. I.3.2), but using the Fourier transformation we can also obtain $w_\eta^{(k)}$

$$w_\eta^{(k)}(y_1, \ldots, y_k) = \frac{1}{(2\pi)^k} \int\limits_{-\infty}^{+\infty} \varphi_\eta^{(k)}(s_1, \ldots, s_k)$$

$$\times \exp\{-i[s_1 y_1 + \ldots + s_k y_k]\}\, ds_1 \ldots ds_k \quad.$$

Substituting (3.34) and considering that

$$\int\limits_{-\infty}^{+\infty} \exp\{is_j[F(x_j) - y_j]\} ds_j = 2\pi\delta[F(x_j) - y_j] \quad,$$

we find

$$w_\eta^{(k)}(y_1, \ldots, y_k) = \int \delta[F(x_1) - y_1]$$

$$\times \ldots \delta[F(x_k) - y_k] w_\xi^{(k)}(x_1, \ldots, x_k)\, dx_1 \ldots dx_k \quad. \qquad (3.35)$$

The delta-function $\delta[F(x_j) - y_j]$ contributes to the integral only at points where $F(x_j) = y_j$, i.e., at $x_j = x_j^{(\nu)} = F_\nu^{-1}(y_j)$, where $F_\nu^{-1}(y)$ is the νth branch ($\nu = 1, 2, \ldots, m$) of the inverse function of $F(t)$. In the immediate neighborhood of each of the points (for the delta-function an arbitrarily small neighborhood is sufficient) we can ssume that

$$F(x_j) - y_j \approx F(x_j^{(\nu)}) + F'(x_j^{(\nu)})(x_j - x_j^{(\nu)}) - y_j = F'(x_j^{(\nu)})(x_j - x_j^{(\nu)})$$

and hence, in the neighborhood of $x_j^{(\nu)}$

$$\delta[F(x_j) - y_j] = \delta[F'(x_j^{(\nu)})(x_j - x_j^{(\nu)})]$$

$$= |F'(x_j^{(\nu)})|^{-1}\delta(x_j - x_j^{(\nu)})$$

$$= |F_\nu^{-1'}(y_j)|\delta[x_j - F_\nu^{-1}(y_j)] \quad.$$

And for all the points $(x_j^{(\nu)})$ ($\nu = 1, 2, \ldots, m$) we have

$$\delta[F(x_j) - y_j] = \sum_{\nu=1}^{m} |F_\nu^{-1'}(y_j)|\delta[x_j - F_\nu^{-1}(y_j)] \quad.$$

We now write the product

$$\prod_{j=1}^{k} \delta[F(x_j) - y_j] = \sum_{\alpha,\beta,\dots,\kappa=1}^{m} |F_\alpha^{-1'}(y_1)F_\beta^{-1'}(y_2)\dots F_\kappa^{-1'}(y_k)|$$
$$\times \delta[x_1 - F_\alpha^{-1}(y_1)]\dots\delta[x_k - F_\kappa^{-1}(y_k)] \quad ,$$

substitute it into (3.35) and arrive at

$$w_\eta^{(k)}(y_1,\dots,y_k)$$
$$= \sum_{\alpha,\dots,\kappa=1}^{m} w_\xi^{(k)}[F_\alpha^{-1}(y_1), F_\beta^{-1}(y_2),\dots,F_\kappa^{-1}(y_k)]$$
$$\times |F_\alpha^{-1'}(y_1)\dots F_\kappa^{-1'}(y_k)| \quad . \tag{3.36}$$

In the special case of the univariate distribution function this yields (3.33). With a one-to-one correspondence, (3.36) reduces to the single-term formula (3.30).

We have seen that a process which is normal at the input of a *linear* system also produces a normal process at the output. Furthermore, for a sufficiently narrow pass-band the linear system could give a normal process with non-Gaussian action at the output. The equations derived indicate that a nonlinear system (at least in the memoryless case under consideration) drastically changes the distribution. In particular, it disturbs any normal distribution of the input process.

Consider several simple examples.

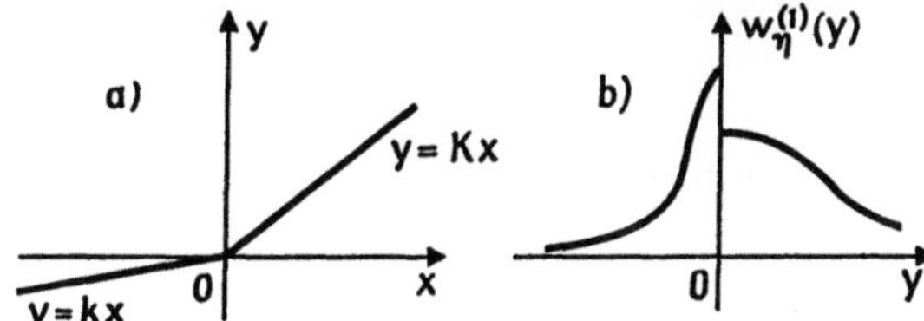

Fig. 3.6. (a) Broken linear characteristic; (b) distribution of an output process due to a Gaussian process at the input

Let the characteristic $y = F(x)$ consist of two rays with slopes K (for $x>0$) and k (for $x<0$) (Fig. 3.6a). If the input process is Gaussian with a standard deviation σ, then it is intuitively clear that the distribution at the output will consist of the two Gaussian laws: for $x>0$ with standard deviation $K\sigma$; and for $x<0$, with standard deviation $k\sigma$ (Fig. 3.6b). As $k \to 0$ (the transition to the so-called linear detector) the left side of $w_\eta^{(1)}(y)$ compresses into the delta-function $\delta(y)/2$: at $k = 0$ any $x<0$ gives at the output $y = 0$.

For the square-law detector

$$y = \beta x^2 \quad ,$$

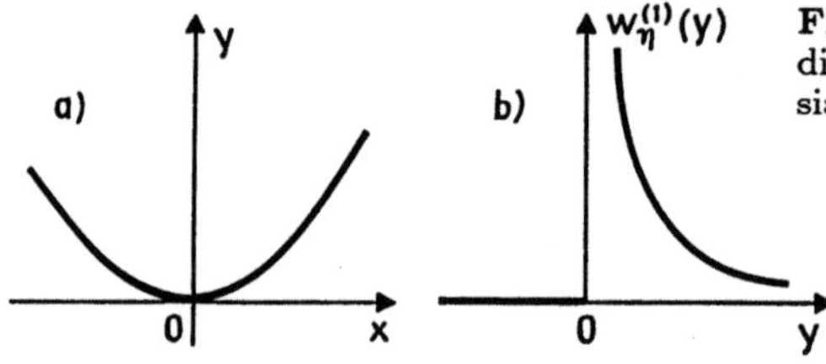

Fig. 3.7. (a) Square-law characteristic; (b) distribution of an output process with a Gaussian process at the input

(Fig. 3.7a) the inverse function is two-valued

$$x = F_{1,2}^{-1}(y) = \pm\sqrt{y/\beta} \quad .$$

By (3.33), we get

$$w_\eta^{(1)}(y) = \begin{cases} \dfrac{1}{2\sqrt{\beta y}}\left\{ w_\xi^{(1)}\left(\sqrt{\dfrac{y}{\beta}}\right) + w_\xi^{(1)}\left(-\sqrt{\dfrac{y}{\beta}}\right)\right\} & \text{for} \quad y>0 \quad , \\ 0 & \text{for} \quad y<0 \quad . \end{cases}$$

If the distribution of ξ is normal, then the distribution for η will be

$$w_\eta^{(1)}(y) = \begin{cases} \dfrac{1}{\sqrt{2\pi\beta y}\,\sigma}e^{-y/2\beta\sigma^2} & \text{for} \quad y>0 \quad , \\ 0 & \text{for} \quad y<0 \quad , \end{cases} \tag{3.37}$$

(Fig. 3.7b), i.e., just as in the first example, it is no longer Gaussian.

Let us turn to the ways in which the mean $\overline{\eta}$ and the moment $\overline{\eta\eta_\tau}$ can be derived in some special cases. According to (2.7), we have

$$\overline{\eta}(t) = \int F(x)w_\xi^{(1)}(x,t)\,dx \quad ,$$

$$\overline{\eta\eta_\tau} = \int F(x_1)F(x_2)w_\xi^{(2)}(x_1,t;x_2,t+\tau)\,dx_1\,dx_2 \quad .$$

Various mathematical tricks are employed to evaluate the double integral in the latter expression: reduction to a contour integral, a variety of expansions of $w_\xi^{(2)}$ into a series, e.g., into a system of orthogonal functions $\varphi_n(x,t)$

$$w_\xi^{(2)}(x_1,t_1;x_2,t_2)$$
$$= w_\xi^{(1)}(x_1,t_1)w_\xi^{(1)}(x_2,t_2) + \sum_{m,a} a_{mn}(t_1,t_2)\varphi_m(x_1,t_1)\varphi_n(x_2,t_2)$$

and other techniques related to specific features of the problem.

Let, for instance, $\xi(t)$ be a *stationary Gaussian* process, such that

$$w_\xi^{(2)}[x_1,x_2,K(\tau)] = \frac{1}{2\pi\sigma^2\sqrt{1-K^2}}\exp\left\{-\frac{x_1^2+x_2^2-2Kx_1x_2}{2\sigma^2(1-K^2)}\right\} \quad . \tag{3.38}$$

Expanding $w_\xi^{(2)}$ in powers of $K(\tau)$, we readily obtain

$$\overline{\eta\eta_\tau} = \sum_{n=0}^{\infty} C_n^2 K^n(\tau) \quad , \qquad \text{where} \tag{3.39}$$

$$C_n = \frac{1}{\sqrt{2\pi n!}} \int\limits_{-\infty}^{+\infty} F(\sigma x) \frac{d^n e^{-x^2/2}}{dx^n} \, dx \quad .$$

Suppose that $K(\tau) \to 0$ as $\tau \to \infty$. Then, from (3.39), we get

$$\overline{\eta\eta_\tau} \to \overline{\eta}^2 = C_0^2 \quad (\tau \to \infty), \qquad \text{hence}$$

$$\psi_\eta(\tau) = \overline{\eta\eta_\tau} - \overline{\eta}^2 = \sum_{n=1}^{\infty} C_n^2 K^n(\tau) \quad . \tag{3.40}$$

For positive frequencies the spectral density will be

$$g_{\eta+}(\omega) = \frac{2}{\pi} \int\limits_0^\infty \psi_\eta(\tau) \cos \omega\tau \, d\tau = \frac{2}{\pi} \sum_{n=1}^{\infty} C_n^2 \int\limits_0^\infty K^n(\tau) \cos \omega\tau \, d\tau \quad . \tag{3.41}$$

The shape of the output spectrum is thus determined both by the correlation coefficient $K(\tau)$ of the input process (i.e., its spectrum), and by the nonlinear characteristic $y = F(x)$, whose form determines the values of the coefficients C_n.

   If the Gaussian process at the input has a narrow spectrum and is (for simplicity) symmetrical with respect to a frequency ω_0, then according to Sect. 2.2

$$K(\tau) = k(\tau) \cos \omega_0\tau \quad , \tag{3.42}$$

where $k(\tau)$ is a slowly-varying function of τ. In that case, the bivariate distribution (3.38) is a periodic function of $\omega_0\tau$ and can be expanded into a Fourier series

$$w_\xi^{(2)}(x_1, x_2, k \cos \omega_0\tau) = \sum_{n=0}^{\infty} \varepsilon_n v_n[x_1, x_2, k(\tau)] \cos n\omega_0\tau \quad ,$$

where

$$\varepsilon_n = \begin{cases} 1 & \text{for} \quad n = 0 \quad , \\ 2 & \text{for} \quad n > 0 \quad . \end{cases}$$

Substitution into the expression for $\overline{\eta\eta_\tau}$ gives

$$\overline{\eta\eta_\tau} = \sum_{n=0}^{\infty} \varepsilon_n B_n(\tau) \cos n\omega_0\tau \quad , \tag{3.43}$$

where the coefficients

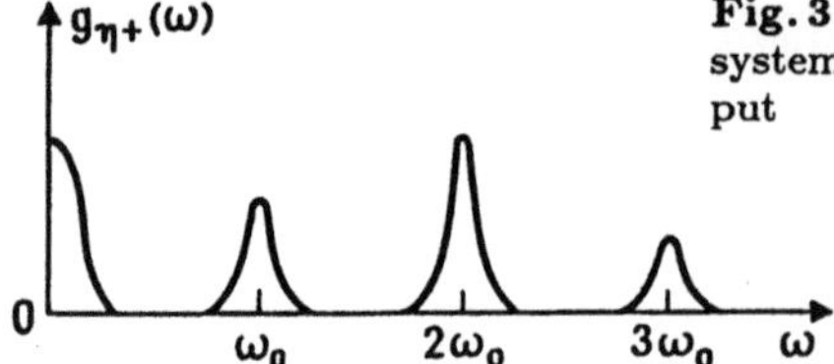

Fig. 3.8. Output spectrum for a nonlinear memoryless system with a narrow-band Gaussian process at the input

$$B_n(\tau) = \int F(x_1)F(x_2)v_n(x_1, x_2, k)\, dx_1\, dx_2$$

are slowly-varying functions of τ. The spectrum $\eta(t)$ therefore consists of narrow bands at frequencies $n\omega_0$ (Fig. 3.8). Of course, if we substitute the expression (3.42) for $K(\tau)$ in (3.39), express the powers of $\cos \omega_0\tau$ in terms of the cosines of multiple arcs and rearrange the terms, we will arrive at (3.43).

For the quasi-monochromatic input process $\xi(t) = A(t) \cos \Phi(t)$, where $\Phi(t) = \omega_0 t + \varphi(t)$, we can calculate $\overline{\eta}$ and $\overline{\eta\eta_\tau}$ in another way as well, namely by using the distribution functions for the envelope A and phase φ. The periodic function $\eta = F(\xi) = F(A \cos \Phi)$ is expanded into a Fourier series in Φ

$$\eta = \sum_{n=0}^{\infty} \varepsilon_n F_n(A) \cos n\Phi \quad , \qquad \text{where} \tag{3.44}$$

$$F_n(A) = \frac{1}{\pi} \int_0^{\pi} F(A \cos \Phi) \cos n\Phi\, d\Phi \quad .$$

Since $\Phi(t) = \omega_0 t + \varphi(t)$, the nth term of the series (3.44) is a quasi-monochromatic process whose spectrum lies in the neighborhood of $n\omega_0$. From Sect. 2.3 we know that the stationary process $\xi(t)$ has for A and φ a joint distribution of the form

$$w(A, \varphi)\, dA\, d\varphi = w_A(A)\, dA \frac{d\varphi}{2\pi}$$

and hence

$$\overline{\eta} = \sum_{n=0}^{\infty} \varepsilon_n \overline{F_n(A)}\,\overline{\cos n\Phi} = \overline{F_0(A)} = \int_0^{\infty} F_0(A) w_A(A)\, dA \quad .$$

To calculate the mean of $\eta(t)$ requires only a knowledge of the distribution of the envelope $w_A(A)$, but to calculate $\overline{\eta\eta_\tau}$ already requires the four-variate distribution of A, A_τ, φ, and φ_τ. If $\xi(t)$ is a Gaussian process, this distribution is (2.27). All the computation can be done analytically.

Cumulant equations (as exemplified by Exercises I.3.7.15, 16 and by 3.6.3, 4) may be of substantial help in evaluating the moments of a process at the output of a memoryless nonlinear circuit element, especially when the input process is not normal.

Results partaining to particular forms of the nonlinear characteristic (3.25) can, of course, be derived from general relationships, but if the characteristic is simple, it is easier to perform calculations directly for this characteristic.

Let, for instance, the square-law detector $\eta = \beta \xi^2$ be exposed to a superposition of the deterministic signal $s(t)$ and the Gaussian noise $n(t)$ with $\overline{n} = 0$, $\overline{n^2} = \sigma^2$ and $\overline{nn_\tau} = \sigma^2 K(\tau)$

$$\xi(t) = s(t) + n(t) \quad .$$

Hence

$$\eta = \beta(n^2 + 2ns + s^2) \quad , \quad \overline{\eta} = \beta(\sigma^2 + s^2) \quad ,$$

$$\eta - \overline{\eta} = \beta(n^2 + 2ns - \sigma^2) \quad ,$$

so that

$$\begin{aligned}
\psi_\eta(t,\tau) &= \overline{(\eta - \overline{\eta})(\eta_\tau - \overline{\eta}_\tau)} = \beta^2 \overline{(n^2 + 2ns - \sigma^2)(n_\tau^2 + 2n_\tau s_\tau - \sigma^2)} \\
&= 2\beta^2 \sigma^2 |\sigma^2 K^2(\tau) + 2ss_\tau K(\tau)| \quad .
\end{aligned} \tag{3.45}$$

This takes into account the fact that, with the normal function $n(t)$, odd moments are zero and $\overline{n^2 n_\tau^2} = \sigma^4(2K^2 + 1)$. The covariance $\psi_\eta(t,\tau)$ depends on t through the product $ss_\tau = s(t)s(t+\tau)$. In the absence of a signal the process $\eta(t)$ at the detector output is stationary and the covariance is

$$\psi_\eta(\tau) = 2\beta^2 \sigma^4 K^2(\tau) \quad . \tag{3.46}$$

This is nothing other than the power expansion (3.40) that now only contains one term, namely $C_2 = \sqrt{2}\beta\sigma^2 \neq 0$.

If the noise is quasi-monochromatic and (3.42) holds, then

$$\psi_\nu(\tau) = \beta^2 \sigma^4 k^2(\tau)(1 + \cos 2\omega_0 \tau) \quad , \tag{3.47}$$

i.e., the initial spectrum localized in the neighborhood of ω_0 produces two lines at $\omega = 0$ and $\omega = 2\omega_0$ at the square-law detector output (see Fig. 3.9). The slowly-varying part lying close to $\omega = 0$ can be separated, and this is actually done in the majority of cases by the subsequent video filter. In general, if a nonlinear element is included between linear circuits (e.g., filters) and one link

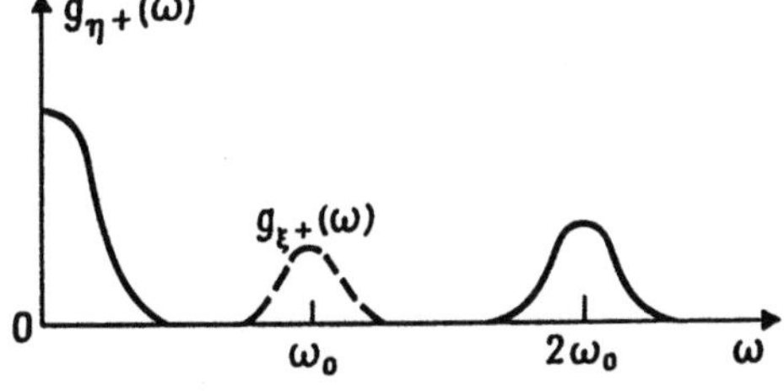

Fig. 3.9. Spectrum $g_{\eta+}(\omega)$ at the output of a square-law detector and spectrum $g_{\xi+}(\omega)$ of the normal noise at the input

of the chain does not affect the preceding one, then the circuit can be designed step by step.

Formula (3.47) is, of course, the Fourier expansion (3.43), where only $B_0(\tau)$ and $B_2(\tau)$ are nonzero.

According to (3.45), the variance in the presence of a signal at the output of the square-law detector is

$$D[\eta(t)] = \psi_\eta(t,0) = 2\beta^2\sigma^2[\sigma^2 + 2s^2(t)] \quad .$$

If the signal and noise are quasi-monochromatic, and the process at the detector output is averaged over the high-frequency period $T_0 = 2\pi/\omega_0$ (such an operation is also carried out by a video filter cutting off frequencies $\omega \geq \omega_0$), then the variance will be

$$\widetilde{D[\eta(t)]}_{T_0} = 2\beta^2\sigma^2[\sigma^2 + 2\widetilde{s^2(t)}_{T_0}] \quad ,$$

the wave line signifying time averaging. The simplest condition for the signal to be *directly* detected against the background of noise is the inequality $2\widetilde{s^2(t)}_{T_0} > \sigma^2$, i.e., a sufficiently large ratio of the mean powers of the signal and noise (or, for short, "signal-to-noise" ratio).

Of course, such a criterion for signal detection in noise is neither general nor universal. The general formulation of the problem is complicated enough, as it should take into account the distinctive features of the signal (its pulse or stationary character; its dependence on various parameters that may be in part random and in part deterministic). It must also give both the probability of correct detection (the probability that in the presence of a signal a certain threshold at the detector output will be exceeded) and the probability of false detection — or to use radar terminology , of false alarm (the probability that the threshold will be exceeded without any signal, due to noise only). Such issues are at present of particular interest for applications of the theory of random functions. They are of importance not only for radio communication and radar, but for measurements in the presence of noise in general. However, on the whole, this domain may be referred to as typical "radiomathematics". As far as the physics is concerned it presents no problems of interest and therefore we will not deal with its inherent issues, which are discussed in a large body of literature (noteworthy are, for instance, the books [3.8–11]).

3.3 Noise Measurement. Radiometers [5]

We will now use the result of the preceding two sections to examine a widely-used circuit consisting of three blocks (Fig. 3.10): a high-pass filter (h.p.f.) with transfer function $k_1(i\omega)$, a square-law detector (s.l.d.) with nonlinear response

---

[5] This section was written in collaboration with N.N. Kalachevsky.

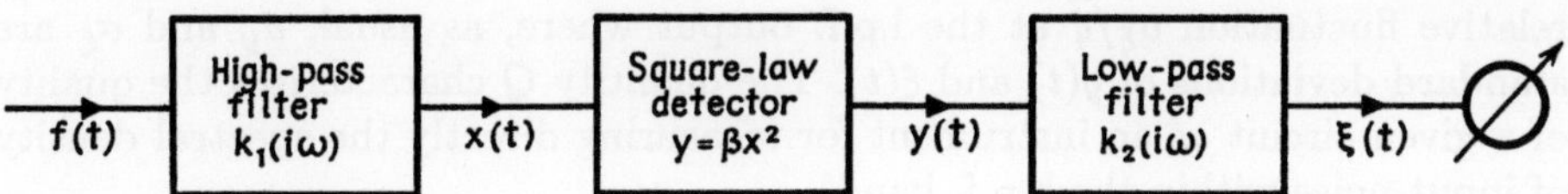

Fig. 3.10. Three-block circuit diagram

$y = \beta x^2$, and a low-pass filter (l.p.f.) with transfer function $k_2(i\omega)$. A measuring device is connected to the l.p.f. output. As an input signal we will use the stationary normal process $f(t)$ with zero mean. Regarding the circuit we assume that:

1. There is no reaction of a block to the preceding one;
2. The filters have narrow pass-bands, i.e., for the h.p.f. the band $\Delta\omega$ that contains a frequency ω_0 is far narrower than both ω_0 and the width of the spectrum for the process $f(t)$; and the l.p.f. band $\Delta\Omega$ (for $\omega>0$) is much narrower than not only ω_0, but also $\Delta\omega$ (Fig. 3.11).

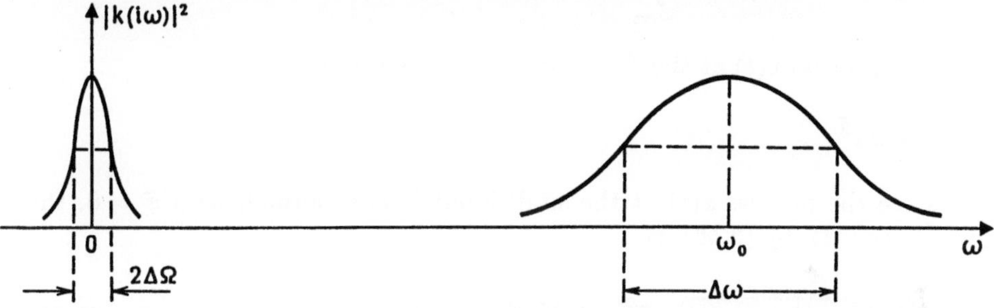

Fig. 3.11. High- and low-pass filter bands

The relationship between the statistical behavior of the input process $f(t)$ and the process $\xi(t)$ measured at the output has been considered in many works, beginning with the classic paper by *Rice* [3.12]. Under assumption (1) problems with varying degrees of generality have been studied: distribution functions for $\xi(t)$ – both w_1 and w_2, and w_n – have been found; moments of $\xi(t)$ have been obtained when there is not only noise at the input but also a deterministic or random signal; the same issues have been studied for a non-Gaussian distribution of $f(t)$, for arbitrary nonlinearity of the detector or, in the absence of constraints (2) on the band width of the filters (see [3.11]). We will confine ourselves, however, to a simple problem, namely that of finding the benefits provided by the smoothing action of the l.p.f. given the above assumptions. Or, more exactly, we will calculate the *smoothing coefficient* for fluctuations Q

$$Q = \frac{\sigma_y/\overline{y}}{\sigma_\xi/\overline{\xi}} \ , \tag{3.48}$$

that is, the ratio of the relative fluctuation $\sigma_y/\overline{y}$ at the detector output to the relative fluctuation $\sigma_\xi/\overline{\xi}$ at the l.p.f. output where, as usual, σ_y and σ_ξ are standard deviations of $y(t)$ and $\xi(t)$. The quantity Q characterizes the quality of a given circuit as an instrument for measuring directly the spectral density of input noise within the h.p.f. band[6].

Since the spectrum of the noise $f(t)$ is assumed to be wide as compared with the h.p.f. band, the spectral density $x(t)$ can be written as

$$g_x(\omega) = g_f(\omega_0)|k_1(\mathrm{i}\omega)|^2 \quad . \tag{3.49}$$

Hence, the covariance of the process $x(t)$ is

$$\psi_x(\tau) = g_f(\omega_0) \int\limits_{-\infty}^{+\infty} |k_1(\mathrm{i}\omega)|^2 e^{\mathrm{i}\omega\tau}\, d\omega \quad ,$$

and its mean square is

$$\overline{x^2} = \psi_x(0) = g_f(\omega_0) \int\limits_{-\infty}^{+\infty} |k_1(\mathrm{i}\omega)|^2\, d\omega \quad . \tag{3.50}$$

The process $y(t)$ at the detector output has the mean

$$\overline{y} = \beta\overline{x^2} \quad ,$$

and since the process $x(t)$ at the s.l.d. input is also normal, with $\overline{x} = 0$, the covariance $y(t)$, by (3.46), is

$$\psi_y(\tau) = \overline{y(t+\tau)y(t)} - \overline{y}^2 = 2\beta^2\psi_x^2(\tau) \quad . \tag{3.51}$$

Thus, $\sigma_y^2 = \psi_y(0) = 2\beta^2\langle x^2\rangle^2$ and the relative fluctuation at the detector output will be

$$\sigma_y/\overline{y} = \sqrt{2} \quad . \tag{3.52}$$

By the theorem on the Fourier transform of a product of two functions (Sect. 1.5) (or in the present case the square of $\psi_x(\tau)$) we obtain from (3.51) an expression for the spectral density $y(t)$ in the form of the convolution of $x(t)$[7]

---

[6] The filters may, of course, contain amplifiers operating in a linear mode. In the general case, "input noise" refers not only to the external noise signal, but also to some equivalent of receiver noise (see below).

[7] In Sect. 1.5 (footnote 11) the theorem was given for the convolution of two functions of t and, accordingly, for the product of their Fourier transforms, i.e., of functions of ω. But the fact that the direct and inverse Fourier transforms are equivalent enables one to "interchange" t and ω, i.e., to formulate a similar theorem for the convolution of two functions of ω and, consequently, of the product of their Fourier transforms, which are functions of t.

$$g_y(\omega) = 2\beta^2 \int\limits_{-\infty}^{+\infty} g_x(\omega')g_x(\omega - \omega')\,d\omega' \quad . \tag{3.53}$$

This formula describes, of course, the *entire* spectrum of $y(t)$: both the high-frequency (band width $2\Delta\omega$ at $\omega = \pm 2\omega_0$), and the low-frequency part (band width $2\Delta\omega$ at $\omega = 0$; see Fig. 3.9). It goes without saying that the l.p.f. does not transmit the high-frequency part.

Knowing $g_y(\omega)$, we can write the spectral density at the l.p.f. output

$$g_\xi(\omega) = g_y(\omega)|k_2(\mathrm{i}\omega)|^2 \quad , \tag{3.54}$$

and hence the expression for the covariance of $\xi(t)$

$$\sigma_\xi^2 = \overline{\xi^2} - \overline{\xi}^2 = \int\limits_{-\infty}^{+\infty} g_\xi(\omega)\,d\omega = \int\limits_{-\infty}^{+\infty} g_y(\omega)|k_2(\mathrm{i}\omega)|^2\,d\omega \quad .$$

Substituting $g_y(\omega)$ from (3.53) and $g_x(\omega)$ from (3.49) gives

$$\sigma_\xi^2 = 2\beta^2 g_f^2(\omega_0) \iint\limits_{-\infty}^{+\infty} |k_2(\mathrm{i}\omega)|^2 |k_1(\mathrm{i}\omega')|^2 |k_1[\mathrm{i}(\omega - \omega')]|^2\,d\omega\,d\omega' \quad . \tag{3.55}$$

Considering the fact that the l.p.f. band $2\Delta\Omega$ is narrow as compared with the width $2\Delta\omega$ of the *low-frequency part* $g_{yl}(\omega)$ of the spectrum $g_y(\omega)$ at the s.l.d. output, we may take the mean value of $\xi(t)$ to be

$$\overline{\xi} = k_2(0)\overline{y} = k_2(0)\beta\overline{x^2} \quad ,$$

or by (3.50),

$$\overline{\xi} = \beta g_f(\omega_0)k_2(0) \int\limits_{-\infty}^{+\infty} |k_1(\mathrm{i}\omega)|^2\,d\omega \quad . \tag{3.56}$$

It should be noted that in (3.54), too, we could write $g_{yl}(\omega)$ instead of $g_y(\omega)$, i.e., we could ignore the high-frequency part $g_y(\omega)$ because $|k_2(\mathrm{i}\omega)|^2 \neq 0$ only in a narrow band at $\omega = 0$.

Substituting (3.52, 55, 56) into (3.48), we obtain for the smoothing coefficient

$$Q = \frac{k_2(0) \displaystyle\int\limits_{-\infty}^{+\infty} |k_1(\mathrm{i}\omega)|^2\,d\omega}{\left[\displaystyle\iint\limits_{-\infty}^{+\infty} |k_2(\mathrm{i}\omega)|^2 |k_1(\mathrm{i}\omega')|^2 |k_1[\mathrm{i}(\omega - \omega')]|^2\,d\omega\,d\omega'\right]^{1/2}} \quad . \tag{3.57}$$

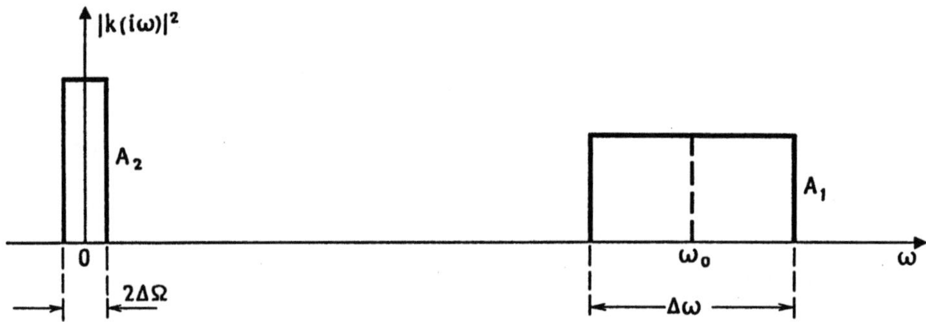

Fig. 3.12. High- and low-pass filters represented schematically by rectangles

In order to extract from this rather involved relationship a simple and pictorial estimate of Q, we replace $|k_1(i\omega)|^2$ and $|k_2(i\omega)|^2$ by rectangles, as shown in Fig. 3.12. Then

$$k_2(0) = \sqrt{A_2} \quad , \qquad \int_{-\infty}^{+\infty} |k_1(i\omega)|^2\, d\omega = 2A_1\,\Delta\omega \quad .$$

Because $|k_2(i\omega)|^2$ is only nonzero in a narrow neighborhood of $\omega = 0$, the double integral in the denominator of (3.57) can be estimated as follows:

$$\int_{-\infty}^{+\infty} |k_2(i\omega)|^2\, d\omega \int_{-\infty}^{+\infty} |k_1(i\omega')|^4\, d\omega' = 2A_2\Delta\Omega \cdot 2A_1^2\Delta\omega \quad .$$

We thus arrive at

$$Q = \sqrt{\Delta\omega/\Delta\Omega} \quad , \tag{3.58}$$

i.e., the fluctuations of $\xi(t)$ about the mean (3.56) are smoothed all the more strongly, the larger the ratio of bands for the h.p.f. and l.p.f., i.e., the longer the averaging time $\tau_\xi \sim 1/\Delta\Omega$ as compared with the correlation time $\tau_x \sim 1/\Delta\omega$ at the h.p.f. output.

In dealing with the above *direct method* of measuring $g_f(\omega_0)$, we made no distinction in the input noise between the noise signal of interest to us and the receiver noise. However, it is often necessary (e.g., in radioastronomy) to measure a weak noise signal against the background of a much more intense noise generated by the device itself (radiometer). The direct method then meets with the following difficulty. The spectral density $g_f(\omega)$ is said to consist of the density of the receiver noise $g_{f_0}(\omega)$ *calculated at the input* and a small correction $\Delta g_f(\omega)$ for the external signal. Accordingly, we have $\overline{\xi} = \overline{\xi}_0 + \Delta\overline{\xi}$, where $\overline{\xi}_0$ and $\Delta\overline{\xi}$ are related by (3.56) to $g_{f_0}(\omega_0)$ and $\Delta g_f(\omega_0)$, respectively. The problem that the deviation $\Delta\overline{\xi}$ of interest to us is far smaller than $\overline{\xi}_0$ is easily circumvented by a *compensation method,* namely by applying a reverse bias

$-\bar{\xi}_0$ to the measuring device and taking a substantially more sensitive device. The problem then is that this sensitive device, indicating now a *large* deviation $\Delta\bar{\xi}$, will respond with the same sensitivity to the fluctuations of $\xi(t)$. The latter are still determined by the *total* spectral density $g_f(\omega_0) = g_{f_0}(\omega_0) + \Delta g_f(\omega_0)$, i.e., essentially by the density of the receiver noise $g_{f_0}(\omega_0) \gg \Delta g_f(\omega_0)$.

The *threshold sensitivity* of a device is the spectral density at the input, $g_{\mathrm{th}}(\omega)$, such that the deviation $\bar{\xi}_{\mathrm{th}}$ is equal to the standard deviation of the fluctuation $\xi(t)$, i.e., $\bar{\xi}_{\mathrm{th}} = \sigma_\xi$. Considering that $\bar{\xi}_{\mathrm{th}}$ is related to $g_{\mathrm{th}}(\omega_0)$ by (3.56), and σ_ξ to $g_{f_0}(\omega_0)$ by (3.55), we obtain from the equality $\bar{\xi}_{\mathrm{th}} = \sigma_\xi$, using (3.57), that

$$g_{\mathrm{th}}(\omega_0) = \frac{\sqrt{2}}{Q} g_{f_0}(\omega_0) \quad .$$

Thus, the smoothing coefficient Q must be large enough for the density $\Delta g_f(\omega_0)$ being measured to exceed the following threshold:

$$\Delta g_f(\omega_0) \geq \frac{\sqrt{2}}{Q} g_{f_0}(\omega_0) \quad . \tag{3.59}$$

Even if by suppressing the level of receiver noise (e.g., by cooling the input stage for thermal noise) and/or by improving the Q value, we satisfy the condition (3.59), there is still one more phenomenon impairing the stability of the output measuring device, namely drifts of the gain (the so-called *zero drift*). For the most part the drift is caused by the input amplifier (the latter is included in the h.p.f.) and is *slow*, i.e., its spectrum is concentrated within an extremely narrow band $2\Delta\omega_d$ at $\omega = 0$. Therefore, after the detector, the low-frequency part $g_{yl}(\omega)$ of the spectrum has the form shown in Fig. 3.13, in which the zero drift spectrum is shown shaded. The l.p.f. transmits these variations of the gain with the result that the pointer of the device "wanders", which is equivalent to impairing the smoothing coefficient Q.

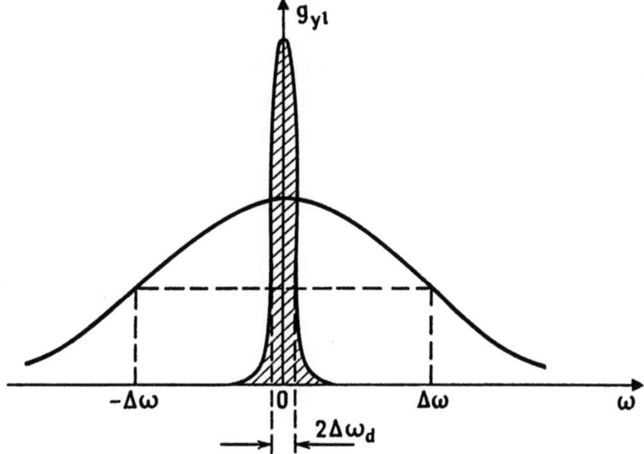

Fig. 3.13. Low-pass filter passes a narrow spectrum of gain drift

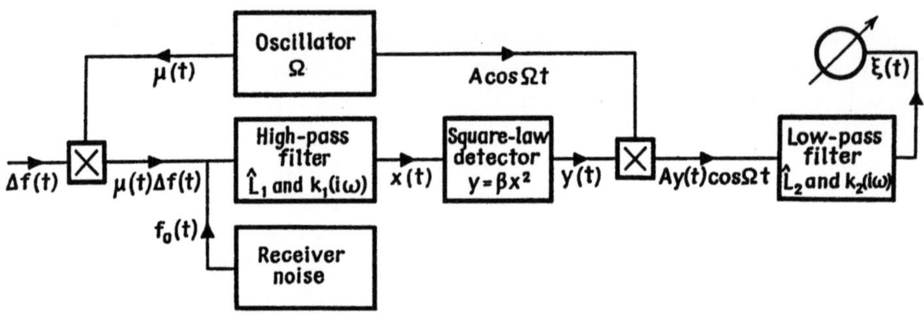

Fig. 3.14. Block diagram of the modulation radiometer

Zero drifts can be eliminated using the modulation technique of measuring. A schematic diagram of the *modulation radiometer* is given in Fig. 3.14. The oscillator Ω is used to modulate by a periodic function $\mu(t)$ with period $2\pi/\Omega$ the signal measured, so that the process at the input of the h.p.f. will be

$$f(t) = f_0(t) + \mu(t)\Delta f(t)$$

where $f_0(t)$ represents the receiver noise. For convenience we take $\mu(t)$ to be unity during one half of the period and zero during the other. Then

$$\mu^2(t) = \mu(t) = \sum_k a_k \cos k\Omega t \quad .$$

Further, the process $y(t)$ at the detector output is multiplied by the fundamental harmonic of the oscillator, and at the l.p.f. will be $Ay(t) \cos \Omega t$. In Fig. 3.14 $\hat{L}_1$ and $\hat{L}_2$ designate linear differential operators describing respectively the action of the h.p.f. and l.p.f., whose respective transfer functions are $k_1(i\omega)$ and $k_2(i\omega)$. We will now see that this arrangement removes the influence of zero drift.

At the h.p.f. output we obtain

$$x(t) = \hat{L}_1\{f_0(t) + \mu(t)\Delta f(t)\} \approx \hat{L}_1\{f_0\} + \mu(t)\hat{L}_1\{\Delta f\} \quad .$$

The period $2\pi/\Omega$ of the function $\mu(t)$ is fairly large as compared with the corerlation time τ_x. Therefore, we placed $\mu(t)$ outside the operator $\hat{L}_1$, retaining only the quasi-static response to $\mu(t)$. At the detector output we thus have

$$y(t) = \beta x^2(t) = \beta([\hat{L}_1\{f_0\}]^2 + \mu(t)[\hat{L}_1\{\Delta f\}]^2 + 2\mu(t)\hat{L}_1\{f_0\}\hat{L}_1\{\Delta f\}) \quad .$$

On averaging the doubled product vanishes, since $f_0(t)$ and Δf are independent, and their means are zero. Hence,

$$\overline{y}(t) = \beta\overline{x^2(t)} = \overline{y}_0 + \mu(t)\Delta\overline{y} \quad , \tag{3.60}$$

where

136

$$\overline{y}_0 = \beta \langle [\hat{L}_1\{f_0\}]^2 \rangle \quad , \quad \Delta \overline{y} = \beta \langle [\hat{L}_1\{\Delta f\}]^2 \rangle \quad .$$

Turning to the fluctuations of $y(t)$, we can, still assuming a weak signal Δf, consider these fluctuations $\tilde{y} = y(t) - \overline{y}(t)$ to be conditioned only by the receiver noise. Hence, for $\psi_y(\tau)$, σ_y^2, and $g_y(\omega)$ the earlier relations, which only include the spectral densities of receiver noise and zero drifts, hold.

At the l.p.f. output we obtain the process

$$\begin{aligned}
\xi(t) &= \hat{L}_2\{Ay(t) \cos \Omega t\} \\
&= A\hat{L}_2\{\overline{y}(t) \cos \Omega t\} + A\hat{L}_2\{\tilde{y}(t) \cos \Omega t\} \quad .
\end{aligned}$$ (3.61)

The first term on the right is the signal $\overline{\xi}(t)$, i.e.,

$$\begin{aligned}
\overline{\xi}(t) &= A[\overline{y}_0 \cos \Omega t + \mu(t)\Delta\overline{y} \cos \Omega t] \\
&= A\left[\overline{y}_0 \cos \Omega t + \frac{\Delta\overline{y}}{2} \sum_k a_k \cos (k\Omega \pm \Omega)t\right] \\
&= A\left\{\overline{y}_0 \cos \Omega t + \frac{\Delta\overline{y}}{2}[2a_0 \cos \Omega t + a_1(1 + \cos 2\Omega t) \right. \\
&\qquad \left. + a_2(\cos \Omega t + \cos 3\Omega t) + \ldots]\right\} \quad .
\end{aligned}$$

The low-pass filter whose band $2\Delta\Omega$ is narrower than the fundamental modulation frequency $(2\Delta\Omega < \Omega)$ will transmit from this Fourier expansion the constant term only, i.e.,

$$\overline{\xi} = \frac{Aa_1}{2}\Delta\overline{y} \quad ,$$

or, using (3.56),

$$\overline{\xi} = \frac{\beta Aa_1}{2}\Delta g_f(\omega_0)k_2(0) \int\limits_{-\infty}^{+\infty} |k_1(i\omega)|^2 \, d\omega \quad .$$ (3.62)

The modulation radiometer thus responds on average to the modulated part of the input noise alone, i.e., to the signal $\Delta f(t)$ that is of interest to us.

The second term in (3.61) describes the fluctuations of $\xi(t)$ about the mean (3.62). To calculate σ_ξ^2 requires the spectrum of the process $\tilde{y}(t) \cos \Omega t$. We will not perform all the computation, confining ourselves to a qualitative picture which is sufficient to illustrate the underlying idea.

It is apparent that the l.p.f. input is dominated by the low-frequency part of the fluctuations of $y(t)$ (Fig. 3.13). Multiplying this low-frequency part $\tilde{y}_l(t)$ by $\cos \Omega t$ implies breaking down the spectrum into two components separated by $\pm \Omega$ from $\omega = 0$ (Fig. 3.15). But the modulation frequency Ω, which is small as compared with the h.p.f. band $2\Delta\omega$, is large as compared both with the l.p.f. band $2\Delta\Omega$ and all the more so with the width $2\Delta\omega_d$ of the zero drift

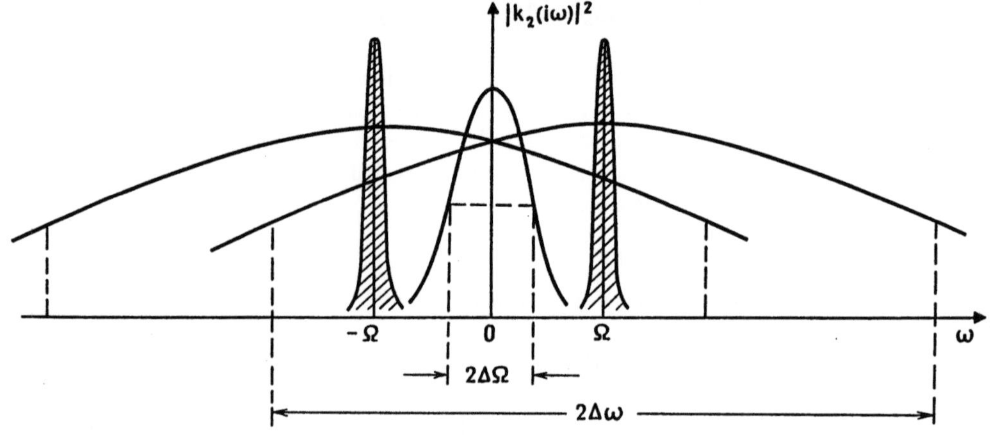

Fig. 3.15. Principle of the modulatioin radiometer

spectrum (in Fig. 3.15 this spectrum is again shaded). Therefore, receiver noise is essentially unaffected within the l.p.f. band, whereas the zero-drift spectrum, owing to the fact that $\Omega > 2\Delta\Omega$ and $\Omega \gg 2\Delta\omega_d$, appears to be removed beyond the l.p.f. band. It follows that the zero drift does not influence the fluctuations of $\xi(t)$, and the latter are determined only be the receiver noise $g_{f_0}(\omega_0)$.

It is easily seen that the smoothing coefficient Q remains the same as in the absence of modulation and zero drift, i.e., it is expressed by (3.58). But the advantage of the modulation technique lies exactly in the fact that zero drift does not reduce this value of Q.

A further improvement is the *zero modulation technique*, in which the radiometer input is periodically (period $2\pi/\Omega$) switched over, using the oscillator (Ω), from the source of the noise measured $\Delta f(t)$ to the noise reference standard and back (Fig. 3.16). If the spectral densities of both sources are the same within the h.p.f. band, then there is no modulation and the measuring device reads zero. This allows purely instrumental errors to be reduced with the help of a more sensitive device.

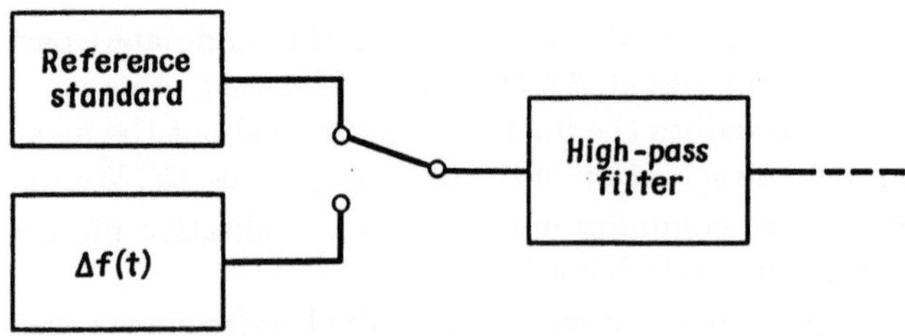

Fig. 3.16. Commutation in zero modulation technique of noise measuring

3.4 Correlation Theory of Fluctuations in the Thomson Oscillator

The Bershtein method discussed in this section, enabled the natural non-monochromaticity of a self-oscillating system to be measured for the first time. It relies heavily on the technical frequency drifts being slow (long correlation times). These slow random variations in the self-oscillatory system are treated as a result of a certain action with a long correlation time (as compared not only with the period, but also with the amplitude settling time). But, as we know, a random process in the system will not be Markovian. Hence it cannot be described using the Einstein-Fokker-Planck equation, as was done in Sect. I.5.8 and justified in Sect. I.6.4. On the other hand, for a sufficiently large amplitude of self-oscillations, the fluctuations are adequately described by *linearized* equations of motion (Sect. I.5.9). We can thus use another mathematical technique that does not require the process to be Markovian, namely stochastic differential equations and correlation theory [3.13]. We will now proceed to discuss this approach.

We will begin by formulating the problem as in Sect. I.5.8. It is based on equation (I.5.98) for the Thomson self-oscillatory system with one degree of freedom

$$\ddot{x} + x = \mu f(x, \dot{x}) + \mu F(t') \quad , \tag{3.63}$$

where $t' = \omega_0 t$. It will be recalled that if the fluctuation "force" $\mu F(t')$ describes the action of the shot current $I_{\mathrm{sh}}(t)$ and thermal noise in an oscillatory circuit [random e.m.f $e(t)$], then, by (I.5.95)

$$\mu F(t') = \frac{1}{I_0}\left[I_{\mathrm{sh}}(t) + \frac{\dot{e}(t)}{L\omega_0^2}\right] \quad . \tag{3.64}$$

Using the slow perturbation method we derived in Sect. I.5.8 in a first approximation the Van der Pol equations (I.5.100) for the amplitude and phase of the oscillation $x(t) = r(\theta) \cos [t' + \varphi(\theta)]$ that depend on the "slow time" $\theta = \mu t'$

$$r' = R(r) - \tfrac{1}{2}F_{\perp}(\theta) \quad , \quad r\varphi' = \Phi(r) - \tfrac{1}{2}F_{\parallel}(\theta) \quad . \tag{3.65}$$

The functions $R(r)$ and $\Phi(r)$ are defined by the equalities (I.3.101). In deriving (3.65) the stationary force $F(t')$ is replaced by the stationary modulated force (I.5.99)

$$\tilde{F}(t') = F_{\parallel}(\theta) \cos [t' + \varphi(\theta)] + F_{\perp}(\theta) \sin [t' + \varphi(\theta)] \quad , \tag{3.66}$$

whose spectrum is bounded by the band $\mu\Omega$ near the dimensionless natural frequency 1 of the circuit. But this narrow band ($\mu\Omega \ll 1$) is still significantly wider than the spectral line of the self-oscillatory system. This means that Ω

substantially exceeds not only the width $2D$ of the line due to phase diffusion, but also the width $2p$ of the "platform" due to amplitude fluctuations, i.e.,

$$\Omega \gg p \gg D \quad ,$$

where p is the dimensionless increment and D is the diffusion coefficient in the slow time θ (Sect. I.5.9, and Sect. 1.5). Within the band $\mu\Omega$ the spectral density of the effective force $\tilde{F}(t')$ is constant and the same as $F(t')$ has at frequency 1.

We denote the spectral density $F(t')$ in terms of the dimensionless frequency $\alpha = \omega/\omega_0$ by $g(\alpha)$

$$\langle F(t'_1)F(t'_2)\rangle = \int\limits_{-\infty}^{+\infty} g(\alpha)\exp\left[i\alpha(t'_1 - t'_2)\right]d\alpha \quad .$$

Let $g(\pm 1) = C/2\pi$. If the spectral density had this value over the entire α axis, then $F(t')$ would be white noise with the covariance

$$\langle F(t'_1)F(t'_2)\rangle = C\delta(t'_1 - t'_2) \quad . \tag{3.67}$$

This very covariance is typical of the shot current, which has been assumed to be independent of the system state (i.e., of $\overline{I}_a$ or v_g); and of the thermal noise (Sect. 3.5).
tuations of x and $\dot{x}$ form a diffusion Markov process, so that all the results of correlation theory must here coincide with the approximate results derived for the linearized problem using the Einstein-Fokker-Planck equation (Sect. I.5.9).

For our further discussion we shall need the covariances of the amplitudes $F_\|(\theta)$ and $F_\perp(\theta)$ that enter into the relation for the effective force (3.66). To derive these it is convenient to replace $\tilde{F}(t')$ by the analytical signal (Sect. 1.1)

$$\mathcal{F}(t') = [F_\|(\theta) - iF_\perp(\theta)]e^{i[t'+\varphi(\theta)]} \quad , \quad \tilde{F}(t') = \mathrm{Re}\left\{\mathcal{F}(t')\right\} \quad , \tag{3.68}$$

and deduce the covariances for this signal. Since the spectrum $\mathcal{F}(t')$ is only nonzero within the band $(1 - \mu\Omega/2,\ 1 + \mu\Omega/2)$ of positive frequencies α, and the spectral density within this band is $4g(1) = 2C/\pi$, the "first" covariance is

$$\langle \mathcal{F}(t'_1)\mathcal{F}^*(t'_2)\rangle = \frac{2C}{\pi} \int\limits_{1-\mu\Omega/2}^{1+\mu\Omega/2} e^{i\alpha(t'_1 - t'_2)}\,d\alpha$$

$$= 4\mu C\frac{\sin\frac{\Omega}{2}(\theta_1 - \theta_2)}{\pi(\theta_1 - \theta_2)}e^{i(t'_1 - t'_2)} \quad .$$

The "second" covariance, owing to $\mathcal{F}(t')$ being stationary, is zero,

$$\langle \mathcal{F}(t'_1)\mathcal{F}(t'_2)\rangle = 0 \quad .$$

Substituting the expression (3.68) for $\mathcal{F}(t')$ into both covariances gives

$$\langle [F_{\|1}F_{\|2} + F_{\perp 1}F_{\perp 2} + \mathrm{i}(F_{\|1}F_{\perp 2} - F_{\perp 1}F_{\|2})]e^{\mathrm{i}(\varphi_1 - \varphi_2)} \rangle$$
$$= 4\mu C \frac{\sin\frac{\Omega}{2}(\theta_1 - \theta_2)}{\pi(\theta_1 - \theta_2)} \quad ,$$

$$\langle [F_{\|1}F_{\|2} - F_{\perp 1}F_{\perp 2} - \mathrm{i}(F_{\|1}F_{\perp 2} + F_{\perp 1}F_{\|2})]e^{\mathrm{i}(\varphi_1 + \varphi)} \rangle = 0 \quad . \tag{3.69}$$

The values of the functions $F_\|$, $F_\perp$, and φ with the arguments $\theta_1 = \mu t'_1$ and $\theta_2 = \mu t'_2$ are simply labelled with subscripts 1 and 2.

Generally speaking, the random phase φ is not independent of $F_\|$ and $F_\perp$. It is clear from the form of (3.69) that the correlation time for $F_\|$ and $F_\perp$ must be of the order of $1/\Omega$, as with the right side, whereas the characteristic diffusion time for φ is of the order of $1/D$, i.e., it is far longer than $1/\Omega$. We may, therefore, assume that within the time intervals $\theta_1 - \theta_2 \sim 1/\Omega$ of interest, the phase is *constant* and *independent* of the fast-varying $F_\|$ and $F_\perp$. Then, for the alone intervals of θ we suppose that $\varphi_1 = \varphi_2$, $\exp[\mathrm{i}(\varphi_1 - \varphi_2)] = 1$, and $\exp[\mathrm{i}(\varphi_1 + \varphi_2)] = \exp(2\mathrm{i}\varphi_y) \approx \mathrm{const}$, with the result that the constant factor in the second equation (3.69) is cancelled. We thus arrive at

$$\overline{F_{\|1}F_{\|2}} + \overline{F_{\perp 1}F_{\perp 2}} + \mathrm{i}(\overline{F_{\|1}F_{\perp 2}} - \overline{F_{\perp 1}F_{\|2}}) = 4\mu C \frac{\sin\frac{\Omega}{2}(\theta_1 - \theta_2)}{\pi(\theta_1 - \theta_2)} \quad ,$$

$$\overline{F_{\|1}F_{\|2}} - \overline{F_{\perp 1}F_{\perp 2}} - \mathrm{i}(\overline{F_{\|1}F_{\perp 2}} + \overline{F_{\perp 1}F_{\|2}}) = 0 \quad .$$

It follows immediately that

$$\overline{F_\|(\theta_1)F_\|(\theta_2)} = \overline{F_\perp(\theta_1)F_\perp(\theta_2)} = 2\mu C \frac{\sin\frac{\Omega}{2}(\theta_1 - \theta_2)}{\pi(\theta_1 - \theta_2)} \quad ,$$

$$\overline{F_\|(\theta_1)F_\perp(\theta_2)} = 0 \quad . \tag{3.70}$$

The very strong inequaltiy $\Omega \gg D$ permits here the formal limiting process $\Omega \to \infty$, i.e., the change

$$\frac{\sin\frac{\Omega}{2}(\theta_1 - \theta_2)}{\pi(\theta_1 - \theta_2)} \to \delta(\theta_1 - \theta_2) \quad ,$$

and then[8]

$$\overline{F_\|(\theta_1)F_\|(\theta_2)} = \overline{F_\perp(\theta_1)F_\perp(\theta_2)} = 2\mu C\delta(\theta_1 - \theta_2) \quad ,$$

$$\overline{F_\|(\theta_1)F_\perp(\theta_2)} = 0 \quad . \tag{3.71}$$

---

[8] We proceeded here from qualitative considerations, but if the "force" $F(t')$ is a normal process, then using the Furutsu-Novikov formula [3.14] we can *prove* the asymptotic (as $D/\Omega \to 0$) validity of the results (3.71). This proof, in essence, justifies the transition to the approximation of the diffusion Markov process in this problem.

Note that for the case of (3.67), i.e., for the delta-correlated force $F(t')$, this result would be immediate.

However, equation (3.63), where $F(t')$ is replaced by $\tilde{F}(t')$, has a wider scope of application than the initial problem of the oscillator subject to shot and thermal noise, when the "force" $F(t')$ can be regarded as delta-correlated. In fact, under steady-state conditions, where in a zero approximation $r = r_0 = $ const, we can write $\tilde{F}(t')$ in the form

$$\tilde{F}(t') = \frac{1}{r_0}\left[F_\parallel(\theta)x - F_\perp(\theta)\frac{dx}{dt'} \right] \quad .$$

It is easily seen that the variation of the coefficient at x with θ, i.e., the function $F_\parallel(\theta)$, expresses the modulation of the circuit frequency, whereas $F_\perp(\theta)$, the coefficient at dx/dt', gives the modulation of the increment of the oscillator. Consequently, in the above approximation, the amplitudes $F_\parallel$ and $F_\perp$ can also describe *technical drifts* caused by random changes of the parameters. If these changes are stationary, the covariances $F_{\parallel\,\text{tech}}$ and $F_{\perp\,\text{tech}}$ will only be dependent on the shift $\theta_1 - \theta_2$, but in that case $\tilde{F}(t')$, generally speaking, will not be stationary. The slowness of the technical drifts suggests that $F_{\parallel\,\text{tech}}$ and $F_{\perp\,\text{tech}}$ possess very long covariance times. From this point of view the case of (3.71) may be thought of as a very fast (delta-correlated) modulation of the oscillator parameters.

In order that the techniques of correlation theory may be effectively used, we will need to linearize equations (3.65) in the vicinity of the undisturbed dynamic mode. Linearization is possible if the oscillator is sufficiently strongly excited (Sect. I.5.9), i.e., linearization presupposes that $F_\parallel$ and $F_\perp$ themselves are very small. Representing then the solutions to (3.65) at $F_\parallel = F_\perp = 0$ in terms of r_0 and φ_0 and the small deviations produced by the random forces in terms of ϱ and χ, we have

$$r_0' = R(r_0) \quad , \quad r_0\varphi_0' = \Phi(r_0) \quad , \tag{3.72}$$

and in the first order in ϱ and χ

$$\varrho' + p\varrho = -\frac{1}{2}F_\perp(\theta) \quad , \quad \chi' + q\varrho = -\frac{1}{2r_0}F_\parallel(\theta) \quad , \tag{3.73}$$

where

$$p = -\frac{\partial R(r_0)}{\partial r_0} \quad , \quad q = -\frac{\partial}{\partial r_0}\left[\frac{\Phi(r_0)}{r_0}\right] \quad . \tag{3.74}$$

Note that in the special case of the oscillator (I.5.96)

$$R(r_0) = \frac{r_0}{2}(p - r_0^2) \quad , \quad \Phi(r_0) = 0 \tag{3.75}$$

so that $q = 0$, and p is the value of the increment for the limiting cycle.

Under steady-state self-oscillatory conditions that are the only ones of interest to us, we have $r_0' = 0$, and the first of (3.72) takes the form

$$R(r_0) = 0$$

i.e., it defines constant (independent of θ) radii of limiting cycles in a zero approximation. If for the limiting cycle $\Phi(r_0) \neq 0$, then from the second of (3.72) we find

$$\varphi_0 = \Delta\theta + \gamma \quad ,$$

where γ is an arbitrary constant, and the quantity

$$\Delta = \varphi_0' = -\Phi(r_0)/r_0 \quad ,$$

gives the first order correction to the self-oscillation frequency. In the interesting special case of (3.75), $r_0 = \sqrt{p}$ and $\Delta = 0$.

Equations (3.73) are precisely the stochastic equations describing the fluctuations of the amplitude and phase in the vicinity of a limiting cycle. For ϱ we have an equation of the relaxation type (for the stable cycle $p>0$). The presence of the quasi-elastic force $-p\varrho$ establishes of steady-state conditions and the existence of a steady-state finite value of $\overline{\varrho^2}$ (assuming that $\sqrt{\overline{\varrho^2}} \ll r_0$).

To calculate the covariance ϱ under steady-state conditions we take the solution to the first of (3.73), the zero initial condition being placed at $\theta = -\infty$

$$\varrho(\theta) = -\frac{e^{-p\theta}}{2} \int\limits_{-\infty}^{\theta} e^{p\theta} F_\perp(\theta)\, d\theta \quad . \tag{3.76}$$

Then

$$\psi_\varrho(\theta - \theta') = \overline{\varrho(\theta)\varrho(\theta')}$$

$$= \frac{1}{4} e^{-p(\theta+\theta')} \int\limits_{-\infty}^{\theta} e^{p\theta}\, d\theta \int\limits_{-\infty}^{\theta'} e^{p\theta'} \overline{F_\perp(\theta)F_\perp(\theta')}\, d\theta' \quad . \tag{3.77}$$

There are no steady-state conditions for phase fluctuations because the system under consideration is closed. With this in mind we take the solution to the second of (3.73). This solution corresponds to the initial condition that $\chi = 0$ at $\theta = 0$

$$\chi(\theta) = -q \int\limits_{0}^{\theta} \varrho(\theta)\, d\theta - \frac{1}{2r_0} \int\limits_{0}^{\theta} F_\parallel(\theta)\, d\theta \quad . \tag{3.78}$$

Of especial interest is the mean square of χ. If $\varrho(\theta)$ and $F_\parallel(\theta)$ are uncorrelated[9],

---

[9] This will obviously occur in the absence of correlation between $F_\parallel$ and $F_\perp$ as, for example, in the cases (3.70, 71), i.e., with stationary actions on the self-oscillatory system. In the general case, $F_\parallel$ and $F_\perp$ may be correlated, but this does not hinder the computation of the covariances and mean squares $\varrho(\theta)$ and $\chi(\theta)$.

then

$$\overline{\chi^2(\theta)} = q^2 \int\limits_0^\theta d\theta \int\limits_0^\theta \psi_\varrho(\theta - \theta')\, d\theta' + \frac{1}{4r_0^2} \int\limits_0^\theta d\theta \int\limits_0^\theta \overline{F_\parallel(\theta) F_\parallel(\theta')}\, d\theta' \quad . \quad (3.79)$$

If we insert into (3.77) the covariance (3.71) for $F_\perp(\theta)$ corresponding to the delta-correlated random action $F(t')$ on the oscillator, we obtain

$$\psi_\varrho(\theta - \theta') = \frac{\mu C}{4p} e^{-p|\theta - \theta'|} \quad , \qquad (3.80)$$

hence, in particular,

$$\overline{\varrho^2} = \psi_\varrho(0) = \mu C/4p \quad . \qquad (3.81)$$

It is to be noted that in the covariances of the random forces $F_\perp(\theta)/2$ and $F_\parallel(\theta)/2$, which enter into the Langevin equations (3.73), the coefficient at the delta-function according to (3.71), is $\mu C/2$. From the general theorem (I.6.46) this quantity must coincide with the coefficient B in the Einstein-Fokker-Planck equation (I.5.112)

$$B = \mu C/2 \quad . \qquad (3.82)$$

By simply comparing (I.5.125) and (3.80), we arrive at exactly the same equality.

In computing the mean fluctuational phase progression $\overline{\chi^2(\theta)}$ from (3.79), we assume that the amplitude fluctuations are already well-established, i.e., we take for ϱ the covariance (3.80). For $F_\parallel(\theta)$ we still take the covariance in the form of (3.71). The result is

$$\overline{\chi^2(\theta)} = \frac{\mu C}{2}\left\{ \frac{q^2}{p^3}(p\theta + e^{-p\theta} - 1) + \frac{\theta}{r_0^2} \right\} \quad . \qquad (3.83)$$

The first term is associated with the amplitude fluctuations, the second with the direct actions of the random force $F_\parallel(\theta)$, or, what is the same, with the delta-correlated frequency modulation. From the very beginning, the second term grows according to the diffusion law, whereas the first only obeys this law for $p\theta \gg 1$. For large θ

$$\overline{\chi^2(\theta)} \approx \frac{\mu C}{2}\left(\frac{q^2}{p^2} + \frac{1}{r_0^2} \right)\theta \quad .$$

If $p\theta \ll 1$, then

$$\overline{\chi^2(\theta)} \approx \frac{\mu C}{2}\left(\frac{q^2 \theta^2}{2p} + \frac{\theta}{r_0^2} \right) \quad .$$

Of course, in the isochronous oscillator (3.75) whose amplitude deviations exert no influence on the frequency ($q = 0$), the first term in (3.83) is absent and the diffusion law holds for all θ

$$\overline{\chi^2(\theta)} = \frac{\mu C}{2r_0^2}\theta = \frac{\mu C}{2p}\theta = 2D\theta \quad . \tag{3.84}$$

By virtue of (3.82) this coincides with (I.5.130).

So far, in dealing with the delta-correlated random "force" $F(t')$ we have not decomposed it in accordance with (3.64) into shot and thermal noises. But now we shall carry out this operation in order to compare the contributions of both kinds of noise.

Since the two random actions are independent, we have

$$\langle F(t_1')F(t_2')\rangle = \frac{1}{I_0^2\mu^2}\left[\overline{I_{\mathrm{sh}}(t_1)I_{\mathrm{sh}}(t_2)} + \frac{\overline{\dot{e}(t_1)\dot{e}(t_2)}}{L^2\omega_0^4}\right] \quad . \tag{3.85}$$

By (2.2), we have

$$\overline{I_{\mathrm{sh}}(t_1)I_{\mathrm{sh}}(t_2)} = e\overline{I}_{\mathrm{a}}\delta(t_1 - t_2)$$

where e is the electron charge, and $\overline{I}_{\mathrm{a}}$ the mean anode current through the tube. If we go over to the dimensionless time $t' = \omega_0 t$, then

$$\overline{I_{\mathrm{sh}}(t_1)I_{\mathrm{sh}}(t_2)} = e\overline{I}_{\mathrm{a}}\omega_0\delta(t_1' - t_2') \quad . \tag{3.86}$$

The covariance of the fluctuating thermo-e.m.f. $e(t)$ is described by (I.6.14)

$$\overline{e(t_1)e(t_2)} = 2kT\,R\delta(t_1 - t_2) = 2kT\,R\omega_0\delta(t_1' - t_2') \quad , \tag{3.87}$$

where k is the Boltzmann's constant, and T is the absolute temperature of the resistance R. But the covariance in (3.85) is not that for $e(t)$, but for $de(t)/dt = \omega_0 de(t')/dt'$, i.e.,

$$\overline{\dot{e}(t_1)\dot{e}(t_2)} = \omega_0^2\overline{\frac{de(t_1')}{dt_1'}\frac{de(t_2')}{dt_2'}} \quad . \tag{3.88}$$

We find this quantity in a zero approximation in the small parameter μ.

Of course, the introduction of the effective modulated force $\tilde{F}(t')$ in place of $F(t')$ calls for a similar change for $e(t')$. We then represent $\tilde{e}(t')$ in the form of the analytical signal

$$\mathcal{E}(t') = \mathcal{E}_0(\theta)\exp\left\{\mathrm{i}[t' + \varphi(\theta)]\right\} \quad .$$

In a zero-order approximation in μ, $d\mathcal{E}(t')/dt' = i\mathcal{E}(t')$, so that

$$\left\langle \frac{d\mathcal{E}(t_1')}{dt_1'}\frac{d\mathcal{E}^*(t_2')}{dt_2'}\right\rangle = \langle\mathcal{E}(t_1')\mathcal{E}^*(t_2')\rangle \quad .$$

It follows from this and (3.87, 88) that

$$\overline{\dot{e}(t_1)\dot{e}(t_2)} = \omega_0^2\overline{e(t_1')e(t_2')} = 2kT\,R\omega_0^3\delta(t_1'-t_2') \quad . \tag{3.89}$$

Substituting (3.86) and (3.89) into (3.85) we arrive at

$$\overline{F(t_1')F(t_2')} = (C + C_1)\delta(t_1'-t_2') \quad , \quad \text{where} \tag{3.90}$$

$$C = \frac{e\overline{I}_a\omega_0}{I_0^2\mu^2} \quad , \quad C_1 = \frac{2kT\,R}{L^2\omega_0 I_0^2\mu^2} \quad . \tag{3.91}$$

Clearly, C and C_1 will be additive in all the bilinear quantities characterizing the correlation and intensity of the fluctuations. So, for instance, for the mean squares of the fluctuations of the amplitude and phase in the isochronous self-oscillatory system we will get the expressions

$$\overline{\varrho^2} = \frac{\mu}{4p}(C + C_1) \quad , \quad \overline{\chi^2} = \frac{\mu}{2p}(C + C_1)\theta \quad .$$

Consequently, the ratio of the intensities of the shot and thermal noises is

$$\frac{C}{C_1} = \frac{e}{2kT}\frac{L^2\omega_0^2\overline{I}_a}{R} = \frac{eU_C}{2kT} \quad , \tag{3.92}$$

where U_C is the voltage amplitude at the capacitor in the circuit, which is equal to the amplitude of the oscillation current

$$I_{\mathrm{amp}} = \frac{L\omega_0}{R}\overline{I}_a$$

divided by $\omega_0 C = 1/\omega_0 L$. According to (3.92), the intensity ratio is the ratio of the work performed in the passage of an electron across the capacitor (at a maximal voltage) to twice the energy of the thermal noise in the capacitor. At $T = 300\,\mathrm{K}$ we have

$$C/C_1 = 17U_C \quad ,$$

where U_C is expressed in volts. Thus, if at room temperature the amplitude of the voltage across the circuit for our oscillator model exceeds $1/17\,\mathrm{V}$, the influence of the shot noise will be stronger than that of the thermal one. It should, however, be borne in mind that this result refers to the oscillator under consideration, with the oscillatory circuit in the anode branch and with no grid current. If non zero grid current is taken into account or if the self-oscillator is different, the situation may differ markedly [3.15].

In the relations for the mean squares of ϱ and χ we go over to the initial (dimensioned) parameters. Since $x = I/I_0$ and $r = r_0 + \varrho$, the current amplitude fluctuations in the oscillatory circuit are $\Delta I = I_0\varrho$. If we only take into consideration the shot noise and use the expressions $\mu = \omega_0 MS$, $\mu p = \omega_0(MS - RC)$ (Sect. I.5.8), we obtain, by (3.81),

$$\overline{(\Delta I)^2} = I_0^2 \overline{\varrho^2} = I_0^2 \frac{\mu C}{4p} = \frac{\overline{I}_a e}{4(MS - RC)} \ .$$

With the help of this expression we can also rewrite the mean square of the fluctuating phase progression (3.84) in terms of dimensioned quantities

$$\overline{\chi^2} = \frac{\mu C}{2p}\theta = 2\frac{\overline{(\Delta I)^2}}{I_0^2}\mu\omega_0 t = \frac{\overline{I}_a e\omega_0^4 M^3 S}{8V^2(MS - RC)}t \ .$$

If we approach the self-excitation threshold ($MS - RC \to 0$), i.e., the stability boundary of the self-oscillatory mode, the "rigidity" of the limiting cycle tends to zero. In the process, the intensity of the amplitude fluctuations and the phase diffusion coefficient grow indefinitely. Of course, in reality the growth of the fluctuations will be bounded, but it is significant enough for the prerequisits for linearizing (3.65) to be violated. Therefore, the question of fluctuations which are so close to the self-excitation boundary that random shifts from the limiting cycle have the same order of magnitude as the radius r_0, can no longer be considered within the framework of correlation theory[1]. If random forces are delta-correlated, we may again turn to the Einstein-Fokker-Planck equation, see [3.16]. In that case, the treatment of the crossing of the self-excitation boundary will present no difficulties (Sect. I.5.8).

Within its scope of application the method of stochastic differential equations and correlation theory offers definite advantages. First, it significantly simplifies the entire statistical scheme, thus enabling problems more involved than that of fluctuations in a closed oscillator with one degree of freedom to be handled. Fluctuations have been considred in an oscillator synchronized by an external harmonic e.m.f.[11]; in an oscillator that is frequency-stabilized by being connected to a high-Q oscillatory circuit (quartz-crystal), i.e., in a system with two degrees of freedom [3.13, 19]; in a Thomson oscillatory system with many degrees of freedom [3.20]; in an oscillator with a grid current and constant or automatic bias [3.15], and so on. A detailed treatment of a variety of problems on fluctuations in self-oscillatory systems, together with a relevant bibliography, is given in [3.21].

Second, this technique enables one to handle *non-Markovian* fluctuations caused by random actions with prolonged correlation, including technical drifts in the self-oscillator. We will now go on to discuss this.

As has already been noted, the "force" $F_{\parallel}(\theta)$ in the second of (3.73) can also be understood as the result of fluctuations of frequency controlling parameters. It is expedient to denote the right side of the second of (3.73) by $\alpha(\theta)$ in order to describe by this random function fluctuations of the dimensionless frequency of oscillations, whatever their origin. We may therefore include

---

[10] This in general applies to all those cases where we approach to bifurcation points of any stationary mode near which fluctuations are studied.

[11] In addition to [3.13], the problem has been studied using the same method in [3.17, 18].

in $\alpha(\theta)$ the term $-q\varrho(\theta)$ transposed from the left side of the equation. This term expresses the frequency fluctuations caused by amplitude fluctuations in a nonisochronous system. The equation will then take the form

$$\chi'(\theta) = \alpha(\theta) = \alpha_n(\theta) + \alpha_t(\theta) \quad ,$$

where $\alpha_n(\theta)$ is the natural frequency fluctuation due to the shot current and thermal noise, and $\alpha_t(\theta)$ is the technical drift associated with random variations of the circuit parameters. Although the source of the fluctuations of the amplitude ϱ is assumed to be delta-correlated natural noise, we will still need to refer the term $-q\varrho$ to $\alpha_t(\tau)$ since ϱ exhibits a rather long correlation time (of the order of $1/p$ in θ). For simplicity we will confine ourselves in the following to the case of the isochronous oscillator with $q = 0$.

Clearly, $\alpha_n(\theta)$ and $\alpha_t(\theta)$ are statistically independent. If we suppose that these random frequency deviations are stationary then we can, without loss of generality, put $\chi(0) = 0$ in calculating the mean square of the phase progression $\chi(\theta) - \chi(0)$ in time θ. Then

$$\chi(\theta) = \int\limits_0^\theta [\alpha_n(\theta) + \alpha_t(\theta)]\, d\theta \quad ,$$

hence, by virtue of the fact that $\alpha_n(\tau)$ and $\alpha_t(\tau)$ are uncorrelated, we get

$$\overline{\chi^2(\theta)} = \int\limits_0^\theta \int\limits_0^\theta [\overline{\alpha_n(\theta)\alpha_n(\theta')} + \overline{\alpha_t(\theta)\alpha_t(\theta')}]\, d\theta\, d\theta' \quad . \tag{3.93}$$

Bearing in mind that the correlation time ϑ_n of natural fluctuations of frequency is much shorter than the period of oscillations (in "slow time" this means that $\vartheta_n \ll 2\pi\mu$), we assume, as earlier, that $\alpha_n(\theta)$ is the delta-correlated function

$$\overline{\alpha_n(\theta)\alpha_n(\theta')} = 2D_n\delta(\vartheta - \vartheta') \quad . \tag{3.94}$$

On the contrary, the correlation time for technical drifts, ϑ_t, is very long even as compared with the amplitude settling time ($\vartheta_t \gg 1/p \gg 2\pi\mu$). It is sufficient for our further treatment to take some "model" of the covariance $\alpha_t(\tau)$, e.g., an exponential one,

$$\overline{\alpha_t(\theta)\alpha_t(\theta')} = \frac{D_t}{\vartheta_t}\mathrm{e}^{-|\theta-\theta'|/\vartheta_t} \quad , \quad D_t/\vartheta_t = \overline{\alpha_t^2} \quad . \tag{3.95}$$

Substituting into (3.93) covariances of this type [already used above in calculating (3.83)] gives

$$\overline{\chi^2(\theta)} = \overline{\chi_n^2(\theta)} + \overline{\chi_t^2(\theta)} = 2D_n\theta + 2D_t[\theta - \vartheta_t(1 - \mathrm{e}^{-\theta/\vartheta_t})] \quad . \tag{3.96}$$

All that was said above regarding (3.83) may be applied to (3.96) with the sole difference that the time for the diffusion law for $\overline{\chi_t^2(\theta)}$, now equals ϑ_t being far longer than the time $1/p$ in (3.83).

For time intervals $\theta \ll \vartheta_t$ the total mean square shift is

$$\overline{\chi^2(\theta)} = 2D_n\theta + \frac{D_t}{\vartheta_t}\theta^2 \quad , \tag{3.97}$$

and for $\theta \gg \vartheta_t$ we arrive at the diffusion law

$$\overline{\chi^2(\theta)} = 2(D_n + D_t)\theta \quad . \tag{3.98}$$

The coefficient of "technical diffusion" $2D_t$ exceeds the coefficient of natural diffusion $2D_n$ (Sect. I.5.9) by many orders of magnitude (thousands of times and more).

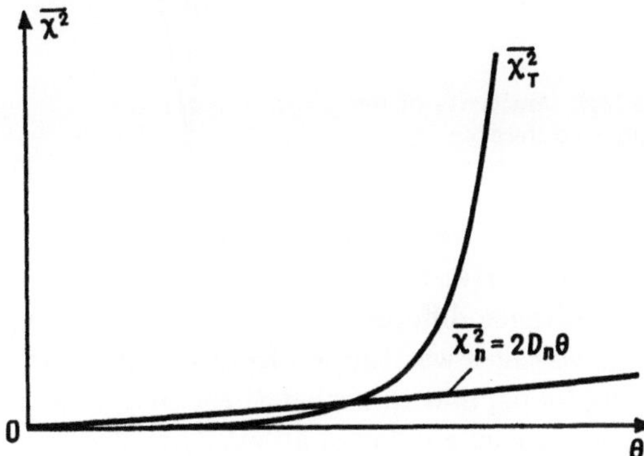

Fig. 3.17. Variation of mean squares of natural $\overline{(\chi_n^2)}$ and technical $\overline{(\chi_t^2)}$ phase progressions with the dimensionless observation time θ

Figure 3.17 shows the behavior of both terms of the function (3.96), with D_n and D_t not given to scale. At large θ the technical shift increases much faster than the natural one, but at sufficiently small θ the growth of $\overline{\chi_t^2}$ obeys a square law, we can always select θ to be sufficiently small i.e., so that $\overline{\chi_n^2}$ will be much larger than $\overline{\chi_t^2}$. Hence, if we are somehow to measure $\overline{\chi^2}$ during just such short intervals, this will amount to measuring the quantity $\overline{\chi_n^2} = 2D_n\theta$, and thereby the *natural* spectral line width of the oscillator. This reasoning is exactly the underlying idea in the method of Bershtein.

How small must the time shifts θ be? An estimate may be made from (3.97), where we require that

$$2D_n\theta \gg \frac{D_t}{\vartheta_t}\theta^2 = \overline{\alpha_t^2}\theta^2 \quad , \qquad \text{whence}$$

$$\theta \ll \frac{2D_n\vartheta_t}{D_t} = \frac{2D_n}{\overline{\alpha_t^2}} \quad . \tag{3.99}$$

Thus, θ must be not only much shorter than the correlation time of technical drifts, ϑ_t, but than a fairly small fraction $2D_n/D_t$ of ϑ_t. If, for example,

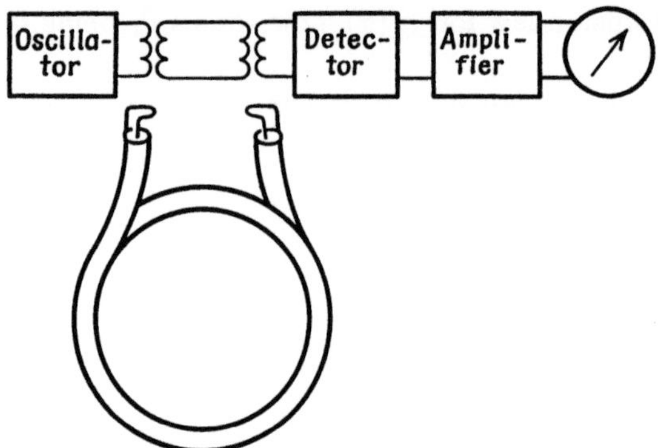

Fig. 3.18. Block diagram of the Bershtein experiment

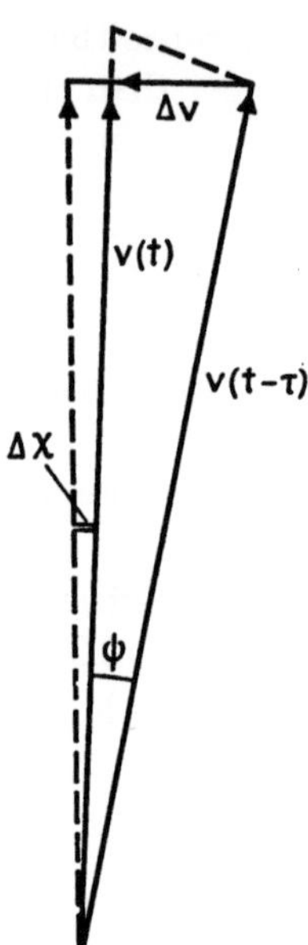

Fig. 3.19. Vector diagram illustrating the high sensitivity of ►
the Bershtein method for measuring the phase difference

$2D_n/D_t \sim 10^{-5}$ and $\tau_t = \vartheta_t/\mu\omega_0 \sim 10\,\text{s}$ (by τ_t we denote the dimensioned correlation time), then the shift must be far shorter than $10^{-4}\,\text{s}$.

In Bershtein's measurements the voltages difference $\Delta v = v(t) - v(t-\tau)$ was applied to the detector; one of the voltages was tapped from the oscillator in question directly through a coupling loops, and the other through a coaxial feeder 500 m in length (Fig. 3.18) that produced a delay of about $3\,\mu\text{s}$. The fluctuations of the amplitude of the voltage difference Δv at the detector input stem primarily from fluctuations of the phase difference $\Delta\chi = \chi(t) - \chi(t-\tau)$ between both voltages. Owing to $\Delta\chi$ being small they are simply proportional to $\Delta\chi$. Using an amplifier and a square-law output device, we measure the quantity $\overline{(\Delta v)^2}$ to find $\overline{(\Delta\chi)^2}$, (i.e., subject to the condition (3.99) we will find the quantity $2D_n\theta = 2\mathcal{D}_n\tau$ ($\mathcal{D}_n = \mu\omega_0 D_n$ and $\tau = \theta/\mu\omega_0$ are the dimensioned diffusion coefficient and delay). It is the possibility of large amplification that made it possible to measure deviations from the coherence as small as $\sqrt{\overline{(\Delta\chi)^2}} \sim 1''$. This value was determined by a geometric path difference as small as 500 m and accounted for only 10^{-12} of the coherent train length.

The vector diagram in Fig. 3.19 illustrates one more significant distinction of the method just described – the possibility of so selecting the mean amplitudes of the voltages and the instrumental phase shift ψ between them that the amplitude of the voltage difference Δv would be strongly dependent on the fluctuations of the phase difference $\Delta\chi$, but weakly on the amplitude fluctuations of $v(t)$.

It will be noted that we did not stick to the actual experimental arrangement, but only explained its principle following *Gorelik* [3.22]. It was not the quantity $\overline{(\Delta\chi)^2}$ that was actually measured, but the *spectrum* of $\Delta\chi$.

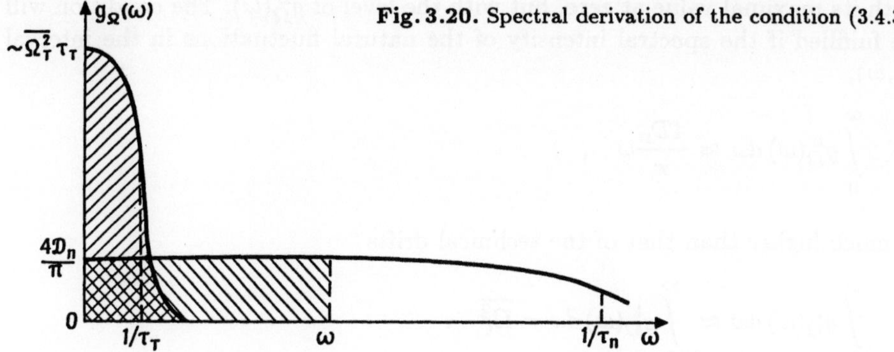

Fig. 3.20. Spectral derivation of the condition (3.4.37)

As was indicated in Sect. 2.4, an examination of the spectrum $g_x(\omega)$ of the oscillation $x(t) = r_0 \cos [\omega_0 t + \varphi(t)]$ in the wings of the line, i.e., for sufficiently large separations $|\omega - \omega_0|$ from the maximum $g_x(\omega)$, allows one to deduce the high-frequency part of the frequency fluctuation spectrum $\Omega(t) = \dot{\varphi}(t)$ (we return here to dimensioned quantities).

In the self-oscillatory system, where $\Omega(t)$ is split up into two uncorrelated components – technical drifts and natural fluctuations – we have

$$g_\Omega(\omega) = g_\Omega^t(\omega) + g_\Omega^n(\omega) \quad .$$

The spectrum $g_\Omega^t(\omega)$ is high and narrow [strong and slow variations of $\Omega_t(t)$] whereas $g_\Omega^n(\omega)$ is low and wide [weak and fast fluctuations of $\Omega_n(t)$], as shown schematically in Fig. 3.20. Accordingly, the mean square phase progression also contains two components

$$\overline{(\Delta\chi)^2} = \overline{(\Delta\chi)_t^2} + \overline{(\Delta\chi)_n^2} \quad .$$

But for the self-oscillation spectrum of the oscillator, i.e., for $g_x(\omega)$, there is no additivity because of an intricate nonlinear relationship between $g_x(\omega)$ and $g_\Omega(\omega)$. Formula (2.64) indicates, however, that measuring $g_x(\omega)$ on the wings of the curve allows one in principle to determine the behavior of $g_\Omega(\omega)$ at high frequencies, i.e., where technical drifts are essentially of no importance and $g_\Omega(\omega) \approx g_\Omega^n(\omega)$. What separation from ω_0 is required for this purpose, i.e., how large must the argument of the function $g_\Omega(\omega)$ be? Here we are looking for the *spectral* formulation of the condition (3.99) which enables natural fluctuations to be detected despite the presence of technical drifts. In terms of the dimensioned quantities, since $\theta = \mu\omega_0\tau$, $\Omega(t) = d\varphi/dt = \mu\omega_0 d\varphi/d\theta = \mu\omega_0\alpha(\theta)$, and $D_n = \mathcal{D}/\mu\omega_0$, this condition can be rewritten as

$$\tau \ll 2\mathcal{D}_n/\overline{\Omega_t^2} \quad . \tag{3.100}$$

A glance at Fig. 3.20 seems to suggest that we should require $\omega \gg 1/\tau_t$. But obviously we want values of ω for which $g_\Omega^t(\omega)$ would be small as compared not

151

with its maximal value at zero, but with the level of $g_\Omega^n(\omega)$. The condition will be fulfilled if the spectral intensity of the natural fluctuations in the interval $(0, \omega)$,

$$\int_0^\omega g_\Omega^n(\omega)\, d\omega \approx \frac{4\mathcal{D}_n}{\pi}\omega \quad ,$$

is much higher than that of the technical drifts

$$\int_0^\omega g_\Omega^t(\omega)\, d\omega \approx \int_0^\infty g_\Omega^t(\omega)\, d\omega = \overline{\Omega_t^2} \quad .$$

We thus arrive at the condition

$$\frac{4\mathcal{D}_n}{\pi}\omega \gg \overline{\Omega_t^2} \quad ,$$

which for $\omega = \pi/2\tau$ coincides with (3.100).

For the wings of the curve to be measured directly it is necessary that $g_x(\omega)$ for the differences $|\omega - \omega_0|$, meeting the condition (3.100), exceed the level of instrumental noise. The Bershtein method is effective when the fluctuational phase drift is not very small, as is the case in conventional vacuum-tube and klystron oscillators. In the molecular oscillator (maser) natural frequency fluctuations are so weak that they are generally measured using other techniques, e.g., by comparing two independent oscillators.

We will finally turn to the following interesting question. The Thomson self-oscillatory system produces quasi-monochromatic oscillations — a spectral line of small width. In the early literature it was sometimes erroneously assumed that fluctuations affecting a self-oscillatory system behave as an extremely selective filter. The self-oscillation spectrum was represented as a superposition of a discrete line and a narrow (filtered) noise background. This is false because the oscillator is a nonlinear system with feedback, whereby its inherent noise sources produce not a superposition of noise and self-oscillations, but a chaotic modulation of the latter, i.e., line broadening.

Now suppose that we have a rather selective filter whose squared modulus of the transfer function, $|k(i\omega)|^2$, reproduces accurately the spectral line shape of the self-oscillatory system. Thus, if we apply a noise with a uniform spectrum to the filter (uniform at least in a certain band that with an adequate margin contains the filter pass-band), then we obtain at the output an oscillation whose spectrum (and hence the covariance) is the same as for the self-oscillatory system. Can we under these conditions distinguish the filtered noise from the self-oscillations simply by examining the oscillations, without access to their sources?

It is only natural to turn to the distribution functions of both processes. Both are stationary quasi-monochromatic processes. Accordingly, the joint dis-

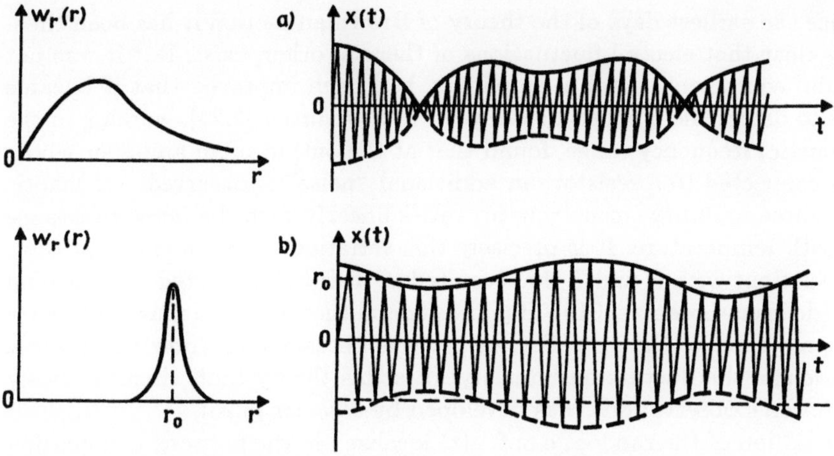

Fig. 3.21a,b. Distributions of amplitude $w_r(r)$ and oscillograms for (**a**) a narrow band noise filtered from white noise and (**b**) oscillations of the self-oscillatory system

tribution of the amplitude and phase of either has the form of (2.35). The phases of both oscillations are distributed uniformly in the interval $(0, 2\pi)$ and the only possibility of differentiating between them is to compare the distributions of the envelope $w_r(r)\, dr$.

The oscillation at the output of a filter with a fairly narrow band will be a Gaussian process with the result that the amplitude r will be distributed following the Rayleigh law (Fig. 3.21a). But in the oscillator, even if the self-excitation boundary is just crossed, the distribution of r, as we saw in Sect. I.5.9, will be Gaussian and centered about the value $\bar{r} = r_0$, i.e., near the radius of a dynamic limiting cycle (Fig. 3.21b). It is quite obvious that this distinction is fairly general in nature. When the oscillation source is inaccessible, it enables one to establish to which type (self-oscillations or filtered noise) the oscillations are closer.

3.5 Thermal Noise in Quasi-Stationary Networks. The Fluctuation-Dissipation Theorem

Fluctuations of electric and magnetic quantities resulting from the thermal motion of microcharges in bodies offer extensive and fascinating possibilities for the use of statistical methods in electrodynamics. This domain also includes those random electromagnetic fields that are produced by the thermal motion of microcharges, including wave fields that are referred to as thermal radiation. These fields will be considered in [3.23]. Here we will confine ourselves to the special case of *quasi-stationary* electric networks, for which radiation as a rule is of no interest. The key element is the fluctuations of *integral* electric quantities (charges, currents, voltages) describing the state of the circuit.

Since the earliest days of the theory of Brownian motion it has been theoretically clear that electric fluctuations of thermal origin exist. But it was not until radio equipment, notably amplifiers, had been improved that it became possible to detect them experimentally. In 1927 *Johnson* [3.24], working in the low (acoustic) frequency range, found that at the output of an amplifier whose input is connected to a resistor, an additional "noise" is observed − a chaotic voltage whose intensity (mean square) varies linearly with the input resistance R and with temperature. It is precisely this variation of the intensity of thermal fluctuations with R and T that resulted in their being treated as Brownian motion (de Haas-Lorentz, 1913). But the general theory had to make allowance for the pass-band of the amplifier, i.e., it was necessary to know the *spectral distribution* of the fluctuation intensity. It is this theory that, simultaneously with Johnson's observations, was developed by *Nyquist* [3.25].

The notion of the random e.m.f. $e(t)$ localized in the network and cuasing fluctuations of currents and voltages has been introduced by analogy with the Langevin random force resulting in the Brownian motion of a particle. Nyquist has given a spectral (in modern language, correlation) theory of this e.m.f. All the macroscopic conditions remaining unchanged, this e.m.f. appears to constitute a stationary random process. Since $\overline{e(t)} = 0$, the principal characteristic of the process is its spectral density $g^{(e)}(\omega)$, which is such that it gives a correct description of electric fluctuations observed in any network. But an intricate network can always be broken down into two-terminal networks to each of which can be assigned its own inherent source of the fluctuational e.m.f. The issue thus reduces to establishing the spectral density $g^{(e)}(\omega)$ of the random e.m.f. localized in any of the two-terminal network. The only electrodynamical characteristic of such a network is its impedance $Z(i\omega)$; accordingly, the relationship between $g^{(e)}(\omega)$ and $Z(i\omega)$ should be found. It is this relationship that is given by a fundamental formula derived by Nyquist and which bears his name. Further development of the theory of thermal fluctuations has lead to a variety of derivations of the formula and to far-reaching generalizations from which the formula follows as a simple special case. But its initial derivation, as given by Nyquist, is still classic from the point of view of clarity and elegance.

Clearly, the desired relationship between $g^{(e)}(\omega)$ and $Z(i\omega)$ is so universal that it is independent of the specific model used to find it. Nyquist considers an ideal two-wire line with a wave impedance $\sqrt{L/C}$ (L and C are the linear inductance and capacitance, respectively), whose ends are connected to resistances $R = \sqrt{L/C}$ (Fig. 3.22) that are matched in frequency ω and have the same temperature T (equilibrium state). To sustain fluctuations in the ideal conductors of which the line is made up requires no internal sources, but it is necessary to include in the resistances R "generators" of the random e.m.f.'s $e_1(t)$ and $e_2(t)$. Energy exchange between the resistances thus occurs by the agency of waves excited in the line by each of these "generators" and undergoing no reflection because the line has been matched to the loads in the spectral interval $(\omega, \ \omega + d\omega)$ under consideration.

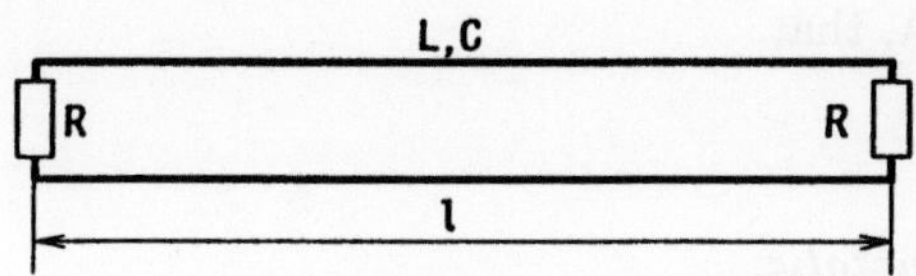

Fig. 3.22. Circuit illustrating part of the derivation of the Nyquist formula

If at a certain instant of time we short-circuit both ends of the line by ideally conducting jumpers, then waves travelling through the line in the opposite directions at various frequencies will be trapped. A system of standing waves with frequencies in the interval $(\omega, \ \omega + d\omega)$ may be thought of as a superposition of those natural oscillations of a given section l of the line, whose frequencies are related to this interval. Natural frequencies are equidistant and given by $\omega_m = m\pi v/l$, where $m = 1, 2, 3, \ldots$ and v is the phase velocity of waves in the line. The natural frequencies are spaced at $\Delta\omega_m = \pi v/l$. Accordingly, the interval of *positive* frequencies $d\omega$ of interest to us will contain

$$dN = \frac{d\omega}{\Delta\omega_m} = \frac{l\,d\omega}{\pi v}$$

natural oscillations (degrees of freedom). By the equipartition theorem, in thermodynamic equilibrium the mean energy per degree of freedom is kT (k is the Boltzmann's constant), so that the energy of the trapped waves in the interval $(\omega, \ \omega + d\omega)$ is

$$dU = kT\,dN = kT\frac{l\,d\omega}{\pi v} \quad . \tag{3.101}$$

When loads R are connected there is no reflection, therefore the expression derived is equal to the energy sent into the line by both resistances during the time required for the waves to cover the distance l, i.e., during the time $\tau = l/v$. It follows that each of the resistances generates in the positive frequency interval $d\omega$ per unit time the energy

$$p(\omega)\,d\omega = \frac{dU}{2\tau} = \frac{kT}{2\pi}\,d\omega \quad . \tag{3.102}$$

A few remarks are in order concerning this result.

First, dN is only proportional to $d\omega$ when the number of natural oscillations dN within $d\omega$ is sufficiently large ($dN \gg 1$). Expressing ω in terms of the wavelength in the line, $\lambda = 2\pi v/\omega$, we rewrite the above condition as

$$\frac{d\lambda}{\lambda} \gg \frac{\lambda}{2l} \quad .$$

On the other hand, the nonmonochromaticity $d\lambda$ should also not be unduly large, so that frequency-dependent quantities [e.g., $R(\omega)$] will vary only slightly over $d\omega$. In any event, it is necessary that $d\omega \ll \omega$, or in terms of λ,

$$1 \gg \frac{d\lambda}{\lambda} \quad .$$

It follows from both conditions for $d\lambda/\lambda$, that

$$l \gg \lambda \quad ,$$

i.e., the result (3.102) is in a sense *asymptotic*.

Second, (3.102) has been derived for a special form of line (ideal two-wire line) and dN has only been calculated for the so-called principal waves (transverse-electric and transverse-magnetic simultaneously), which have no dispersion of the velocity v. It can however be proven [Ref. 3.26, Sect. 17] that (3.102) holds for a line (one-dimensional channel) of an arbitrary type and for any waves possible in such a line, and not only for the principal ones, which in the general case may be nonexistent (e.g., in a waveguide). We should here take into account the dispersion of $v = v(\omega)$ and, accordingly, the difference between the phase and group velocities (in the presence of dispersion the transit time τ is determined by the group velocity).

Thirdly, the equipartition theorem confines us to the nonquantum region $\hbar\omega \ll kT$, which is natural enough for the quasi-stationary range of interest. If, however, we do not confine ourselves to sufficiently low frequencies, we will have to employ the quantum form of the energy partition theorem. In that case, in (3.101, 102) instead of kT, we would have the mean energy of the quantum oscillator

$$\Theta(\omega, T) = \frac{\hbar\omega}{2} + \frac{\hbar\omega}{\exp(\hbar\omega/kT) - 1} = \frac{\hbar\omega}{2} \coth\left(\frac{\hbar\omega}{2kT}\right) \quad . \tag{3.103}$$

But let us return to the Nyquist formula.

Now, let identical resistances R be connected through a purely *reactive* filter X that only passes the positive frequency band $(\omega, \ \omega + d\omega)$, within which it is absolutely "transparent" (Fig. 3.23). Such a filter, which functions either as a short circuit, or as a circuit breaker, introduces no noise of its own. Therefore, the current through the network at a frequency ω, i.e., the spectral amplitude density $\tilde{I}(\omega)$, is expressed in terms of the analogous amplitude densities $\tilde{e}_1(\omega)$ and $\tilde{e}_2(\omega)$ by

$$\tilde{I}(\omega) = \frac{\tilde{e}_1(\omega) + \tilde{e}_2(\omega)}{2R(\omega)} \quad . \tag{3.104}$$

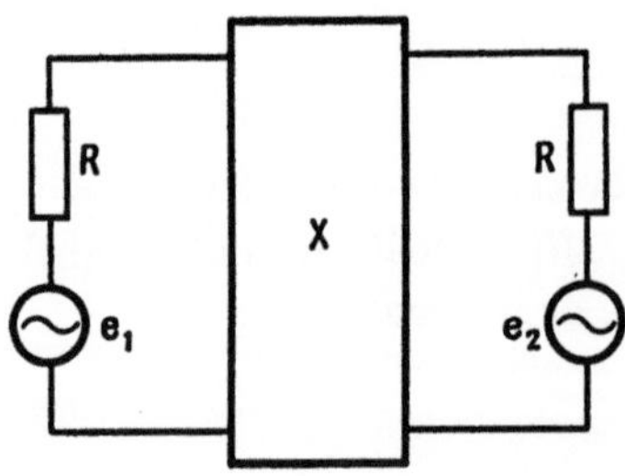

Fig. 3.23. Scheme referring to the derivation of the Nyquist formula

For the stationary processes in question

$$\langle \tilde{I}(\omega)\tilde{I}^*(\omega')\rangle = g^{(I)}(\omega)\delta(\omega - \omega') \quad ,$$

and similar relations take place for $\tilde{e}_1(\omega)$ and $\tilde{e}_2(\omega)$. Note that $g_1^{(e)}(\omega) = g_2^{(e)}(\omega) \equiv g_e(\omega)/2$ where, to simplify the notation, we denote the spectral density in *positive* frequencies by $g_e(\omega)$ instead of $g^{(e)+}(\omega)$. Assume further that $\tilde{e}_1(\omega)$ and $\tilde{e}_2(\omega)$ are uncorrelated, which is natural enough if we remember that the true source of each of these e.m.f.'s is the thermal motion of microcharges in an appropriate resistor. However, below we give a more general justification for the assumption.

We now multiply the expressions (3.104) for $\tilde{I}(\omega)$ and $\tilde{I}^*(\omega)$ and average the result, taking into account the fact that $\tilde{e}_1(\omega)$ and $\tilde{e}_2(\omega)$ are uncorrelated. Having discarded $\delta(\omega - \omega')$ on either side of the equality, we find

$$R(\omega)g^{(I)}(\omega) = \frac{g_1^{(e)}(\omega) + g_2^{(e)}(\omega)}{4R(\omega)} \quad . \tag{3.105}$$

Of course, this equality, written for the spectral densities ω in the interval $(-\infty, \infty)$, is also valid for the spectral densities in positive frequencies $g_I(\omega) = 2g^{(I)}(\omega)$, and so forth. According to (3.105) the power (in the interval $d\omega$ of *positive* frequencies) coming from the second to the first resistance is given by

$$2\cdot\frac{g_2^{(e)}(\omega)}{4R(\omega)}\, d\omega = \frac{g_e(\omega)}{4R(\omega)}\, d\omega \quad .$$

Exactly the same power comes from the first resistance to the second. This power must, by (3.102), be $kT d\omega/2\pi$, which leads to the Nyquist formula for the spectral density (in positive ω) of the random thermal e.m.f. $e(t)$

$$g_e(\omega) = \frac{2}{\pi}kTR(\omega) = \frac{2}{\pi}kT\,\mathrm{Re}\{Z(i\omega)\} \quad . \tag{3.106}$$

It follows from (3.106) that systems that contain no resistance, also do not include any source of thermal noise. We already discussed such a connection between thermal fluctuations and energy dissipation in a system earlier when considering Brownian motion (Sect. I.6.3.). The assumption that purely reactive systems are not noisy was used above. A further assumption is the absence of mutual correlation of e.m.f.'s in different two-terminal networks. This property can be deduced from the general and obvious requirement that a series and a parallel connection of these two-terminal networks (Z_1, e_1 and Z_2, e_2) be equivalent to one two-terminal network with an impedance Z equal respectively to either $Z_1 + Z_2$, or $Z_1 Z_2/(Z_1 + Z_2)$, and with an e.m.f. e equal to either $e_1 + e_2$ or $(Z_2 e_1 + Z_1 e_2)/(Z_1 + Z_2)$ (Fig. 3.24). Again, e and Z must obey the Nyquist formula.

What is more, in writing the Kirchhoff equations for a given network, the use of local, uncorrelated e.m.f.'s e enables the possible temperature difference

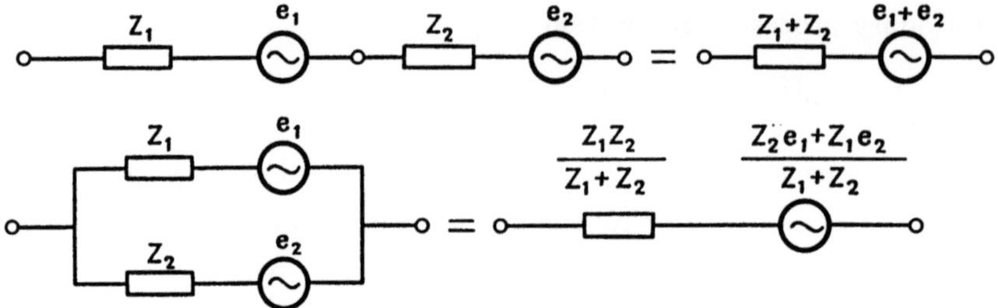

Fig. 3.24. Series and parallel connection of two noisy two-terminal networks

between individual two-terminal networks to be conveniently taken into account (see Exercises 3.6.7, 8).

According to (3.106), the spectral density of the Nyquist e.m.f. increases with resistance, and as $R \to \infty$ (circuit breakage) also tends to infinity. This apparent paradox only emerges because (3.106) with $\mathrm{Re}\{Z\} = R$ is solely applicable to sufficiently good conductors, i.e., to those *inside* which we can neglect, for a given frequency ω, the displacement current as compared with the conduction current. If this condition is not met, we will have to include the dielectric permittivity ε of the conductor as well as the conductivity σ. Take, for example, a length of wire (Fig. 3.25). If, besides the resistance $R = l/\sigma S$, we introduce the internal capacitance between the end faces $C = \varepsilon S/4\pi l$ (S is the cross-sectional area), we will obtain, as will be shown in later volumes, the following relation for the spectral density of the thermal e.m.f.:

$$g_e(\omega) = \frac{2}{\pi} kT \frac{R}{1 + \omega^2 C^2 R^2} \quad . \tag{3.107}$$

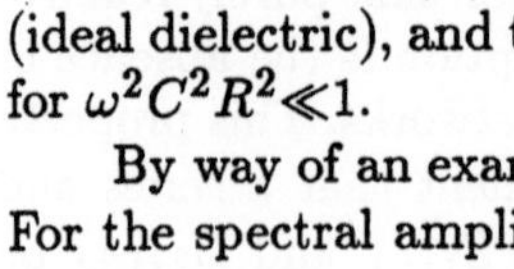

Fig. 3.25. To the derivation of the Nyquist formula including internal capacitance

It follows that $g_e(\omega)$ vanishes both at $R = 0$ (ideal conductor), and $R = \infty$ (ideal dielectric), and that the conventional relationship (3.106) is only suitable for $\omega^2 C^2 R^2 \ll 1$.

By way of an example, consider the thermal noise in an oscillatory circuit. For the spectral amplitudes of the current and e.m.f. $[\tilde{I} = \tilde{I}(\omega)$ and $\tilde{e} = \tilde{e}(\omega)]$ we have

$$\tilde{I} = \frac{\tilde{e}}{Z(i\omega)} = \frac{\tilde{e}}{R + i(\omega L - 1/\omega C)} \quad .$$

The spectral current density in positive frequencies is

$$g_I(\omega) = \frac{g_e(\omega)}{|Z|^2} = \frac{2}{\pi} kT R \frac{1}{R^2 + (\omega L - 1/\omega C)^2} \quad .$$

We find that the mean total magnetic energy of the noise in the circuit is

$$\overline{U}_m = \frac{L}{2}\overline{I^2} = \frac{L}{2}\int\limits_0^\infty g_I(\omega)\,d\omega = \frac{kT\,RL}{\pi}\int\limits_0^\infty \frac{d\omega}{R^2 + (\omega L - 1/\omega C)^2} \quad .$$

Because of the skin-effect, R and L are, at fairly high frequencies, dependent on ω (in a 1-mm dia copper conductor this dependence will already be noticeable beginning with $\omega \sim 10^5$). Suppose that the natural frequency of the circuit is far lower than these frequencies and that we can, to sufficient accuracy, assume the circuit parameters to be independent of ω.

We now introduce the integration variable $x = \omega\sqrt{LC}$ and denote the inverse of the quality by γ ($\gamma = 1/Q = R/\sqrt{L/C}$, whereby the expression for $\overline{U}_m$ becomes

$$\overline{U}_m = \frac{\gamma kT}{2\pi}\int\limits_{-\infty}^{+\infty} \frac{x^2\,dx}{\gamma^2 x^2 + (x^2 - 1)^2} \quad .$$

We then close the integration path in the upper half-plane and take the residues at the poles $x_{1,2} = i\gamma/2 \pm \sqrt{1 - \gamma^2/4}$ to obtain

$$\overline{U}_m = kT/2 \quad ,$$

as is to be expected by virtue of the equipartition theorem and the relation $\overline{U}_m = \overline{E}/2$, where $\overline{E}$ is the total mean energy of the linear oscillator.

Similarly, for the mean total electric energy of the oscillatory circuit, which depends on the spectral intensity of the charge at the capacitor $g_q(\omega) = g_I(\omega)/\omega^2$, we arrive at

$$\overline{U}_e = \frac{\overline{q^2}}{2C} = \frac{1}{2C}\int\limits_0^\infty g_I(\omega)\frac{d\omega}{\omega^2}$$

$$= \frac{\gamma kT}{2\pi}\int\limits_{-\infty}^{+\infty} \frac{dx}{\gamma^2 x^2 + (x^2 - 1)^2} = \frac{kT}{2} \quad .$$

If the capacitance C is connected to an arbitrary two-terminal network (Fig. 3.26) with impedance

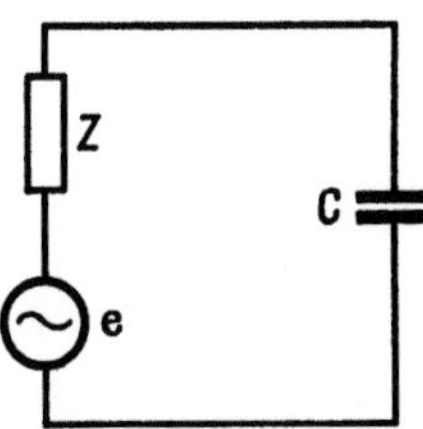

Fig. 3.26. To the issue of the total energy of thermal fluctuations of the charge at the capacitor

$$Z(\mathrm{i}\omega) = R(\omega) + \mathrm{i}X(\omega) \quad ,$$

then the mean electric energy of the capacitor will, of course, no longer be equal to $kT/2$: kT is the energy per degree of freedom (per each normal oscillation of the system) and not per capacitance and inductance. According to Ohm's law and Nyquist's formula, the spectral intensity of the current through the network Z and C is given by

$$g_I(\omega) = \frac{2}{\pi}kT\frac{\mathrm{Re}\,\{Z\}}{|Z + 1/\mathrm{i}\omega C|^2} = \frac{2}{\pi}kT\frac{R(\omega)}{R^2(\omega) + [X(\omega) - 1/(\omega C)]^2} \quad .$$

Again, taking into account that $g_q(\omega) = g_I(\omega)/\omega^2$, we obtain

$$\overline{U}_e = \frac{CkT}{2\pi}\int\limits_{-\infty}^{+\infty} \frac{R(\omega)\,d\omega}{\omega^2 C^2 R^2(\omega) + [\omega C X(\omega) - 1]^2} \quad ,$$

where, since $R(\omega)$ is an even function, we extend the integration to include the entire real axis ω. If

$$Z = \frac{R}{1 + \mathrm{i}\omega c R}$$

i.e., the two-terminal network consists of a capacitance c and a resistance R connected in parallel, then the mean energy $\overline{U}_e = kT/2$ will refer to the total capacitance $C+c$, and the capacitor C will only account for the part $C/(C+c)$ of $kT/2$.

It is customary in radio engineering and radiophysics to express the intensity of noise of any origin (noise in vacuum tubes and semiconductor devices, cosmic radiation noise, etc.) in terms of the so-called *equivalent noise temperature*. Let a source (two-terminal network) give to a matched load in the (positive) frequency band $\Delta\omega$ the power

$$P = \int\limits_{\omega}^{\omega+\Delta\omega} p(\omega)\,d\omega = \overline{p}(\omega)\Delta\omega \quad ,$$

where, clearly, $\overline{p}(\omega)$ is the mean power density in the band under consideration. According to (3.102), the resistance (at a temperature T_s) matched to the same load produces thermal noise of power $kT_s\Delta\omega/2\pi$. If T_s is selected so that the powers are equal

$$\frac{kT_s}{2\pi}\Delta\omega = P = \overline{p}(\omega)\Delta\omega \quad ,$$

then T_s is termed the *equivalent* (noise) temperature of the source. It is often convenient to divide T_s by the "room" temperature $T_0 = 290\,\mathrm{K}$, i.e., to introduce the *relative* equivalent temperature of the source

$$n_T(\omega) = \frac{T_{\rm s}}{T_0} = \frac{2\pi \overline{p}(\omega)}{kT_0} \quad .$$

Moreover, based on the Nyquist formula (3.106), the idea of *equivalent noise resistance* is often used. We will consider this with reference to the shot noise of the tube $(I_a = \overline{I}_a + I_{\rm sh})$, whose spectral density (for $\omega > 0$) is $g_{\rm sh}(\omega) = e\overline{I}_a/\pi$.

Suppose that the tube is not noisy, but that a resistance $R_{\rm s}(\omega)$ to its grid which is the source of the Nyquist e.m.f. $e(t)$ is connected. The spectral density of this e.m.f. at room temperature T_0 is

$$g_e(\omega) = \frac{2}{\pi}kT_0 R_{\rm s}(\omega) \quad .$$

In the anode current, $e(t)$ produes a thermal noise with the spectral density $g_I(\omega) = S^2 g_e(\omega) = 2kT_0 S^2 R_{\rm s}(\omega)/\pi$, where S is the slope of the anode characteristic. If $R_{\rm s}(\omega)$ is chosen so that $g_I(\omega) = g_{\rm sh}(\omega)$, i.e.,

$$R_{\rm s}(\omega) = \frac{e\overline{I}_a}{2kT_0 S^2} \quad ,$$

then this will be the equivalent noise resistance of the shot current. Both noises here are white, whereby $R_{\rm s}(\omega)$ does not vary with frequency.

In the general case of a linear network with n degrees of freedom the behavior of the system is described by $2n$ variables: either, as is generally the case in electric and radio engineering, by n currents and n voltages, or, in terms of the Lagrange-Maxwell equations, by n charges (generalized coordinates) q_j and n currents (generalized velocities) $\dot{q}_j = I_j$. The linear differential equations for $q_j(t)$ give for the spectral amplitudes of the currents I_j and e.m.f.'s $\tilde{\mathcal{E}}_j$ the generalized Kirchhoff equations

$$\tilde{I}_j = \sum_k Y_{jk}(\mathrm{i}\omega)\tilde{\mathcal{E}}_k \quad , \quad \tilde{\mathcal{E}}_j = \sum_k Z_{jk}(\mathrm{i}\omega)\tilde{I}_k \quad . \tag{3.108}$$

here $\boldsymbol{Y} = \{Y_{jk}\}$ and $\boldsymbol{Z} = \{Z_{jk}\}$ are the mutually inverse matrices of admittance and impedance $[\boldsymbol{YZ} = \boldsymbol{E}$, where $\boldsymbol{E} = \{\delta_{jk}\}$ is the unit matrix].

The e.m.f.'s $\tilde{\mathcal{E}}_j(\omega)$ in Kirchhoff's laws (3.108) are no longer local e.m.f.'s included into corresponding two-terminal networks, into which we can divide our complicated network. In (3.108) they are the spectral amplitudes of the *generalized* "forces" $\mathcal{E}_j(t)$ that, in terms of the Lagrange equations, correspond to the generalized "coordinates" $q_j(t) = \int I_j(t)\,dt$. In other words, the work done on the system in unit time by all the e.m.f.'s $\mathcal{E}_j(t)$ (i.e., the instantaneous power absorbed by the system) will be

$$Q(t) = \sum_j \mathcal{E}_j(t)\dot{q}_j(t) = \sum_j \mathcal{E}_j(t)I_j(t) \quad .$$

If $\mathcal{E}_j(t)$ are thermal fluctuational e.m.f.'s, which at a constant temperature T are stationary and stationarily related random processes with $\overline{\mathcal{E}_j(t)} = 0$ then, in terms of correlation theory, their statistical behavior is described by the correlation matrix $\langle \mathcal{E}_j(t)\mathcal{E}_k(t')\rangle$ or the corresponding matrix of the spectral densities $g_{jk}^{(\mathcal{E})}(\omega)$

$$\langle \tilde{\mathcal{E}}_j(\omega)\tilde{\mathcal{E}}_k^*(\omega')\rangle = g_{jk}^{(\mathcal{E})}(\omega)\delta(\omega - \omega') \quad .$$

Generalization of the Nyquist formula (3.106) to systems with many degrees of freedom yields a connection between the matrix $g_{jk}^{(\mathcal{E})}(\omega)$ and the impedance matrix $Z_{jk}(i\omega)$ of the network. Namely, if the entire network has the same temperature T (equilibrium state), the Nyquist formula, together with the Kirchhoff equations (3.108) and the Thévenin theorem, gives for $\omega > 0$ the spectral densities (see [3.27]):

$$g_{\mathcal{E}jk}(\omega) = \frac{kT}{\pi}(Z_{jk} + Z_{kj}^*) \quad . \tag{3.109}$$

For the nonbranching network ($j = k = 1$, $Z_{11} \equiv Z$) this clearly gives the Nyquist formula (3.106).

Note that the generalization (3.109) of the Nyquist formula to branched networks also holds when the network does not satisfy the reciprocity principle. If this principle is satisfied, the matrices Z and Y will be symmetrical, and then

$$g_{\mathcal{E}jk}(\omega) = \frac{kT}{\pi}(Z_{jk} + Z_{jk}^*) = \frac{2}{\pi}kT\,\mathrm{Re}\,\{Z_{jk}\} \quad . \tag{3.110}$$

From (3.109) we can readily derive the expressions for the matrix elements of the spectral currents densities

$$\langle \tilde{I}_j(\omega)\tilde{I}_k^*(\omega')\rangle = g_{jk}^{(I)}(\omega)\delta(\omega - \omega').$$

By the first of (3.108), we obtain

$$\langle \tilde{I}_j(\omega)\tilde{I}_k^*(\omega')\rangle = \sum_{l,m} Y_{jl}(i\omega)Y_{km}^*(i\omega')\langle \tilde{\mathcal{E}}_l(\omega)\tilde{\mathcal{E}}_m^*(\omega')\rangle \quad .$$

If we substitute the covariances, $\delta(\omega - \omega')$ cancels and we can go over to densities for $\omega > 0$,

$$g_{Ijk}(\omega) = \sum_{l,m} Y_{jl}(i\omega)Y_{km}^*(i\omega)g_{\mathcal{E}lm}(\omega) \quad .$$

Inserting (3.109) gives

$$g_{Ijk}(\omega) = \frac{kT}{\pi}\sum_{l,m} Y_{jl}Y_{km}^*(Z_{lm} + Z_{ml}^*) \quad .$$

But

$$\sum_l Y_{jl} Z_{lm} = \delta_{jm} \quad \text{and} \quad \sum_m Y_{km}^* Z_{ml}^* = \delta_{kl} \quad ,$$

therefore,

$$g_{Ijk}(\omega) = \frac{kT}{\pi}(Y_{jk} + Y_{kj}^*) \quad . \tag{3.111}$$

Unlike the thermal e.m.f.'s $e_l(t)$ (localized at individual two-terminal networks), which are suitable for those cases where the entire structure of the network (and not only its impedance matrix) is known, the e.m.f.'s $\mathcal{E}_j$ which appear in (3.108) are, generally speaking, mutually correlated. This is because the generalized e.m.f.'s $\mathcal{E}_j$ are linearly expressed in terms of local, uncorrelated e.m.f.'s e_l, with the result that different $\mathcal{E}_j$ may contain the same e_l (see Exercise 3.7.6). When e_l can be used (the network structure is known), they are convenient not only due to the fact that they are mutually uncorrelated, but also due to the ease with which possible differences in temperature of individual two-terminal networks can be taken into account. Several examples of such nonequilibrium conditions are considered in Exercises 3.6.7, 8.

Equations (3.109) and (3.111) are just special cases of a much more general theorem proved by *Callen* et al. [3.28–30] in 1951. This is the so-called *fluctuation-dissipation theorem* (FDT) to be discussed below.

Let the state of a system with n degrees of freedom and constant (time-independent) parameters be described by n generalized coordinates $q_j(t)$. The system has a thermostat at a temperature T, and thermal fluctuations occur, i.e., $q_j(t)$ are random processes. We will denote the ensemble-average, i.e., macroscopic, values of the coordinates by $x_j(t) = \langle q_j(t) \rangle$.

There are also generalized forces $F_j(t)$ that are conjugate to $x_j(t)$ in the sense that the mean instantaneous power imparted to the system by these *deterministic* forces is given by

$$Q(t) = \dot{x}_j(t) F_j(t) \quad . \tag{3.112}$$

Here (and below) the summation is over the double subscript from 1 to n. It should be stressed that $x_j(t)$ in (3.112) are the means derived in the presence of the forces $F_j(t)$, i.e., over the *nonequilibrium* ensemble.

The mean response $\boldsymbol{x}(t) = \{x_1(t), \ldots, x_n(t)\}$ of the generally *nonlinear* system to the forces $\boldsymbol{F}(t) = \{F_1(t), \ldots, F_n(t)\}$ is described by certain nonlinear functionals of $\boldsymbol{F}(t)$

$$x_j(t) = \hat{\Phi}_j\{\boldsymbol{F}(t)\} \quad . \tag{3.113}$$

Without loss of generality, we can suppose that in the absence of $F_j(t)$, i.e., at a thermodynamic equilibrium, the mean response is zero:

$$x_j(t)|_{\boldsymbol{F}=0} = \hat{\Phi}_j(0) = 0 \quad . \tag{3.114}$$

We now expand the functionals $\hat{\Phi}_j$ into functional series in $\boldsymbol{F}(t)$. According to (3.114), these series will begin with terms linear in $\boldsymbol{F}(t)$, i.e., terms describing the response of a linear dynamic system with constant parameters (harmonic system)

$$x_j(t) = \int\limits_{-\infty}^{+\infty} a_{jk}(t-\tau)F_k(\tau)\,d\tau + \ldots \quad , \tag{3.115}$$

where $a_{jk}(t-\tau)$ is the pulse response (at t) of the jth coordinate to the delta-impulse (at τ) of the kth force. If $\tilde{x}_j(\omega)$, $\tilde{F}_k(\omega)$ and $\alpha_{jk}(\omega) \equiv \tilde{a}_{jk}(\omega)$ are the spectral amplitudes of $x_j(t)$, $F_k(t)$ and $a_{jk}(t)$, then the spectral representation of (3.115) will be

$$\tilde{x}_j(\omega) = \alpha_{jk}(\omega)\tilde{F}_k(\omega) \quad , \quad \text{or} \quad \tilde{F}_j(\omega) = \alpha_{jk}^{-1}(\omega)\tilde{x}_k(\omega) \quad . \tag{3.116}$$

The matrix $\boldsymbol{\alpha} = \{\alpha_{jk}(\omega)\}$ is called the *generalized susceptibility matrix* for the system, and $\boldsymbol{a}^{-1}$ is its inverse. The generalized susceptibility $\boldsymbol{\alpha}$ characterizes, by definition, the mean (macroscopic) *linearized* response of a dynamic system, i.e., the response to the deterministic actions $F_k(t)$ which are so weak that we can retain only the linear terms in (3.115).

We will now turn to the *second* moments of the fluctuation coordinates $q_j(t)$. In the absence of any deterministic forces, $F_j(t)$, $q_j(t)$ fluctuate about the state of thermodynamic equilibrium, the fluctuations being *stationary* random processes. Therefore, we can introduce the spectral density matrix $g_{jk}^{(q)}(\omega)$ for these fluctuations (the frequency ω varying from $-\infty$ to $+\infty$) in a conventional way:

$$\langle \tilde{q}_j(\omega)\tilde{q}_k^*(\omega') \rangle = g_{jk}^{(q)}(\omega)\delta(\omega-\omega') \quad , \tag{3.117}$$

where the averaging is now over the *equilibrium* ensemble, i.e., $F_j(t)$ are not present.

FDT states that the *spectral density matrix for thermal fluctuations about the state of thermodynamic equilibrium at a temperature T is expressed in terms of the skew-Hermitian part of the generalized susceptibility of the linearized* (in the above sense) *dynamic system* in the following way:

$$g_{jk}^{(q)}(\omega) = \frac{i\Theta(\omega,T)}{2\pi\omega}[\alpha_{jk}(\omega) - \alpha_{kj}^*(\omega)] \quad , \tag{3.118}$$

where $\Theta(\omega,T)$ is the mean energy of the quantum oscillator (3.103). This general and fundamental result is valid for any frequencies ω, the quantum region $\hbar|\omega| \gtrsim kT$ included, and is applicable to macroscopic systems of any physical nature.

It should be stressed that the quantum-mechanical proof of the theorem (3.118) *does not assume that the fluctuations q_j are small*[12]. This proof consists of the exact calculation of (1) the spectral density $g_{jk}^{(q)}(\omega)$ and (2) the susceptibility matrix $\alpha_{jk}(\omega)$ as defined above. Simple comparison of both expressions shows that the equilibrium fluctuations obey the equality (3.118). Thus, the dynamic behavior of the system subjected to strong external actions, when we can no longer limit ourselves to expressions (3.116) (i.e., to terms that are linear in the deterministic actions $F_k(t)$), has nothing to do with the fluctuations about the state of thermodynamic equilibrium. This crucial point, clarified by *Bernard* and *Callen* [3.33] and independently by *Bunkin* [3.34], implies that any corrections to the FDT due to the dynamic equations of a macroscopic system being nonlinear are simply out of the question. Also out of the question are, in particular, various paradoxes associated with macroscopic nonlinearity, e.g., one involving a detector connected into a passive circuit in a thermostat. The detector would rectify thermal fluctuations, i.e., produce a direct current, thereby suggesting that some work may be done by a system in thermodynamic equilibrium.

The above in no way implies that the susceptibility α is insensitive to the nonlinearity of the *stochastic* equations of the system in question. In the general case, the equations for the *random* coordinates $q_j(t)$ are

$$\hat{L}_j\{q(t)\} = F_j(t) + f_j(t) \quad , \tag{3.119}$$

where $\hat{L}_j$ are nonlinear operators, $F_j(t)$ are deterministic forces, and $f_j(t)$ are random forces introduced together with $F_j(t)$ and describing the thermostat effect ($\langle f_j(t)\rangle = 0$). In the general case, determining α is a rather complicated exercise involving us with solving (3.119), i.e., finding

$$q_j(t) = \hat{L}_j^{-1}\{F(t) + f(t)\} \quad .$$

We then average the solution over a nonequilibrium ensemble to arrive at the mean response (3.113)

$$x_j(t) = \langle q_j(t)\rangle = \langle \hat{L}_j^{-1}\{F(t) + f(t)\}\rangle \equiv \hat{\Phi}_j\{F(t)\} \quad .$$

Finally, we have to separate out the part of $x_j(t)$ that is linear in $F(t)$. When we proceed from *macroscopic* equations, i.e., equations for $x_j(t)$, in order to find α it is only necessary to linearize *these* equations.

If the nonlinear terms in (3.119) are small, the solution can be reached by the perturbation method. The problem is, however, greatly simplified when the stochastic equations are linear [$\hat{L}_j$ in (3.119) are linear operators]. The equations for the spectral amplitudes $\tilde{q}_k(\omega)$ will then have the form

$$C_{jk}(\omega)\tilde{q}_k(\omega) = \tilde{F}_j(\omega) + \tilde{f}_j(\omega) \tag{3.120}$$

---

[12] See [Ref. 3.31, Sects. 87, 88], [Ref. 3.32, Sects. 126, 127].

and their averaging will yield (3.116)

$$C_{jk}(\omega)x_k(\omega) = \tilde{F}_j(\omega) \quad ,$$

i.e., $\alpha_{jk}^{-1}(\omega) = C_{jk}(\omega)$ and $\alpha_{jk}(\omega) = C_{jk}^{-1}(\omega)$. Thus, in this particular, but important, case the stochastic and macroscopic equations will have the same coefficients.

This is exactly the case with the linear electric networks described by the charges $q_j(t)$ $[\tilde{q}_j(\omega) = \tilde{I}_j(\omega)/i\omega]$. If $\tilde{F}_j(\omega) = \mathcal{E}_j(\omega)$ are the spectral amplitudes of the conjugate e.m.f.'s, then, by Kirchhoff's laws (3.108), $\alpha_{jk}(\omega) = Y_{jk}(\omega)/i\omega$, where $Y = \{Y_{jk}\}$ is the admittance matrix for the network. If we confine ourselves to the nonquantum region $\hbar|\omega| \ll kT$ [when $\Theta(\omega, T) = kT$], the general formula (3.118), in terms of the spectral densities for frequencies will lead to (3.111).

At equilibrium $[\tilde{F}_j(\omega) = 0]$, the stochastic equations (3.120) take the form

$$\alpha_{jk}^{-1}(\omega)\tilde{q}_k(\omega) = \tilde{f}_j(\omega) \quad . \tag{3.121}$$

From this we can easily obtain another common form of FDT, namely the spectral density matrix $f_{jk}^{(f)}(\omega)$ for the random Langevin forces $f_j(t)$

$$\langle \tilde{f}_j(\omega)\tilde{f}_k^*(\omega')\rangle = g_{jk}^{(f)}(\omega)\delta(\omega - \omega') \quad .$$

In fact, from (3.121) and (3.117) we have

$$g_{jk}^{(f)}(\omega) = \alpha_{jm}^{-1}(\omega)\alpha_{kn}^{-1*}(\omega)g_{mn}^{(q)}(\omega) \quad .$$

Substituting (3.118) gives

$$g_{jk}^{(f)}(\omega) = \frac{i\Theta(\omega, T)}{2\pi\omega}\alpha_{jm}^{-1}\alpha_{kn}^{-1*}(\alpha_{mn} - \alpha_{nm}^*) \quad .$$

But since $\alpha_{jm}^{-1}\alpha_{mn} = \delta_{jn}$ and $\alpha_{kn}^{-1*}\alpha_{nm}^* = \delta_{km}$, we thus obtain

$$g_{jk}^{(f)}(\omega) = \frac{i\Theta(\omega, T)}{2\pi\omega}[\alpha_{kj}^{-1*}(\omega) - \alpha_{jk}^{-1}(\omega)] \quad , \tag{3.122}$$

which is a generalization of the electrodynamic formula (3.110).

3.6 Exercises

3.6.1 Express the transfer function $k(i\omega)$ of a filter matched to a pulse $F(t)$ in terms of the spectral amplitude of the pulse, $\tilde{F}(\omega)$, and find the spectral density $g_\xi(\omega)$ of the Poisson process at the output of the matched filter.

Solution. We have

$$F(t) = \int\limits_{-\infty}^{+\infty} \tilde{F}(\omega)e^{i\omega t}\,d\omega \quad ,$$

where the spectral amplitude for the real function $F(t)$ meets the condition $\tilde{F}(-\omega) = \tilde{F}^*(\omega)$. By (3.18) and (3.21), the pulse response of the matched filter is

$$H(t) = \frac{1}{2\pi}\int\limits_{-\infty}^{+\infty} k(i\omega)e^{i\omega t}\,d\omega = F(t_0 - t)$$

$$= \int\limits_{-\infty}^{+\infty} \tilde{F}(\omega)e^{i\omega(t_0-t)}\,d\omega = \int\limits_{-\infty}^{+\infty} \tilde{F}^*(\omega)e^{i\omega(t-t_0)}\,d\omega \quad ,$$

so that

$$k(i\omega) = 2\pi\tilde{F}^*(\omega)e^{-i\omega t_0} \quad . \tag{3.123}$$

Consequently, the "covariance" $H_2(\chi)$ of the pulse response [see (3.6) and (3.20)] has the spectral amplitude

$$\frac{1}{2\pi}|k(i\omega)|^2 = 2\pi|\tilde{F}(\omega)|^2. \tag{3.124}$$

But according to (1.66), the same spectral amplitude is also peculiar to the "covariance" of the initial pulse $F(t)$

$$\Psi_F(\tau) = \int\limits_{-\infty}^{+\infty} F(t)F(t+\tau)\,dt \quad ,$$

hence $H_2(\tau) = \Psi_F(\tau)$.

The spectral density of the Poisson process $f(t) = \sum_\nu a_\nu F(t - t_\nu)$ at the filter input is

$$g_f(\omega) = 2\pi n_1\overline{a^2}|\tilde{F}(\omega)|^2 \quad , \tag{3.125}$$

hence, by (3.10) and (3.125), the spectral density at the output is

$$g_\xi(\omega) = |k(i\omega)|^2 g_f(\omega) = 8\pi^3 n_1\overline{a^2}|\tilde{F}(\omega)|^4 \quad . \tag{3.126}$$

[Recall that we put $H(t) = F(t_0 - t)$ without introducing a proportionality coefficient.] We can, of course, obtain the result (3.126) directly from (3.25) if we take into account that the spectral amplitude of the pulses $F_\xi(t - t_\nu) = \Psi_F(t - t_0 - t_\nu)$ at the filter output is given by (3.124), and $g_\xi(\omega)$ is proportional (with the coefficient $2\pi n_1 \overline{a^2}$) to the squared modulus of the spectral amplitude of the pulse.

Clearly, all the functions (3.124–126) have the width $\Delta\omega$ of the order of $1/\tau_{F_c}$.

3.6.2 Compute the spectral densities (3.125, 126) for the pulse

$$F(t) = e^{-2t^2/\vartheta^2} \, \cos\left(\omega_0 t + \frac{\alpha t^2}{2}\right)$$

(see Exercise 1.6.6).

Solution. We find the spectral amplitude of the pulse:

$$\tilde{F}(\omega) = \frac{1}{2\pi} \int\limits_{-\infty}^{+\infty} e^{-2t^2/\vartheta^2} \, \cos\left(\omega_0 t + \frac{\alpha t^2}{2}\right) e^{-i\omega t} \, dt$$

$$= \tilde{F}(\omega, \omega_0, \alpha) + \tilde{F}(\omega, -\omega_0, -\alpha) \quad,$$

where

$$\tilde{F}(\omega, \omega_0, \alpha) = \frac{1}{4\pi} \int\limits_{-\infty}^{+\infty} \exp\left\{-\frac{2t^2}{\vartheta^2} + i(\omega_0 - \omega)t + i\frac{\alpha t^2}{2}\right\} dt \quad .$$

Using the relationship

$$\int\limits_{-\infty}^{+\infty} \exp\left\{-\alpha x^2 + i(px^2 + 2qx + r)\right\} dx$$

$$= \frac{\sqrt{\pi}}{\sqrt[4]{a^2 + p^2}} \exp\left\{-\frac{aq^2}{a^2 + p^2} + i\left(\frac{1}{2}\arctan\frac{p}{q} + r - \frac{pq^2}{a^2 + p^2}\right)\right\} \quad,$$

we get

$$\tilde{F}(\omega) = \frac{\vartheta}{4\sqrt{2\pi}\sqrt[4]{s}}\{\exp\left(-A_- + iB_-\right) + \exp\left(-A_+ - iB_+\right)\} \quad,$$

where

$$A_\pm = \frac{(\omega \pm \omega_0)^2 \vartheta^2}{8s} \quad,$$

$$B_\pm = \frac{1}{2}\arctan\frac{\alpha\vartheta^2}{4} - \frac{\alpha\vartheta^2}{4}A_\pm \quad,$$

$$s = 1 + \frac{\alpha^2\vartheta^4}{16} \quad .$$

Hence

$$|\tilde{F}(\omega)|^2 = \frac{\vartheta^2}{32\pi\sqrt{s}}\Big\{\exp\big[-(\omega-\omega_0)^2\vartheta^2/4s\big] + \exp\big[-(\omega+\omega_0)^2\vartheta^2/4s\big]$$

$$+ 2\exp\big[-(\omega^2+\omega_0^2)\vartheta^2/4s\big]\cos\Big(\arctan\frac{\alpha\vartheta^2}{4} - \frac{\alpha(\omega^2+\omega_0^2)\vartheta^4}{8s}\Big)\Big\} \quad .$$

It is clear that the third term in the braces is always small as compared with unity, i.e., with the maxima of the first and second terms at $\omega = \pm\omega_0$, respectively. The width of these maxima at the $1/e$ level is

$$\Delta\omega = 4\sqrt{s}/\vartheta = 8/\tau_{F_c} \quad ,$$

where τ_{F_c} is the effective "correlation time" of the pulse under consideration (see Exercise 1.6.6). If $\alpha\vartheta \gg 1/\vartheta$ (intricate pulse-shapes), so that $s \approx \alpha^2\vartheta^4/16 \gg 1$, then $\Delta\omega \approx \alpha\vartheta$, i.e., the width of the maxima $|\tilde{F}(\omega)|^2$ is equal to the deviation of frequency during the pulse duration ϑ. The spectral densities of the input and output processes are thus approximately expressed by the relationships

$$g_f(\omega) = \frac{n_1\overline{a^2}\vartheta^2}{16\sqrt{s}}\big\{\exp\big[-4(\omega-\omega_0)^2/(\Delta\omega)^2\big]$$

$$+ \exp\big[-4(\omega+\omega_0)^2/(\Delta\omega)^2\big]\big\} \quad ,$$

$$g_\xi(\omega) = \frac{\pi n_1\overline{a^2}\vartheta^4}{128s}\big\{\exp\big[-4(\omega-\omega_0)^2/(\Delta\omega)^2\big]$$

$$+ \exp\big[-4(\omega+\omega_0)^2/(\Delta\omega)^2\big]\big\}^2 \quad .$$

3.6.3 Let the components $x_1(t)$ and $x_2(t)$ of a bivariate stationary random process $\boldsymbol{\xi}(t)$ be subjected to a memoryless nonlinear transformation of the form $y_1(t) = f_1[x_1(t)]$, $y_2(t) = f_2[x_2(t)]$, resulting in the stationary process $\boldsymbol{\eta}(t) = \{y_1(t), y_2(t)\}$. Using equation (I.3.75), obtain the equation relating the moments

$$\alpha_{r+s}^\eta = \langle y_1^r(t)y_2^s(t+\tau)\rangle = \alpha_{r+s}^\eta \underbrace{(0,\ldots,0}_{r\text{ times}};\ \underbrace{\tau,\ldots\tau)}_{s\text{ times}}$$

of the process $\boldsymbol{\eta}(t)$ with the cumulants

$$\lambda_{p+q}^\xi \underbrace{(0,\ldots0}_{p\text{ times}};\ \underbrace{\tau,\ldots\tau)}_{q\text{ times}}$$

of the initial process $\boldsymbol{\xi}(t)$ [3.35].

Solution. Setting into (I.3.75) $x = x_1(t)$, $y = x_2(t+\tau) \equiv x_{2\tau}$, $f(x,y) = f_1^r(x_1)f_2^s(x_{2\tau})$, gives

$$\frac{\partial^m a^\eta_{r+s}}{(\partial \lambda^\xi_{p+q})^m} = \frac{1}{(p!q!)^m} \left\langle \frac{d^{mp} f_1^r(x_1)}{dx_1^{mp}} \frac{d^{mq} f_2^s(x_{2\tau})}{dx_{2\tau}^{mq}} \right\rangle \quad . \tag{3.127}$$

In particular, if $x_1(t) = x_2(t) = x(t)$, then

$$\frac{\partial^m \alpha^\eta_{r+s}}{(\partial \lambda^\xi_{p+q})^m} = \frac{1}{(p!q!)^m} \left\langle \frac{d^{mp} f_1^r(x)}{dx^{mp}} \frac{d^{mq} f_2^s(x_\tau)}{dx_\tau^{mq}} \right\rangle \quad , \tag{3.128}$$

where

$$\alpha^\eta_{r+s} = \langle y^r(t) y^s(t+\tau) \rangle \quad .$$

For $r = s = 1$ and $p = q = 1$, i.e., for the mixed second moment $\alpha^\eta_{1+1} = \langle y(t)y(t+\tau) \rangle = B_\eta(\tau)$ of the process $\eta(t)$ and the covariance $\lambda^\xi_{1+1} = \overline{x x_\tau} - \overline{x}^2 \equiv \psi_\xi(\tau)$ of the initial process $\xi(t)$, we obtain from (3.128)

$$\frac{\partial^m B_\eta(\tau)}{[\partial \psi_\xi(\tau)]^m} = \left\langle \frac{d^m f_1(x)}{dx^m} \frac{d^m f_2(x_\tau)}{dx_\tau^m} \right\rangle \quad . \tag{3.129}$$

It follows from the derivation that this equation holds for any distribution $\{x, x_\tau\}$, and not only for the normal one, as was initially proved [3.36].

It should be emphasized that the cumulant equations derived in Exercises I.3.2.15, 16 for random variables are also valid for random processes, since the variation of the moments and cumulants with time introduces no change into the equations (they contain no time derivatives).

3.6.4 Consider the process $\eta(t) = \beta \xi^2(t)$ at the output of a square-law detector. Find the expression for the mixed moment $B_\eta(\tau) = \langle y(t)y(t+\tau) \rangle$ in terms of the cumulants of the process $\xi(t)$ at the detector input [3.35].

Solution. The detector transforms the quantities $x = x(t)$ and $x_\tau = x(t+\tau)$ similarly, i.e., by the equations of the preceding problem

$$f_1(x) = f_2(x) = \beta x^2 \quad .$$

Therefore, at $r = s = 1$ and $m = 1$ (3.128) takes the form

$$\frac{\partial^2 B_\eta(\tau)}{\partial \lambda^\xi_{p+q}} = \frac{\beta^2}{p!q!} \left\langle \frac{d^p x^2}{dx^p} \frac{d^q x_\tau^2}{dx_\tau^q} \right\rangle \quad , \tag{3.130}$$

whence it is clear that the values of p and q should not exceed 2.

Since the moment $B_\eta = \langle y y_\tau \rangle = \langle x^2 x_\tau^2 \rangle$ is expressed in terms of the cumulants x and x_τ of not higher than fourth order ($1 \leq p+q \leq 4$ and is, clearly, symmetrical about x and x_τ, its general form is as follows:

$$B_\eta(\tau) = a\lambda_{2+2}^\xi + b(\lambda_{1+0}^\xi + \lambda_{0+1}^\xi)(\lambda_{2+1}^\xi + \lambda_{1+2}^\xi)$$
$$+ (\lambda_{1+0}^\xi + \lambda_{0+1}^\xi)^2[c(\lambda_{2+0}^\xi + \lambda_{0+2}^\xi) + d\lambda_{1+1}^\xi]$$
$$+ e(\lambda_{1+1}^\xi)^2 + f(\lambda_{2+0}^\xi + \lambda_{0+2}^\xi)^2 \quad .$$

Equating the expressions for the derivatives of $B_\eta(\tau)$ with respect to λ_{p+q}^ξ resulting from this expression, and from (3.127), we get

$$\frac{\partial B_\eta}{\partial \lambda_{2+2}^\xi} = a = \beta^2 \quad , \quad \frac{\partial B_\eta}{\partial \lambda_{2+1}^\xi} = \frac{\partial B_\eta}{\partial \lambda_{1+2}^\xi} = b(\lambda_{1+0}^\xi + \lambda_{0+1}^\xi) = 2\beta^2\overline{x} \quad ,$$

$$\frac{\partial B_\eta}{\partial \lambda_{2+0}^\xi} = \frac{\partial B_\eta}{\partial \lambda_{0+2}^\xi} = c(\lambda_{1+0}^\xi + \lambda_{0+1}^\xi)^2 + 2f(\lambda_{2+0}^\xi + \lambda_{0+2}^\xi) = \beta^2\overline{x^2} \quad ,$$

$$\frac{\partial B_\eta}{\partial \lambda_{1+1}^\xi} = d(\lambda_{1+0}^\xi + \lambda_{0+1}^\xi)^2 + 2e\lambda_{1+1}^\xi = 4\beta^2\overline{xx_\tau} \quad .$$

But

$$\lambda_{1+0}^\xi = \lambda_{0+1}^\xi = \overline{x}, \quad \lambda_{2+0}^\xi = \lambda_{0+2}^\xi = \overline{x^2}, \quad \lambda_{1+1}^\xi = \overline{xx_\tau} - \overline{x}^2 = \psi_\xi(\tau) \quad .$$

With this in mind we obtain

$$a = b = d = \beta^2 \quad , \quad c = 0 \quad , \quad f = \beta^2/4 \quad , \quad e = 2\beta^2 \quad ,$$

and finally

$$B_\eta(\tau) = \beta^2[\lambda_{2+2}^\xi + 2\overline{x}(\lambda_{2+1}^\xi + \lambda_{1+2}^\xi) + 4\overline{x}^2\psi_\xi(\tau) + 2\psi_\xi^2(\tau) + (\overline{x^2})^2] \quad .$$

If the process $\xi(t)$ at the detector input is Gaussian, then $\lambda_3^\xi = 0$, $\lambda_4^\xi = 0$. Denoting $\overline{x^2} = \sigma^2$ and introducing the correlation coefficient $[\psi_\xi(\tau) = \sigma^2 K(\tau)]$, we find

$$B_\eta(\tau) = \beta^2[4\overline{x}^2\sigma^2 K(\tau) + 2\sigma^4 K^2(\tau) + \sigma^4] \quad .$$

If, additionally, $\overline{x} = 0$, then we arrive at (3.46)

$$\psi_\eta(\tau) = B_\eta(\tau) - \beta^2\sigma^4 = 2\beta^2\sigma^4 K^2(\tau) \quad .$$

3.6.5 Give the Nyquist formula for the case where a two-terminal network includes a generator not of the external e.m.f. $e(t)$, but of the *external current* $i(t)$.

Solution. The current source is connected to the two-terminal network not in series, but in parallel, so that the current $i(t)$ flows through the open two-terminal network and, accordingly, $\tilde{i}(\omega)Z = \tilde{e}(\omega)$. Hence

$$g^{(i)}(\omega) = g^{(e)}(\omega)/|Z|^2 \quad .$$

By the Nyquist formula,

$$g^{(e)}(\omega) = \frac{2}{\pi}kTR = \frac{2}{\pi}kT\,\mathrm{Re}\,\{Z\} \quad ,$$

and thus

$$g^{(i)}(\omega) = \frac{2}{\pi}kT\,\mathrm{Re}\,\{Y\} \quad ,$$

where $Y = 1/Z$ is the admittance of the two-terminal network.

3.6.6 The state of a system consisting of two connected oscillatory circuits (Fig. 3.27) is described by the currents $I_j(t) = \dot{q}_j(t)$ $(j = 1, 2)$. Find the connection between the generalized thermal e.m.f.'s $\mathcal{E}_j$, conjugate according to Lagrange to q_j and to the local e.m.f.'s e_1, e_2, and e, connected into the branches with resistances R_1, R_2, and R, respectively.

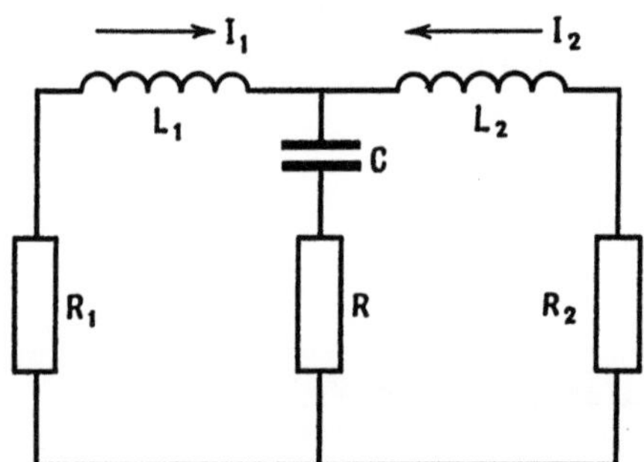

Fig. 3.27. System with two connected oscillatory circuits

Solution. In terms of the spectral amplitudes of the currents and e.m.f.'s, the equality of voltages in the three parallel branches of the network, with connected e.m.f.'s e_1, e_2, and e, is written in the form

$$\tilde{e}_1 + (\mathrm{i}\omega L_1 + R_1)\tilde{I}_1 = -\tilde{e} + \left(\frac{1}{\mathrm{i}\omega C} + R\right)(\tilde{I}_1 + \tilde{I}_2)$$
$$= \tilde{e}_2 - (\mathrm{i}\omega L_2 + R_2)\tilde{I}_2 \quad .$$

Hence the Kirchhoff equations are

$$Z_{11}\tilde{I}_1 + Z_{12}\tilde{I}_2 = \tilde{e} + \tilde{e}_1 = \tilde{\mathcal{E}}_1 \quad , \quad Z_{21}\tilde{I}_1 + Z_{22}\tilde{I}_2 = \tilde{e} + \tilde{e}_2 = \tilde{\mathcal{E}}_2 \quad ,$$

where

$$Z_{jj} = \mathrm{i}\omega L_j + R_j + Z \quad (j = 1, 2) \quad , \quad Z_{12} = Z_{21} = Z = \frac{1}{\mathrm{i}\omega C} + R \quad .$$
$$\tag{3.131}$$

The local thermal e.m.f.'s $\tilde{e}_1$, $\tilde{e}_2$, and $\tilde{e}$ are mutually uncorrelated, and their spectral densities, by the Nyquist formula (3.106), are

$$g_j^{(e)}(\omega) = \frac{2}{\pi}kTR_j \quad (j=1,2) \quad , \quad g^{(e)}(\omega) = \frac{2}{\pi}kTR \quad .$$

Accordingly, the spectral densities of the e.m.f.'s $\tilde{\mathcal{E}}_1 = \tilde{e} + \tilde{e}_1$ and $\tilde{\mathcal{E}}_2 = \tilde{e} + \tilde{e}_2$ are

$$g_j^{(\mathcal{E})}(\omega) = g^{(e)}(\omega) + g_j^{(e)}(\omega) = \frac{2}{\pi}kT(R+R_j) \quad (j=1,2) \quad ,$$

$$g_{12}^{(\mathcal{E})}(\omega) = g^{(e)}(\omega) = \frac{2}{\pi}kTR. \tag{3.132}$$

Correlation between $\tilde{\mathcal{E}}_1$ and $\tilde{\mathcal{E}}_2$ is thus conditioned by the fact that both contain the local e.m.f. $\tilde{e}$. Clearly, expressions (3.132) also follow immediately from the general formula (3.110) and the value (3.131) for Z_{jk}.

3.6.7 For the network described in the preceding problem (Fig. 3.27), find the spectral densities of the generalized e.m.f.'s $\tilde{\mathcal{E}}_1$ and $\tilde{\mathcal{E}}_2$ for the case where the resistances R_1, R_2 and R are at different temperatures – T_1, T_2 and T, respectively.

Solution. For the local e.m.f.'s e_1, e_2 and e, the Nyquist formula will now give

$$g_j^{(\mathcal{E})}(\omega) = g^{(e)}(\omega) + g_j^{(e)}(\omega) = \frac{2}{\pi}k(TR + T_jR_j) \quad (j=1,2) \quad ,$$

$$g_{12}^{(\mathcal{E})}(\omega) = g^{(e)}(\omega) = \frac{2}{\pi}kTR \quad .$$

These relationships are no longer obtainable from the "equilibrium theorem" (3.110).

3.6.8 In the RC-network in Fig. 3.28 the temperatures of the resistances R_1 and R_2 are T_1 and T_2, respectively. Find the spectral density of the energy in the capacitors C_1, C_2 and C. Find the total energies in these capacitors and show that at a thermal equilibrium we have equipartition of energy.

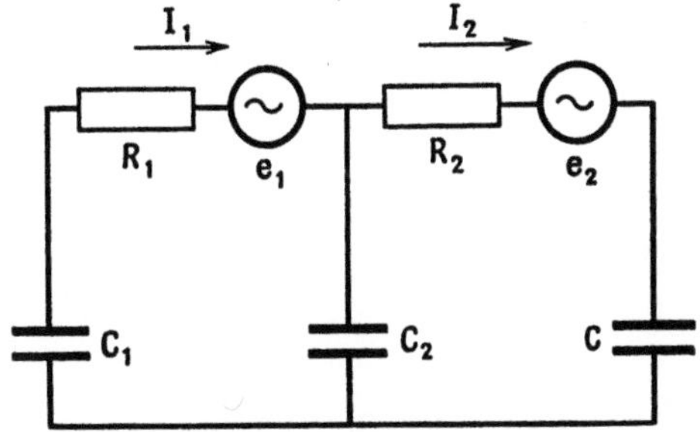

Fig. 3.28. RC-network

Solution. From the Kirchhoff equations we obtain the spectral amplitudes of the currents I_1 and I_2 in terms of the spectral amplitudes of e_1 and e_2 :

$$\tilde{I}_1 = \frac{i\omega}{\Delta}[(i\omega R_2 + p_2)\tilde{e}_1 + p\tilde{e}_2] \quad,$$

$$\tilde{I}_2 = \frac{i\omega}{\Delta}[p\tilde{e}_1 + (i\omega R_1 + p_1)\tilde{e}_2] \quad, \tag{3.133}$$

where

$$\begin{aligned}
\Delta &= (i\omega R_1 + p_1)(i\omega R_2 + p_2) - p^2 \\
&= p_1 p_2 - p^2 - \omega^2 R_1 R_2 + i\omega(R_1 p_2 + R_2 p_1),
\end{aligned}$$

$$p_1 = \frac{1}{C_1} + \frac{1}{C_2} \quad, \quad p_2 = \frac{1}{C} + \frac{1}{C_2} \quad, \quad p = \frac{1}{C_2} \quad.$$

We obtain, by (3.133), the spectral densities of I_1 and I_2, and their difference $I = I_1 - I_2$

$$g_1^{(I)}(\omega) = \frac{\omega^2}{|\Delta|^2}[(\omega^2 R_2^2 + p_2^2)g_1^{(e)}(\omega) + p^2 g_2^{(e)}(\omega)] \quad,$$

$$g_2^{(I)}(\omega) = \frac{\omega^2}{|\Delta|^2}[p^2 g_1^{(e)}(\omega) + (\omega^2 R_1^2 + p_1^2)g_2^{(e)}(\omega)] \quad,$$

$$\begin{aligned}
g^{(I)}(\omega) = \frac{\omega^2}{|\Delta|^2}\{&[\omega^2 R_2^2 + (p_2 - p)^2]g_1^{(e)}(\omega) \\
&+ [\omega^2 R_1^2 + (p_1 - p)^2]g_2^{(e)}(\omega)\} \quad.
\end{aligned}$$

Substituting here the spectral densities of the e.m.f. (for frequencies $\omega \gtrless 0$)

$$g_j^{(e)}(\omega) = \frac{1}{\pi}kT_j R_j \quad (j = 1,2) \quad,$$

we find the spectral densities of the electrical energy at the capacitors C_1, C_2 and C

$$U_1(\omega) = \frac{1}{2C_1\omega^2}g_1^{(I)}(\omega) = \frac{k}{2\pi C_1|\Delta|^2}[(\omega^2 R_2^2 + p_2^2)R_1 T_1 + p^2 R_2 T_2] \quad,$$

$$\begin{aligned}
U_2(\omega) &= \frac{1}{2C_2\omega^2}g^{(I)}(\omega) \\
&= \frac{k}{2\pi C_2|\Delta|^2}\{[\omega^2 R_2^2 + (p_2 - p)^2]R_1 T_1 + [\omega^2 R_1^2 + (p_1 - p)^2]R_2 T_2\},
\end{aligned}$$

$$U(\omega) = \frac{1}{2C\omega^2}g_2^{(I)}(\omega) = \frac{k}{2\pi C|\Delta|^2}[p^2 R_1 T_1 + (\omega^2 R_1^2 + p_1^2)R_2 T_2] \quad.$$

The integrals of the form $U = \int_{-\infty}^{\infty} U(\omega)\, d\omega$ are readily evaluated by taking the residues at the appropriate poles (at zeros of Δ):

$$U_1 = \frac{A}{C_1}\{[(p_1 p_2 - p^2)R_2 + p_2^2 R_1]T_1 + p^2 R_2 T_2\} \quad,$$

$$U_2 = \frac{A}{C_2}\{[(p_1 p_2 - p^2)R_2 + (p_2 - p)^2 R_1]T_1$$
$$+ [(p_1 p_2 - p^2)R_1 + (p_1 - p)^2 R_2]T_2\} \quad, \tag{3.134}$$

$$U = \frac{A}{C}\{p^2 R_1 T_1 + [(p_1 p_2 - p^2)R_1 + p_1^2 R_2]T_2\} \quad,$$

$$A = k/\{2(p_1 p_2 - p^2)(R_1 p_2 + R_2 p_1)\} \quad.$$

If we sum all these expressions together, we obtain the total electrical energy of the fluctuations in the network U_Σ, namely

$$U_\Sigma = U_1 + U_2 + U = k(T_1 + T_2)/2 \quad.$$

At thermal equilibrium ($T_1 = T_2 = T$) this gives $U_\Sigma = kT$, as was to be expected for the system with one degree of freedom under consideration. It should be noted that at equilibrium the energies U_1, U_2, and U are independent of resistances

$$U_1 = \frac{kT p_2}{2C_1(p_1 p_2 - p^2)} = \frac{kT}{2}\frac{C + C_2}{C_1 + C_2 + C} \quad,$$

$$U_2 = \frac{kT(p_1 + p_2 - 2p)}{2C_2(p_1 p_2 - p^2)} = \frac{kT}{2}\frac{C + C_1}{C_1 + C_2 + C} \quad,$$

$$U = \frac{kT p}{2C(p_1 p_2 - p^2)} = \frac{kT}{2}\frac{C_1 + C_2}{C_1 + C_2 + C} \quad.$$

As $C_2 \to \infty$, the network breaks down into two independent RC-cells with $U_1 = kT_1/2$ and $U = kT_2/2$. If we "freeze out" half a degree of freedom, say by setting $C_2 = 0$, then, by (3.134) we have

$$U_1 = \frac{k(R_1 T_1 + R_2 T_2)}{2(R_1 + R_2)}\frac{C}{C + C_1} \quad, \quad U_2 = 0 \quad,$$

$$U = \frac{k(R_1 T_1 + R_2 T_2)}{2(R_1 + R_2)}\frac{C_1}{C + C_1} \quad.$$

The total energy is determined by the effective temperature

$$U_{\Sigma} = U + U_1 = \frac{kT_e}{2} \quad , \quad T_e = \frac{R_1 T_1 + R_2 T_2}{R_1 + R_2}$$

and is shared between the series-connected capacitors C and C_1 inversely to their values. At thermal equilibrium we now obtain $U_{\Sigma} = kT/2$.

4. Certain Kinds of Nonstationary Processes

Nonstationary processes, as opposed to their stationary counterparts, do not constitute a class of random processes since they are singled out in a purely negative way: all the processes that do not fall into the stationary class are nonstationary. Of course, this gives rise to an infinite variety of types of nonstationary processes. This chapter is confined to those types which are most common in practice.

Section 4.1 is devoted to flicker noise. This noise is interesting because it is extremely common, and also because it raises the question of whether it is possible to distinguish stationarity from nonstationarity in the presence of *extremely* slow fluctuations when the duration of measurements is much shorter than the characteristic times of those fluctuations.

The introduction of random processes with stationary increments (RPSI) made it possible to *circumvent* the above ambiguity and was an important contribution to the theory of turbulence. For RPSI's we may still apply the notion of spectral density (in the same sense as for stationary processes); however, the spectral expansion itself now takes another form, and a convenient second-order moment is the so-called *structure* function, not the covariance (Sect. 4.2).

With nonstationary processes, second-order moments are, in the general case, nonlocalized in frequency. This makes the spectral density dependent not on the coordinate along the ω axis, but on coordinates in the plane of the two frequencies ω_1 and ω_2. For stationary processes the spectral density lies on the bisector $\omega_1 = \omega_2$. If the density is nonzero within a rather narrow band in the vicinity of the bisector, then the process is *quasi-stationary* (Sect. 4.3). The two-dimensionality of spectral density allows us to approach the issues of filtration of nonstationary processes from another angle (Sect. 4.4).

The chapter concludes with Sect. 4.5, which treats *periodically nonstationary* processes, whose spectral density lies not only on the bisector $\omega_1 = \omega_2$, but also on equidistant straight lines parallel to the bisector.

4.1 The Flicker Effect

Among the fluctuational noises encountered in radio engineering, a somewhat special place is occupied by the flicker effect, which has already been mentioned in Sects. I.3.4 and I.5.8. In vacuum tubes this kind of noise (discovered by Johnson in 1925) is associated with *slow* local fluctuations of the emissivity of the cathode, so that the corresponding spectral density to be added to the uniform spectrum of the shot noise is concentrated in the low-frequency range

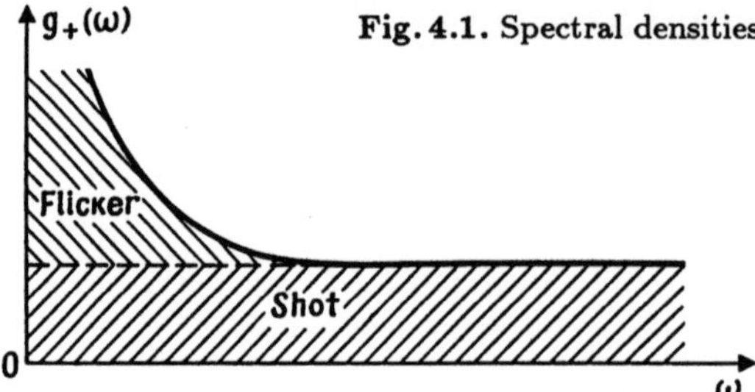

Fig. 4.1. Spectral densities of shot noise and flicker effect

(Fig. 4.1). The effect has the distinctive feature that the growth of spectral density with decreasing frequency shows no tendency to slow down or stop. Experiment shows that within a wide frequency range, $g_+(\omega)$ is satisfactorily described by

$$g_+(\omega) = K\frac{\overline{I}^m}{\omega^n} \quad , \tag{4.1}$$

where $\overline{I}$ is the mean emission current, and the numbers K, m and n depend on the tube properties (type, cathode material and surface finish, degree of vacuum, etc.) and on its operating mode. For the most part, m is close to 2, and n to 1 but it can vary from 0.6 to 2. The frequency spectrum of the flicker effect in vacuum tubes lies below 5–10 kHz. The intensity grows as the frequency decreases to the lowest values at which it is still measureable (to 10^{-1} Hz and below).

Noise with such a spectrum is observed not only in tubes, but also in a number of other conductors: in grained resistors (within the frequency range from fractional hertz to several megahertzs), at contacts, in semiconductor devices, e.g., in germanium and silicon detectors, photoresistors, contact photocells, thermistors (the so-called *excess noise* in semiconductors), in gas-discharge devices and electrolytes (dry cells and accumulator batteries). Thus, low-frequency noise is found in widely dissimilar conductors. Its spectral density within a wide range is proportional to ω^{-n}, where $0.6 < n < 2$, and the current behavior follows the law $\overline{I}^m$, where $m > 1$ and is mostly (in particular, in non-wire-wound resistors with a linear volt-ampere characteristic) close to 2, which with good reason might be attributed to resistance fluctuations.

In fact, if the volt-ampere characteristic is $v = RI + \alpha I^2 + \ldots$, then at a constant e.m.f. $\mathcal{E}$ the fluctuations of the parameters $R, \alpha, \ldots$ will cause the current fluctuations $\Delta I = I - \overline{I}$, which enter into the relation

$$\mathcal{E} = R\overline{I} + \alpha\overline{I}^2 + \ldots = (R + \Delta R)(\overline{I} + \Delta I) + (\alpha + \Delta\alpha)(\overline{I} + \Delta I)^2 + \ldots \quad .$$

Hence

$$\Delta I = -\overline{I}\frac{\Delta R + \overline{I}\Delta\alpha + \ldots}{R + 2\alpha\overline{I} + \ldots}$$

i.e.,

$$\overline{(\Delta I)^2} = \overline{I}^2 \frac{\overline{(\Delta R)^2} + 2\overline{\Delta R \cdot \Delta \alpha} \cdot \overline{I} + \dots}{R^2 + 4R\alpha \overline{I} + \dots} \ .$$

In consequence, for a linear characteristic we have $\overline{(\Delta I)^2} \sim \overline{I}^2$, and any deviations from this law are due to deviations of the characteristic from linearity.

The key feature in the empirical relation (4.1) is the frequency behavior of the spectral density. If the flicker noise is a *stationary* process, then for the integral

$$\overline{(\Delta I)^2} = \int\limits_0^\infty g_+(\omega)\, d\omega \tag{4.2}$$

to be finite requires that n be more than unity as $\omega \to \infty$ (which always seems to be the case), and less then unity as $\omega \to 0$.

As is seen from

$$g_+(\omega) = \frac{2}{\pi} \int\limits_0^\infty \psi(\tau) \cos \omega\tau \, d\tau \ ,$$

the behavior of $g_+(\omega)$ as $\omega \to 0$ is dependent on the rate at which $\psi(\tau)$ falls off as $\tau \to \infty$. The slower this rate, the larger the slope of $g_+(\omega)$ at zero. For instance, if the covariance $|\psi(\tau)| < M$ for $\tau < \tau_0$, and beginning with $\tau = \tau_0$, it falls off following the law $A\tau^{-\nu}$ ($\nu > 0$). then

$$g_+(\omega) = \frac{2}{\pi} \left\{ \int\limits_0^{\tau_0} \psi(\tau) \cos \omega\tau \, d\tau + \int\limits_{\tau_0}^\infty \frac{\cos \omega\tau}{\tau^\nu} \, d\tau \right\} \ .$$

The first term at all ω does not exceed, in absolute value, $2M\tau_0/\pi$. The second is finite as $\omega \to 0$, if $\nu > 1$, and increases as $\ln \omega\tau_0$ at $\nu = 1$ and as $\omega^{\nu-1}$ for $\nu < 1$.

As the flicker noise spectrum is measured over longer time periods, i.e., as the spectral density is measured at increasingly lower frequencies, the region in which the growth of $g_+(\omega)$ is expected to slow down as compared with $1/\omega$, is shifted ever lower, down to 10^{-5} Hz. This corresponds to a correlation time of the order of 5 h. If the covariance of the flicker noise has a *set* of various time-scales (associated, for instance, with different residence times of impurity atoms on the cathode surface or with different life-times of charge carriers in semiconductors), then such long time periods can hardly be accounted for by some reasonable physical causes.

On the other hand, the longer the measurement, the more difficult it is to sustain at a constant level all the conditions which affect the stationarity of the observed effect. Apart from external factors (the stability of power supplies, temperature fluctuations in circuit elements), in the vacuum tube such phenomena come into play as the drift of the cathode emission (frequencies

10^{-2}–10^{-4} Hz) and tube ageing (less than 10^{-4} Hz). In other words, it is quite possible that the fluctuations observed are *not stationary* (so that the requirement that $g_+(\omega)$ be integrable no longer holds), but their statistical characteristics vary fairly slowly with time.

This issue lies outside the framework of the phenomenon discussed. Of interest here is only whether it makes sense to distinguish the slow fluctuations of a stationary random function and the slow variation in time of moments of a nonstationary function, if we only have at our disposal realizations of limited duration. Let, for instance, the stationary function $\xi(t)$ vary with two distinct correlation times: ϑ_1 and $\vartheta_2 \gg \vartheta_1$ (Fig. 4.2 where the time ϑ_2 is not shown as it lies far beyond the confines of the drawing). An experiment involving the averaging of fast ($\sim \vartheta_1$) fluctuations over the time $T \ll \vartheta_2$, tells us nothing about whether the slow variation of $\xi(t)$ is stationary or nonstationary. Given the results of these experiments alone, we may consider that $\xi(t)$ has the correlation time ϑ_1 and is nonstationary, i.e., the mean $\overline{\xi(t)}$ is time-dependent, but changes insignificantly over times of the order of ϑ_1. Such an interpretation is possible in addition to the initial one when the function $\xi(t)$ is taken to be stationary, but showing, apart from ϑ_1, correlation time(s) $\vartheta_2 \gg T$.

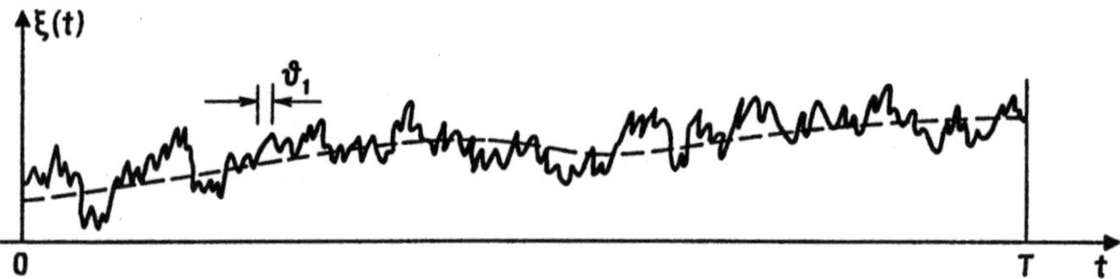

Fig. 4.2. Stationary process with two characteristic time-scales $-\vartheta_1 \ll T$ and $\vartheta_2 \gg T$, where T is measurement time

When the finite duration of a measurement permits of such an ambiguous interpretation, it is only natural to conceive of an approach that would not be concerned with $\xi(t)$ being stationary or nonstationary, but would simply exclude from consideration those variations of the function which are too slow for an experiment of a given duration. In the next section we will take up this formulation of the problem. Here we will discuss, with reference to a particular example of the pulse process, the transition from stationarity to nonstationarity as the pulses fall off more slowly, i.e., as the covariance decreases more slowly with τ.

When, in Sect. I.3.1, we considered the Poisson process of the form

$$\xi(t) = \sum_{\nu} a_\nu F(t - t_\nu) \ , \tag{4.3}$$

we assumed that the interval $(-T/2, T/2)$ of averaging in t_ν was very large as compared with the pulse duration. This allowed us to ignore the boundary effects – the trimming of parts of pulses by the interval boundaries – since the fraction of "damaged" pulses could be made, as $T \to \infty$, arbitrarily small. If, however, the pulse $F(\theta)$ falls off as $\theta \to \infty$ insufficiently fast, then the replacement of the finite interval T of integration with respect to t_ν by the finite one may be unjustified.

For simplicity, we will make the following assumptions. Let the pulses in (4.3) be semi-infinite $[F(\theta) = 0$ for $\theta < 0]$. Moreover, we select the range of t_ν to be $(0, T)$ rather than $(-T/2, T/2)$, so that the initial point of the interval does not shift as T varies. Then the characteristic function of the pulse Poisson process computed in Sect. I.3.3 (before the limiting process as $T \to \infty$) will be written in the form

$$\varphi_\xi(u, t) = \exp\left\{ n_1 \int w_a(a)\, da \int\limits_{t-T}^{t} [e^{iuaF(\theta)} - 1]\, d\theta \right\}$$

and the mth cumulant will be

$$\lambda_m(t) = n_1 \overline{a^m} \int\limits_{t-T}^{t} F^m(\theta)\, d\theta \quad .$$

For $T \to \infty$ and any finite t we may (beginning with $T > t$) replace the lower limit by zero. If $\overline{a} = 0$, the first nonzero cumulant will be the variance

$$\lambda_2(t) = D[\xi(t)] = n_1 \overline{a^2} \int\limits_{0}^{t} F^2(\theta)\, d\theta \quad .$$

Since $D[\xi]$ varies with t, the process (4.3) is nonstationary. But if $F(\theta)$ decreases sufficiently fast with increasing θ, the variance will tend to a constant value, i.e., for large enough t the process (4.3) will be approximately stationary.

Suppose now that, in the interval $(0, \theta_0)$ the function $F(\theta)$ is limited, $|F(\theta)| < M$, and for $\theta > \theta_0$ falls off as $A\theta^{-\nu}$ $(\nu \geq 0)$. Then we have for $t > \theta_0$

$$D[\xi(t)] = \begin{cases} a_1 + b_1 (t/\theta_0)^{1-2\nu} & \text{for} \quad \nu \neq 1/2 \quad , \\ a_2 + b_2 \ln(t/\theta_0) & \text{for} \quad \nu = 1/2 \quad . \end{cases}$$

Thus, for $\nu > 1/2$ stationarity is achieved with increasing t, and for $\nu \leq 1/2$ the variance grows indefinitely. The case of $\nu = 1/2$ corresponds to very slow (logarithmic) growth of $D[\xi]$, and the case of $\nu = 0$ (nondecreasing pulses), leads to the diffusion law.

4.2 Random Functions with Stationary Increments. The Structure Function

If we are not *a priori* sure that a random process is stationary (e.g., in Fig. 4.2 the finite duration of the experiment tells us nothing about the character of the slow variation of the "local" mean), then this does not always suggest that the theory of nonstationary processes of the most general kind should be applied. There exist a large class of nonstationary processes that include stationary processes as a particular case. At the same time they are so special as to allow an appropriate theory to be advanced further than is possible in the most general case. Suffice it to indicate, anticipating events somewhat, that these processes, which are termed *random processes* (functions) *with stationary increments* (RPSI), still retain frequency localization for the second-order moments, i.e., they retain the notion of spectral intensity. Random functions with stationary increments, together with adequate methods for describing them, were first proposed by Kolmogorov in 1940 [4.1, 2][1].

For a nonstationary process $\xi(t)$ we write the increment in the time interval $(t, t + T)$ of an arbitrary, but fixed, duration T

$$\eta_T(t) = \xi(t + T) - \xi(t) \quad . \tag{4.4}$$

Clearly, slow variations of $\xi(t)$ will only slightly affect $\eta_T(t)$; and the more slightly these are affected, the slower they are. If $\xi(t)$ contains a constant component, then this component cancels out. It might appear, because the components with very large periods are suppressed, that the *increment* $\xi(t)$, i.e., $\eta_T(t)$, is *stationary*. In that case, $\xi(t)$ is called a random process with stationary *first* increments.

By definition,

$$\overline{\eta_T(t)} = \overline{\xi(t + T)} - \overline{\xi(t)} = \text{const} \quad .$$

It follows that the mean of $\xi(t)$ can only be a linear function of t

$$\overline{\xi(t)} = at + b \quad . \tag{4.5}$$

This, of course, opens up new possibilities as compared with the stationary process, for which $a = 0$, i.e., $\overline{\xi(t)} = b = \text{const}$. It is not difficult to surmise that for the process $\xi(t)$ with stationary *second* increments, i.e., with the stationary difference $\eta_T(t + T_1) - \eta_T(t)$, the mean will be polynomial of no higher than second degree

$$\overline{\xi(t)} = at^2 + bt + c \quad ,$$

and so forth. We will confine our discussion to stationary *first* increments, and therefore in the following we will omit the word "first", speaking only about RPSI.

The mean of the form (4.5) is peculiar, for example, to the process

$$\xi(t) = \alpha t + \zeta(t) \quad , \tag{4.6}$$

where α is random variable with $\overline{\alpha} = a$, and $\zeta(t)$ is a stationary process with

---

[1] Among later treatments of RPSI theory special mention must be made of [4.3].

$\overline{\zeta(t)} = b$. But the fact that $\eta_T(t)$ is stationary also implies that the mixed moment $B_{\eta_T} = \overline{\eta_T(t+\tau)\eta_T^*(t)}$ is a function of τ only. Having written for (4.6) the difference $\eta_T = \alpha T + \zeta(t+T) - \zeta(t)$, we can readily work out that

$$B_{\eta_T}(\tau) = \overline{|\alpha|^2}T^2 + [\overline{\alpha\zeta^*(t-T)} - \overline{\alpha\zeta^*(t)} + \text{complex conjugate}]T$$
$$+ 2B_\zeta(\tau) - B_\zeta(\tau+T) - B_\zeta(\tau-T) \quad .$$

It is seen from this that for uncorrelated α and $\zeta^*(t)$, when $\overline{\alpha\zeta^*(t)} = \overline{\alpha} \cdot \overline{\zeta^*(t)} = ab^*$, the second term will disappear and the condition that B_{η_T} be independent of t will be met.

For simplicity, we will confine ourselves to real random processes. Consider the *fluctuation* of an RPSI $\xi(t)$

$$\tilde{\xi}(t) = \xi(t) - \overline{\xi(t)} = \xi(t) - at - b \quad , \tag{4.7}$$

and introduce the so-called *structure fucntion* of $\xi(t)$, defined as the mean square of the *increment* of the fluctuation $\tilde{\xi}(t)$ in the interval (t_1, t_2)

$$D_\xi(t_1, t_2) = \langle[\tilde{\xi}(t_2) - \tilde{\xi}(t_1)]^2\rangle = \langle[\xi(t_2) - \xi(t_1) - a(t_2 - t_1)]^2\rangle \quad . \tag{4.8}$$

As regards the second moments of RPSI's, the structure function is as much the principal characteristic as is the covariance for stationary processes. Clearly, D_ξ is a measure of the intensity of those fluctuations of $\xi(t)$ whose periods are not very close to $(t_2 - t_1)/n$, where $n = 1, 2, \ldots$, and are also not too large as compared with $t_2 - t_1$: i.e., D_ξ "does not feel" very slow variations of $\xi(t)$.

Using the identity

$$(a - b)(c - d) = \tfrac{1}{2}[(a-b)^2 + (b-c)^2 - (a-c)^2 - (b-d)^2] \quad ,$$

we can readily express the covariance for the stationary increment (4.4) through the structure function of the RPSI $\xi(t)$ itself

$$\begin{aligned}
\psi_\eta(t_2 - t_1) &= \langle[\eta_T(t_1) - \overline{\eta_T(t_1)}][\eta_T(t_2) - \overline{\eta_T(t_2)}]\rangle \\
&= \langle[\tilde{\xi}(t_1 + T) - \tilde{\xi}(t_1)][\tilde{\xi}(t_2 + T) - \tilde{\xi}(t_2)]\rangle \\
&= \tfrac{1}{2}\{\langle[\tilde{\xi}(t_1 + T) - \tilde{\xi}(t_2)]^2\rangle + \langle[\tilde{\xi}(t_1) - \tilde{\xi}(t_2 + T)]^2\rangle \\
&\quad - \langle[\tilde{\xi}(t_1 + T) - \tilde{\xi}(t_2 + T)]^2\rangle - \langle[\tilde{\xi}(t_1) - \tilde{\xi}(t_2)]^2\rangle\} \\
&= \tfrac{1}{2}\{D_\xi(t_2, t_1 + T) + D_\xi(t_2 + T, t_1) \\
&\quad - D_\xi(t_2 + T, t_1 + T) - D_\xi(t_2, t_1)\} \quad .
\end{aligned} \tag{4.9}$$

It follows that the right-hand-side will only be dependent on $t_2 - t_1$ under the condition that the structure function is homogeneous,

$$D_\xi(t_1, t_2) = D_\xi(t_2 - t_1) \quad . \tag{4.10}$$

In that case, denoting $t_2 - t_1 = \tau$, we obtain from (4.9)

$$\psi_\eta(\tau) = \tfrac{1}{2}[D_\xi(\tau - T) + D_\xi(\tau + T) - 2D_\xi(\tau)] \quad . \tag{4.11}$$

Of course, a structure function can also be set up for a stationary process $\xi(t)$. Since here $\overline{\xi(t)} = b = \text{const}$ and $a = 0$, then, according to (4.8) and using (4.10), we get

$$D_\xi(\tau) = 2[B_\xi(0) - B_\xi(\tau)] = 2[\psi_\xi(0) - \psi_\xi(\tau)] \quad , \tag{4.12}$$

where, as usual,

$$\psi_\xi(\tau) = \langle \xi(t + \tau)\xi(t) \rangle - \langle \xi(t) \rangle^2 = B_\xi(\tau) - b^2 \quad .$$

If for $\xi(t)$ the sufficient ergodicity condition is met, i.e., $\psi_\xi(\infty) = 0$, then

$$D_\xi(\infty) = 2\psi_\xi(0) = 2[\langle \xi^2(t) \rangle - b^2]$$

and (4.12) can be rewritten as

$$\psi_\xi(\tau) = \tfrac{1}{2}[D_\xi(\infty) - D_\xi(\tau)] \quad . \tag{4.13}$$

For stationary ergodic process we can thus use both the covariance and the structure function. The latter has the advantage that it can be more widely applied: it is suitable for describing the behavior of not only stationary processes, but also of RPSI's. Furthermore, $D_\xi(\tau)$ is unaffected by possible unertainties in measuring $\overline{\xi(t)}$. If it turns out that $D_\xi(\infty) = 2\overline{\xi^2(t)}$ is satisfied, then, by (4.13), we immediately calculate $\psi_\xi(\tau)$.

Now we turn to spectral representations of RPSI and their structure functions.

The derivative of RPSI is, by the very definition of this kind of random process, stationary and can thus be represented by the Fourier-Stieltjes integral

$$\dot{\xi}(t) = \alpha + \int\limits_{-\infty}^{+\infty} e^{i\omega t}\, d\tilde{C}(\omega) \quad , \tag{4.14}$$

where $\alpha = \overline{\dot{\xi}(t)} = \text{const}$, $\overline{d\tilde{C}(\omega)} = 0$ and

$$\langle d\tilde{C}(\omega) d\tilde{C}^*(\omega') \rangle = \delta(\omega - \omega')\, d\omega'\, d\tilde{G}(\omega) \quad . \tag{4.15}$$

Integrating (4.14) from 0 to t gives

$$\xi(t) = \xi(0) + \alpha t + \int\limits_{-\infty}^{+\infty} \frac{e^{i\omega t} - 1}{i\omega}\, d\tilde{C}(\omega) \quad , \tag{4.16}$$

where $\xi(0)$ is a random variable. The fluctuation of $\xi(t)$ will be

$$\tilde{\xi}(t) = \xi(t) - \overline{\xi(t)} = \xi(0) - \overline{\xi(0)} + \int\limits_{-\infty}^{+\infty} \frac{e^{i\omega t} - 1}{i\omega}\, d\tilde{C}(\omega) \quad . \tag{4.17}$$

The structure function of $\xi(t)$, by (4.8) and (4.10), will thus be given by

$$\begin{aligned}
D_\xi(\tau) &= \langle [\tilde{\xi}(t+\tau) - \tilde{\xi}(t)]^2 \rangle \\
&\doteq \iint\limits_{-\infty}^{+\infty} \frac{(e^{i\omega\tau} - 1)e^{i\omega t}(e^{-i\omega'\tau} - 1)e^{-i\omega't}}{\omega\omega'} \langle d\tilde{C}(\omega)d\tilde{C}^*(\omega') \rangle \quad .
\end{aligned}$$

Substituting (4.15), we obtain

$$D_\xi(\tau) = 2 \int\limits_{-\infty}^{+\infty} \frac{1 - \cos\omega\tau}{\omega^2}\, d\tilde{G}(\omega) \quad .$$

If the process $\xi(t)$ were stationary, then the increment $dG(\omega) = d\tilde{G}(\omega)/\omega^2$ would represent its spectral intensity in the interval $(\omega, \omega + d\omega)$, and its covariance would be expressed in the form

$$\psi_\xi(\tau) = \int\limits_{-\infty}^{+\infty} e^{i\omega\tau}\, dG(\omega) = \int\limits_{-\infty}^{+\infty} \cos\omega\tau\, dG(\omega) \quad . \tag{4.18}$$

Using the spectral intensity $G(\omega)$, we can eventually write the structure function of the RPSI $\xi(t)$ in the spectral form

$$D_\xi(\tau) = 2 \int\limits_{-\infty}^{+\infty} (1 - \cos\omega\tau)\, dG(\omega) \quad . \tag{4.19}$$

Equations (4.16) and (4.19) are to be understood as spectral representations of the RPSI itself and of its structure function, respectively. Of course, instead of $d\tilde{C}(\omega)$ we can introduce into (4.16, 17) the spectral amplitude $dC(\omega) = d\tilde{C}(\omega)/i\omega$, which, according to (4.15), has the covariance

$$\langle dC(\omega)dC^*(\omega') \rangle = \delta(\omega - \omega')\, d\omega'\, dG(\omega).$$

Here $dG(\omega)$ has the same sense as for stationary processes. It is the spectral intensity in the frequency interval $(\omega, \omega + d\omega)$. Similarly, in the case of a continuous spectrum, the derivative $g(\omega) = dG(\omega)/d\omega$ has for RPSI and stationary processes the sense of intensity per unit frequency range. Using the spectral treatment, we need not be concerned beforehand with whether the process is stationary or RPSI. However, in these two cases the constraints imposed on $dG(\omega)$ are markedly different: the integral (4.19) exists under less stringent conditions than (4.18).

For both integrals to converge at infinity, the same condition must hold

$$\lim_{|\omega|\to\infty} \{\omega\, dG(\omega)\} = 0 \quad , \tag{4.20}$$

but for the integral (4.19), which represents the structure function, to converge at zero, it is necessary that the requirement

$$\lim_{|\omega|\to 0} \{\omega^3\, dG(\omega)\} = 0 \tag{4.21}$$

be met, whereas for the integral (4.18) (representing the covariance) to converge, the more severe condition

$$\lim_{|\omega|\to 0} \{\omega\, dG(\omega)\} = 0 \tag{4.22}$$

must be satisfied. The structure function thus permits at zero of the singularity $dG(\omega) \sim |\omega^{-\nu}|$ for $\nu<3$, whereas the covariance only exists for $\nu<1$, i.e., for the finite total intensity $G(\infty) = \overline{\xi^2(t)}$. If $1\leq\nu<3$, then $D_\xi(\tau)$ exists, but diverges as $|\tau|\to\infty$, so that $\psi_\xi(\tau)$ does not make sense any more, see (4.13).

Using the same expedient we can readily verify that for the process with stationary *second* increments, we can have already at $\omega = 0$ the singularity $dG(\omega)$ with $\nu<5$. Generally, for the stationary nth increments, it is permissible that $\nu<2n + 1$. It was stressed in Sect. 1.4 that for $\xi(t)$ to be wide-sense stationary, *two* requirements must be met: first, the spectral "mass" $\gamma(\omega,\omega')$ must be concentrated on the bisector $\omega' = \omega$, and second, the linear density of this "mass", $g(\omega)$, must be integrable with respect to ω in the entire interval $(-\infty, +\infty)$. It is this last condition that is violated with RPSI. But the first requirement, i.e., that the spectral "mass" be concentrated on the line $\omega' = \omega$, is satisfied with RPSI.

Thus, RPSI's enable us to embrace nonstationary processes $\xi(t)$ whose mean $\overline{\xi(t)}$ may be a linear function of t, and whose spectral intensity may be infinitely large since $dG(\omega)$ grows fast at low frequencies. According as to whether or not the singularity $dG(\omega)$ is integrable at zero, we can go over respectively either to $\psi_\xi(\tau)$ using (4.18), or to $D_\xi(\tau)$ using (4.19).

These remarks apply to the flicker effect discussed in the previous section. To interpret this fluctuational process we need not assume that it is stationary, which would impose extraneous and unjustified constraints on the spectral density behavior at small ω. The fact that the density in any case grows with decreasing ω more slowly than $1/\omega^3$ enables the flicker effect to be treated as an RPSI.

Again, it is clear that if the frequency $\Omega(t)$ of an oscillation is a stationary process, then the phase $\varphi(t)$ of this oscillation, i.e., the integral of $\Omega(t)$, will

be an RPSI. It is then obvious that the mean square of the phase shift —
the quantity $F(\tau) = \langle [\varphi(t-\tau) - \varphi(t)]^2 \rangle$ introduced in Sect. 1.8 — is nothing
other than the structure function of $\varphi(t)$, and (2.44) is the spectral representation of this structure function. The spectral density of the RPSI $\varphi(t)$ is

$$g_\varphi(\omega) = g_\Omega(\omega)/\omega^2 \quad ,$$

where $g_\Omega(\omega)$ is the spectral density of the stationary process $\Omega(t)$.

We will now derive equations enabling the spectral density $g(\omega) = dG(\omega)/d\omega$
(continuous spectrum) to be found from a given structure function. Differentiating (4.19) with respect to τ, we obtain

$$D'_\xi(\tau) = 2 \int\limits_{-\infty}^{+\infty} \omega g(\omega) \sin \omega\tau \, d\omega \quad , \quad D''_\xi(\tau) = 2 \int\limits_{-\infty}^{+\infty} \omega^2 g(\omega) \cos \omega\tau \, d\omega \quad .$$

Inversion of these Fourier integrals gives

$$g(\omega) = \frac{1}{2\pi\omega} \int\limits_0^\infty \sin \omega\tau D'_\xi(\tau) \, d\tau \quad , \tag{4.23}$$

$$g(\omega) = \frac{1}{2\pi\omega^2} \int\limits_0^\infty \cos \omega\tau D''_\xi(\tau) \, d\tau \quad . \tag{4.24}$$

Equation (4.23) is valid if

$$\lim_{\tau \to \infty} D'_\xi(\tau) = 0 \quad , \quad \lim_{\tau \to 0} \tau^2 D'_\xi(\tau) = 0 \quad , \tag{4.23a}$$

and (4.24) if *other* conditions are met:

$$\lim_{\tau \to \infty} D''_\xi(\tau) = 0 \quad , \quad \lim_{\tau \to 0} \tau D''_\xi(\tau) = 0 \quad . \tag{4.24a}$$

Given an expression for $D_\xi(\tau)$, we can verify which of the above conditions are
satisfied, and accordingly make use of (4.23) or (4.24) to find $g(\omega)$.

Finally, let us return to the pulse Poisson process

$$\xi(t) = \sum a_\nu F(t - t_\nu) \quad ,$$

and see which constraints shuld be imposed on $F(t)$, if $\xi(t)$ is to be an RPSI.
Let $\overline{\xi(t)} = 0$, so that the structure function $\xi(t)$ is

$$D_\xi(t,\tau) = \langle [\xi(t+\tau) - \xi(t)]^2 \rangle \quad .$$

Assuming that $F(t) = 0$ for $t<0$, or falls off fast enough as $t \to -\infty$, it is a
straightforward exercise to see in a conventional way (see Sect. I.3.3) that

$$D_\xi(t,\tau) = n_1\overline{a^2} \int\limits_{-\infty}^{t} [F(\theta+\tau) - F(\theta)]^2 \, d\theta \quad .$$

For a steady-state process to exist, i.e., for the upper limit of t to tend to infinity, now requires that not $F(\theta)$ but the *difference* $F(\theta+\tau) - F(\theta)$ drop sufficiently fast. Assuming, as in Sect. 4.1, that for $\theta > \theta_0$ the function $F(\theta)$ has the form $A\theta^{-\nu}$, it is easily seen that for a steady state to come about it is necessary that $\nu > -1/2$, i.e., $F(\theta)$ may even grow with increasing θ, but more slowly than $\sqrt{\theta}$. Then, $\xi(t)$ will become an RPSI with increasing t and will have the structure function

$$D_\xi(\tau) = n_1\overline{a^2} \int\limits_{-\infty}^{+\infty} [F(\theta+\tau) - F(\theta)]^2 \, d\theta \quad .$$

Let the derivative $\dot{F}(\theta)$ be representable as a Fourier integral

$$\dot{F}(\theta) = \int\limits_{-\infty}^{+\infty} \varphi(\omega)e^{i\omega\theta} \, d\omega \quad .$$

Then

$$F(\theta+\tau) - F(\theta) = \int\limits_{0}^{\theta+\tau} \dot{F}(\theta)\, d\theta = \int\limits_{-\infty}^{+\infty} \frac{\varphi(\omega)}{i\omega}(e^{i\omega\tau} - 1)e^{i\omega\theta} \, d\omega \quad ,$$

hence

$$\int\limits_{-\infty}^{+\infty} [F(\theta+\tau) - F(\theta)]^2 \, d\theta$$

$$= \iint\limits_{-\infty}^{+\infty} \frac{\varphi(\omega)\varphi^*(\omega')}{\omega\omega'}(e^{i\omega\tau} - 1)(e^{-i\omega'\tau} - 1) \, d\omega \, d\omega' \int\limits_{-\infty}^{+\infty} e^{i(\omega-\omega')\theta} \, d\theta$$

$$= 4\pi \int\limits_{-\infty}^{+\infty} \frac{|\varphi(\omega)|^2}{\omega^2}(1 - \cos \omega\tau) \, d\omega \quad .$$

As a result, the structure function will take the form

$$D_\xi(\tau) = 4\pi n_1\overline{a^2} \int\limits_{-\infty}^{+\infty} \frac{|\varphi(\omega)|^2}{\omega^2}(1 - \cos \tau) \, d\omega \quad .$$

A comparison of this relationship with (4.19) shows that

$$g(\omega) = 2\pi n_1\overline{a^2}\frac{|\varphi(\omega)|^2}{\omega^2}$$

is the spectral density of the RPSI $\xi(t)$.

The condition that $\xi(t)$ be stationary, in which case the covariance $\psi_\xi(\tau)$ exists, requires that $|\varphi(\omega)|^2/\omega^2$ grow as $\omega \to 0$ more slowly than $1/\omega$, i.e., that $|\varphi(\omega)|/\omega$ grow more slowly than $1/\sqrt{\omega}$. It is not difficult to surmise that this is the condition that the very pulse $F(\theta)$, whose spectral amplitude in this case is $f(\omega) = \varphi(\omega)/i\omega$, be representable as a Fourier integral.

4.3 Spectra of Nonstationary Processes. Quasi-Stationary Processes

The question of the spectral analysis of nonstationary processes has attracted attention for many years. The interest stems both from the needs of measurement technology, and from the need to deal here with more sophisticated concepts than in the case of stationary processes. This is, of course, associated with the fact that nonstationary processes are nonlocalized in frequency (Sect. 1.3), i.e., there is a correlation between the harmonic components $\exp{(i\omega t)}\,dC(\omega)$ (having *different frequencies*) of the nonstationary and (supposedly) harmonizable process $\zeta(t)$. Because of this correlation, the contributions these components make to "energy-related" parameters are not additive, and the moments of second order are not representable in terms of single Fourier expansions with a certain constant (independent of time) spectral intensity, analogous to the intensity $G(\omega)$ of stationary processes. The spectral "mass" of the nonstationary process is *two-dimensional* and the average bilinear (in particular, quadratic) quantities can only be expressed in the form of *double* Fourier-Stieltjes integrals (Sect. 1.3).

It is common knowledge, however, that not infrequently (in science, and especially in engineering practice) one is tempted to extend conventional concepts and techniques beyond their legitimate scope. Spectral representations of bilinear ("energy-related") characteristics of nonstationary processes are no exception. Many works have appeared, and are still appearing which seek to find a certain equivalent of the univariate and nonnegative spectral intensity $G(\omega)$. As it has been clear from the outset that in the general case the problem has no exact solution, all efforts have been directed at finding an approximate solution and delineating the boundaries of applicability of appropriate approximations. The notions of the finite-time averaged power spectrum, short-term spectrum, "physical spectrum", and so forth thus emerged. We will touch upon these issues in the following section, but here we wish to turn to exact (two-dimensional) spectral representations of nonstationary processes.

Suppose that a process $\zeta(t)$, which is generally complex, is harmonizable and, for simplicity, that it has a zero mean $[\overline{\zeta(t)} = 0]$ and a purely continuous spectrum. Accordingly, the Fourier expansion has the form (1.27)

$$\zeta(t) = \int\limits_{-\infty}^{+\infty} e^{i\omega t} c(\omega)\, d\omega \quad , \tag{4.25}$$

such that $\overline{c(\omega)} = 0$. The mixed moment $\zeta(t)$ is

$$B(t_1, t_2) = \langle \zeta(t_1)\zeta^*(t_2)\rangle = \int\!\!\!\int_{-\infty}^{+\infty} \gamma(\omega_1, \omega_2)\exp\left[i(\omega_1 t_1 - \omega_2 t_2)\right] d\omega_1\, d\omega_2 \quad, \tag{4.26}$$

where $\gamma(\omega_1, \omega_2)$ is the bivariate density of the complex spectral "mass" (1.33)

$$d^2\Gamma(\omega_1, \omega_2) = \gamma(\omega_1, \omega_2)\, d\omega_1\, d\omega_2 = \langle c(\omega_1)c^*(\omega_2)\rangle\, d\omega_1\, d\omega_2 \quad. \tag{4.27}$$

We already know (Sect. 1.3) that the fact that $\zeta(t)$ is harmonizable implies that the integral (4.25) exists in the mean square, for which purpose it is necessary and sufficient that at all t_1 and t_2 the integral (4.26) be finite. Specifically, the mean square of the modulus of $\zeta(t)$ is also finite, i.e., the instantaneous mean power of the process[2] will be

$$\frac{1}{2}B(t, t) = \frac{1}{2}\overline{|\zeta(t)|^2} = \frac{1}{2}\int\!\!\!\int_{-\infty}^{+\infty} \gamma(\omega_1, \omega_2)e^{i(\omega_1 - \omega_2)t}\, d\omega_1\, d\omega_2 \quad. \tag{4.28}$$

It is often convenient to utilize the variables

$$t = (t_1 + t_2)/2 \quad, \quad \tau = t_1 - t_2 \quad,$$

instead of t_1 and t_2, so that

$$t_1 = t + \frac{\tau}{2} \quad, \quad t_2 = t - \frac{\tau}{2} \quad, \quad dt_1\, dt_2 = -dt\, d\tau \quad. \tag{4.29}$$

We denote the mixed moment, treated as a function of t and τ, by $B_2(t, \tau)$

$$B\left(t + \frac{\tau}{2} \quad, \quad t - \frac{\tau}{2}\right) = \left\langle \zeta\left(t + \frac{\tau}{2}\right)\zeta^*\left(t - \frac{\tau}{2}\right)\right\rangle \equiv B_2(t, \tau) \quad.$$

Note that the symmetry of t_1 and t_2 about t results in B_2 being Hermitian under shift of τ

$$B_2(t, \tau) = B_2^*(t, -\tau) \quad, \tag{4.30}$$

which in turn implies that B_2 is even in τ for the real process $\zeta(t)$.

According to (4.26) and (4.29),

$$B_2(t, \tau) = \int\!\!\!\int_{-\infty}^{+\infty} \gamma(\omega_1, \omega_2)\exp\left\{i[(\omega_1 - \omega_2)t + \tfrac{1}{2}(\omega_1 + \omega_2)\tau]\right\} d\omega_1\, d\omega_2 \quad.$$

---

[2] In this and later sections we will often for the sake of convenience use the "energy-related" terms for mean bilinear variables.

It is thus clear that it is expedient in the (ω_1,ω_2) plane also to go over to variables similar to (4.29), namely to

$$\omega = \tfrac{1}{2}(\omega_1 + \omega_2) \quad , \quad \Omega = \omega_2 - \omega_1 \ ,$$

or to

$$\omega_1 = \omega - \frac{\Omega}{2} \quad , \quad \omega_2 = \omega + \frac{\Omega}{2} \quad , \quad d\omega_1\,d\omega_2 = d\Omega\,d\omega \quad . \tag{4.31}$$

We denote the two-dimensional spectral density, viewed as a function of Ω and ω, by $g_2(\Omega,\omega)$

$$\gamma\left(\omega - \frac{\Omega}{2},\ \omega + \frac{\Omega}{2}\right) = \left\langle c\left(\omega - \frac{\Omega}{2}\right)c^*\left(\omega + \frac{\Omega}{2}\right)\right\rangle \equiv g_2(\Omega,\omega), \tag{4.32}$$

Then, (4.26) will take the form

$$B_2(t,\tau) = \int\!\!\!\int\limits_{-\infty}^{+\infty} g_2(\Omega,\omega)\mathrm{e}^{-\mathrm{i}(\Omega t - \omega\tau)}\,d\Omega\,d\omega \quad . \tag{4.33a}$$

Hence the inverse will be

$$g_2(\Omega,\omega) = \frac{1}{4\pi^2} \int\!\!\!\int\limits_{-\infty}^{+\infty} B_2(t,\tau)\mathrm{e}^{\mathrm{i}(\Omega t - \omega\tau)}\,dt\,d\tau \quad . \tag{4.33b}$$

By virtue of (4.32), the density g_2 is Hermitian in Ω

$$g_2(\Omega,\omega) = g_2^*(-\Omega,\omega) \quad , \tag{4.34}$$

and if $\zeta(t)$ is a real process, so that $c(\omega) = c^*(-\omega)$, then g_2 is even in ω

$$g_2(\Omega,\omega) = g_2(\Omega,-\omega) \quad . \tag{4.35}$$

In the general case, the complex spectral "mass" is distributed over the entire plane (Ω,ω), but if $\zeta(t)$ is a nonstationary *analytical signal*, then, by (1.51), its bivariate spectral density is nonzero only in the first quadrant of the plane (ω_1,ω_2).

Also, all the correlation properties of $\zeta(t)$ are somehow reflected in the distribution of the complex spectral "mass". In the general case, the variation of B_2 with time t and shift τ can, of course, have many scales in both variables. Let t_c and τ_c be the *smallest* typical time-scales in t and τ, respectively. The bivariate spectral density $g_2(\Omega,\omega)$ in that case has the two *largest* scales of the order of $1/t_c$ in Ω and of $1/\tau_c$ in ω. This means that the complex spectral "mass" is distributed in a region with such scales along the Ω and ω axes.

Let the moment $B_2(t,\tau)$ vary as a function of t very much more *slowly* than as a function of τ. In other words, typical time-scales of variation with

t and τ are markedly different: if T_c is the _largest_ scale in τ, then $t_c \gg T_c$. The mean instantaneous power $B_2(t,0)/2 = \overline{|\zeta(t)|^2}/2$ must, clearly, also be a slowly-varying function in the above sense, i.e., it must vary only slightly in the intervals of t of the order of T_c. For simplicity, let $\zeta(t) \equiv \xi(t)$ be a real process. This then implies that the univariate probability density $w_1(t,x)$ deduced from the bivariate density $w_2(t_1,x_1;\ t_2,x_2)$ by integrating over the whole region of the possible values with respect to either x_1 or x_2, is also a slowly varying function of t. In consequence, the mean

$$\overline{\xi(t)} = \int x w_1(t,x)\,dx$$

will be a slowly-varying function as well.

A random process exhibiting such behavior, i.e., with the moments $\overline{\xi(t)}$, $\overline{\xi^2(t)}$, $D[\xi(t)]$, and $B_2(t,\tau)$ varying slowly with t in the largest characteristic time interval of the fluctuations, T_c, can be called _quasi-stationary_ (in the wide sense)[3]. It is obvious that if the fluctuations in a system are stationary with unchanged parameters and conditions, they often appear to be quasi-stationary when these parameters and conditions vary slowly enough. For instance, shot noise in a vacuum tube or a semiconductor device will be quasi-stationary as the mean current varies if, during the longest correlation time of the noise, T_c, the mean current changes only slightly. The same is true of thermal noise in an electrical network with slowly varying resistance and/or temperature. If the types of noise just described are treated as white noise ($T_c = 0$), there is no limitation on the rate of variation of the system parameters. However, given _finite variance_ of the process, it is necessary to allow for nonzero T_c, and it is expedient to include the latter condition in the definition of stationarity in the wide sense, and as a condition for the process to be harmonizable. Then, of course, the nonstationary white noise $[B_2(t,\tau) = b(t)\delta(\tau)]$ will, strictly speaking, be excluded from the class of wide-sense quasi-stationary processes, just as the stationary white noise $[b(t) = \text{const}]$ will not be classed with wide-sense stationary processes.

It is easy to see how the quasi-stationarity of a process manifests itself in the two-dimensional distribution of its spectral "mass". The characteristic scales of the spectral density $g_2(\Omega,\omega)$ – the largest is of the order of $1/t_c$ in Ω and the smallest, of the order of $1/T_c$ in ω – satisfy the inequality $1/t_c \ll 1/T_c$. The spectral "mass" is thus concentrated in the region that stretches along the bisector $\Omega = \omega_2 - \omega_1 = 0$ (at least, over the interval of ω of the order of $1/t_c$).

---

[3] These processes are introduced in [4.4] and termed _locally stationary_ in the wide sense. This term, however, is not satisfactory as it invites a parallel with _locally-homogeneous_ fields (see subsequent volumes) which represent random functions with _homogeneous increments_, i.e., they are a space analog of processes with _stationary increments_ (Sect. 4.2).

In addition, in [4.4] the notion of local stationarity presupposes the separability of the moment B_2, i.e., its factorization: $B_2(t,\tau) = b(t)K(\tau)$, where $K(\tau)$ is the correlation coefficient of a _stationary_ process. Thus, $b(t) = B_2(t,0)$ is the mean instantaneous power. Admittedly, this factorization is quite possible, but it is superfluous to include it among the _necessary_ conditions of quasi-stationarity.

There are thus two wide-sense stationarity conditions: (1) that the spectral "mass" be concentrated exactly on the bisector $\Omega = 0$ and (2) that the total "mass" be finite. It is worth mentioning that quasi-stationary processes violate the first condition to some extent, whereas processes with stationary increments and white noise violate the second.

Clearly, stationary processes are a limiting case as the time-scale t_c increases indefinitely. In the limit the moment $B_2(t, \tau)$ becomes independent of t, and the spectral "mass" concentrates on the bisector $\Omega = \omega_2 - \omega_1 = 0$, as the largest scale of $1/t_c$ in Ω tends to zero. We thus obtain $g_2(\Omega, \omega) = g(\omega)\delta(\Omega)$, where $g(\omega) \geq 0$ is a real univariate spectral density of the stationary process.

Among the operations to which the nonstationary process $\zeta(t)$ itself or its second moments are subjected, we will need the operation of *smoothing* the moment $B_2(t, \tau)$ or the mean instantaneous power $B_2(t, 0)/2$ over a time interval T[4]. According to (4.33a),

$$\widetilde{B_2(t, \tau)} \equiv \frac{1}{T} \int\limits_{t-T/2}^{t+T/2} B_2(t, \tau)\, dt$$

$$= \iint\limits_{-\infty}^{+\infty} g_2(\Omega, \omega) \frac{\sin(\Omega T/2)}{\Omega T/2} e^{-i(\Omega t - \omega \tau)}\, d\Omega\, d\omega \quad,$$

hence

$$\frac{1}{2}\widetilde{\langle |\zeta(t)|^2 \rangle} = \frac{1}{2} \iint\limits_{-\infty}^{+\infty} g_2(\Omega, \omega) \frac{\sin(\Omega T/2)}{\Omega T/2} e^{-i\Omega t}\, d\Omega\, d\omega \quad.$$

The main maximum of the factor $\sin(\Omega T/2)/(\Omega T/2)$, equal to unity on the bisector $\Omega = 0$, occupies a strip bounded by the values $\Omega = \pm 2\pi/T$, the spectrum $g_2(\Omega, \omega)$ of the process $\zeta(t)$ being essentially suppressed elsewhere. Consequently, the moments $B_2(t, \tau)$ and $\langle |\zeta(t)|^2 \rangle$, smoothed over a sufficiently large interval T (larger than the longest correlation time T_c) vary with time as if these were the moments of the quasi-stationary process with the smallest scale $1/T$ in Ω. Of course, passing over to the limit $T \to \infty$ presupposes either the presence in $\zeta(t)$ of a stationary component having a finite "mass" on the bisector $\Omega = 0$, or the substitution of the *energy accumulated* during the time from $t - T/2$ to $t + T/2$ for the instantaneous smoothed power.

We will now take a further example – the bivariate spectrum of the real modulated process $\xi(t) = A(t) \cos[\omega_0 t + \varphi(t)]$, where $A(t)$ and $\varphi(t)$ are functions varying slowly as compared with $\cos \omega_0 t$. The mixed moment of the appropriate analytical signal

$$\zeta(t) = A(t) \exp\{i[\omega_0 t + \varphi(t)]\} = A(t) e^{i\omega_0 t}$$

---

4 We will refer to averaging over time or frequency intervals as "smoothing", in contrast to statistical averaging over an ensemble of realizations.

is

$$B_2(t,\tau) = \left\langle A\left(t+\frac{\tau}{2}\right)A^*\left(t-\frac{\tau}{2}\right)\right\rangle e^{i\omega_0\tau} = B_{2A}(t,\tau)e^{i\omega_0\tau} \quad .$$

The bivariate spectral density $g_{2A}(\Omega,\omega)$ of the complex amplitude $A(t)$ is localized in the vicinity of the origin in the plane (Ω,ω) in the region whose linear dimensions are small as compared with ω_0 (in Fig. 4.3 this region is delineated by the dashed line and vertical shading). Multiplying by $\exp(i\omega_0\tau)$ shifts the center of the region along the ω axis to the point $\omega = \omega_0$.

The "second" mixed moment of the analytical signal (Sect. 1.1) is

$$\tilde{B}_2(t,\tau) = \left\langle A\left(t+\frac{\tau}{2}\right)A\left(t-\frac{\tau}{2}\right)\right\rangle e^{2i\omega_0 t} = \tilde{B}_{2A}(t,\tau)e^{2i\omega_0 t} \quad .$$

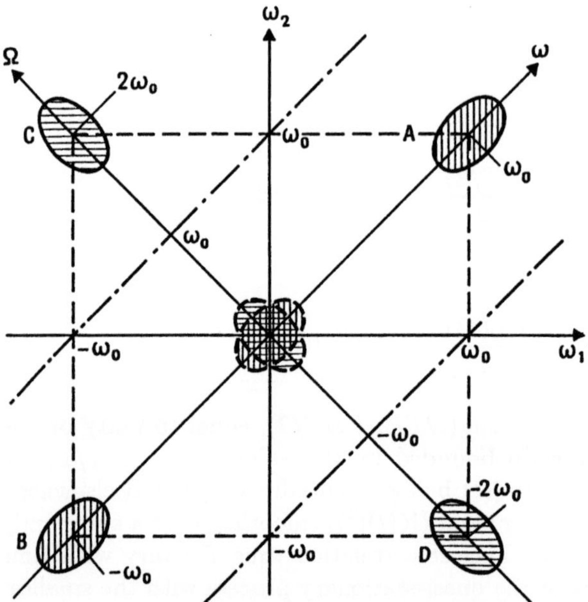

Fig. 4.3. Two-dimensional spectrum of a nonstationary real modulated signal lies in the regions A, B, C, and D of the plane (ω,Ω)

The "second" spectral density $\tilde{g}_{2A}(\Omega,\omega)$ is likewise nonzero only in a small vicinity of the origin (in Fig. 4.3 this region is delineated by the dashed line and horizontal shading). Multiplying $\tilde{B}_{2A}(t,\tau)$ by $\exp(2i\omega_0 t)$ shifts the center of the region along the Ω axis to the point $\Omega = 2\omega_0$. But, according to (1.3), the moment $B_{2\xi}(t,\tau)$ of the *real* process $\xi(t) = \mathrm{Re}\{\zeta(t)\}$ is expressed in terms of the "first" and "second" moments B_2 and $\tilde{B}_2$ of the analytical signal

$$\begin{aligned}
B_{2\xi}(t,\tau) &= \tfrac{1}{2}\mathrm{Re}\{B_2(t,\tau)+\tilde{B}_2(t,\tau)\}\\
&= \tfrac{1}{4}\{B_{2A}(t,\tau)e^{i\omega_0\tau} + \tilde{B}_{2A}(t,\tau)e^{i2\omega_0 t} + \text{complex conjugate}\} \quad .
\end{aligned}$$

$$(4.36)$$

Accordingly, the bivariate spectrum of the nonstationary real modulated process $\xi(t)$ lies in the four regions, indicated in Fig. 4.3 by A, B, C, and D. The first term in (4.36) with its complex conjugate (regions A and B in Fig. 4.3) gives a component that varies slowly with time t, and the second term with its complex conjugate (regions C and D in Fig. 4.3) gives a component that varies with frequencies close to $2\omega_0$.

If the complex amplitude $A(t)$ approaches the stationary random process, so that $B_{2A}(t,\tau)$ no longer depends on t, then the spectral density $\tilde{g}_{2A}(\Omega,\omega) \to g_A(\omega)\delta(\Omega)$. The same occurs in the case of the "second" spectral density $\tilde{g}_2(\Omega,\omega)$, if it does not vanish completely[5]. Regions A and B in Fig. 4.3 thus contract into sections on the ω axis, and C and D into sections parallel to the ω axis. This results in a modulated process belonging to the so-called *periodically nonstationary* processes (Sect. 4.5). If $\tilde{B}_{2A}(t,\tau) \to 0$ as $A(t)$ tends to stationarity, then in the limit the process $\xi(t)$ also becomes a *stationary* modulated process.

As $B_{2\xi}(t,\tau)$ is smoothed over the period $T_0 = 2\pi/\omega_0$, only those sections of the bivariate spectrum are transmitted that lie within the band $2\omega_0$ and Ω (in the figure this band is bounded by a dotted and dashed line), i.e., regions C and D have been truncated. As a result, the smoothed moment $B_{2\xi}(t,\tau)$ is approximately

$$\overset{\sim\sim\sim}{B_{2\xi}(t,\tau)} = \frac{1}{T_0} \int\limits_{t-T_0/2}^{t+T_0/2} B_{2\xi}(t,\tau)\, dt$$

$$\approx \tfrac{1}{4}\{B_{2A}(t,\tau)\mathrm{e}^{\mathrm{i}\omega_0\tau} + B_{2A}^*(t,\tau)\mathrm{e}^{-\mathrm{i}\omega_0\tau}\} \quad .$$

Since it is a low-frequency component of $B_{2\xi}(t,\tau)$, the smoothed moment corresponds to a certain random process that is quasi-stationary or stationary depending on which of these properties are inherent in the complex amplitude $A(t)$.

Besides the double Fourier transforms (4.33), we can also consider quantities related by single transforms. We can thus write (4.33a) in the form

$$B_2(t,\tau) = \int\limits_{-\infty}^{+\infty} g(t,\omega)\mathrm{e}^{\mathrm{i}\omega\tau}\, d\omega \quad , \quad \text{where} \tag{4.37a}$$

$$g(t,\omega) = \frac{1}{2\pi} \int\limits_{-\infty}^{+\infty} B_2(t,\tau)\mathrm{e}^{-\mathrm{i}\omega\tau}\, d\tau \quad . \tag{4.37b}$$

Accordingly, the mean instantaneous power is

---

[5] This is possible because $A(t)$ is not an analytical signal. For example, if the real amplitude $A(t)$ alone is a random process, and the phase is fixed [$\varphi(t) = \varphi_0 = \text{const}$], then

$$B_{2A}(t,\tau) = B_{2A}(t,\tau) \quad , \quad \tilde{B}_{2A}(t,\tau) = B_{2A}(t,\tau)\mathrm{e}^{2\mathrm{i}\varphi_0} \quad ,$$

The spectral densities g_{2A} and $\tilde{g}_{2A}$ also differ only in that the latter has the factor $\exp(2\mathrm{i}\varphi_0)$.

$$\frac{1}{2}B_2(t,0) = \frac{1}{2}\langle|\zeta(t)|^2\rangle = \frac{1}{2}\int\limits_{-\infty}^{+\infty} g(t,\omega)\,d\omega \quad . \tag{4.38}$$

If B_2 is independent of t [the process $\xi(t)$ is stationary], the function $g(t,\omega)$ goes over into the spectral density $g(\omega)$ of the stationary process, and (4.37a) becomes the Khintchine formula. It would seem that expressions (4.37a–38), containing t as a parameter, allow us to treat $g(t,\omega)$ as an "instantaneous" spectral density. This *term* is widely used, and we will also apply it, but it must be realized that the frequency-temporal "hybrid" $g(t,\omega)$ cannot, without some limitations, have the same sense as the univariate density $g(\omega)$ for stationary processes (the power density in frequency).

The density $g(t,\omega)$ is always *real*. According to (4.30),

$$g(t,\omega) = \frac{1}{\pi}\,\mathrm{Re}\left\{\int\limits_0^\infty B_2(t,\tau)e^{i\omega\tau}\,d\tau\right\} \quad .$$

If the process $\zeta(t)$ itself is real, then

$$g(t,\omega) = \frac{1}{\pi}\int\limits_0^\infty B_2(t,\tau)\,\cos\omega\tau\,d\tau \quad , \tag{4.39a}$$

i.e., in addition, according to (4.36), $g(t,\omega)$ here is even in ω. In turn (4.37a) takes the form

$$B_2(t,\tau) = 2\int\limits_0^\infty g(t,\omega)\,\cos\omega\tau\,d\omega \quad . \tag{4.39b}$$

But being a real variable, the instantaneous density $g(t,\omega)$ is *not bound to be nonnegative in the entire plane* (t,ω), and the energy-related interpretation is thereby excluded despite the expression (4.38) for the mean total instantaneous power.

From (4.33, 37) we derive the following relationships between the instantaneous and bivariate spectral densities:

$$g(t,\omega) = \int\limits_{-\infty}^{+\infty} g_2(\Omega,\omega)e^{-i\Omega t}\,d\Omega \quad , \tag{4.40a}$$

$$g_2(\Omega,\omega) = \frac{1}{2\pi}\int\limits_{-\infty}^{+\infty} g(t,\omega)e^{i\Omega t}\,dt \quad . \tag{4.40b}$$

4.4 Filtration of Nonstationary Processes. The Mean Power Spectrum

We now turn to the filtration of nonstationary processes. We are concerned with the relationships between second moments of the nonstationary real process $f(t)$ at the input of a *harmonic* filter and the process $\xi(t)$ at its output. We will still consider that $\overline{f(t)} = 0$, and hence $\overline{\xi(t)} = 0$. Variables referring to the input and output of the filter will be marked by the subscripts f and ξ, respectively.

An arbitrary linear filter is known to be characterized by its pulse response (Green's function) $H(t, \theta)$, where θ is the time of the action of the delta-impulse at the input, and $t \geq 0$ that of the observation of the response. For the passive filter the *causality condition* must be met: the response cannot precede the action, so that

$$H(t, \theta) = 0 \quad \text{for} \quad t < \theta \quad . \tag{4.41}$$

For harmonic filters, i.e., linear systems with *constant* parameters, the response is dependent only on the time *span* $t - \theta$ (Sect. 3.1)

$$H(t, \theta) = H(t - \theta) \quad ,$$

and hence $H(t - \theta) = 0$ for $\theta > t$. Such an asymmetry of the response $H(t - \theta)$ enables us to remove the upper limit to $+\infty$ in all the integrals with respect to θ containing $H(t - \theta)$ in the integrand. It should, however, be emphasized that the integral relations with infinite time limits, which are derived below, are valid *irrespective of whether or not the causality condition is met*. This is to be expected since the function $H(t - \theta)$ can describe not only the response of the passive filter, but also, for instance, a "time window" whose shape can be defined in an arbitrary manner (see below).

To describe the filter, it would be advisable to introduce certain functions that in essence add nothing to its basic characteristic $H(t)$, but render all the relationships between the input and output of the filter symmetric [4.5]. We now write the symmetrized product of the filter responses at $t + \tau/2$ and $t - \tau/2$[6]

$$B_{2H}(t, \tau) = H\left(t + \frac{\tau}{2}\right) H\left(t - \frac{\tau}{2}\right) \quad .$$

---

[6] The integral of $B_{2H}(\theta, \tau)$ with respect to θ

$$\Psi(\tau) = \int\limits_{-\infty}^{+\infty} H\left(\theta + \frac{\tau}{2}\right) H\left(\theta - \frac{\tau}{2}\right) d\theta$$

can arbitrarily be called the "covariance" of the pulse response, in much the same way as in Sect. 3.1 we introduced the "covariance" of the elementary pulse in dealing with pulse processes.

If the causality condition is satisfied, then $B_{2H} \neq 0$ only in the region $t > |\tau|/2$ in the plane (t,τ).

For the pulse response of the filter we define the instantaneous spectral density $g_H(t,\omega)$ and the bivariate density $g_{2H}(\Omega,\omega)$, assuming

$$B_{2H}(t,\tau) = \frac{1}{2\pi} \int\limits_{-\infty}^{+\infty} g_H(t,\omega) e^{i\omega\tau} \, d\omega \quad , \tag{4.42a}$$

$$g_H(t,\omega) = \int\limits_{-\infty}^{+\infty} B_{2H}(t,\tau) e^{-i\omega\tau} \, d\tau \quad , \tag{4.42b}$$

$$B_{2H}(t,\tau) = \frac{1}{4\pi^2} \iint\limits_{-\infty}^{+\infty} g_{2H}(\Omega,\omega) e^{-i(\Omega t - \omega\tau)} \, d\Omega \, d\omega \quad , \tag{4.43a}$$

$$g_{2H}(\Omega,\omega) = \iint\limits_{-\infty}^{+\infty} B_{2H}(t,\tau) e^{i(\Omega t - \omega\tau)} \, dt \, d\tau \quad . \tag{4.43b}$$

It follows that both spectral densities are related by

$$g_H(t,\omega) = \frac{1}{2\pi} \int\limits_{-\infty}^{+\infty} g_{2H}(\Omega,\omega) e^{-i\Omega t} \, d\Omega \quad , \tag{4.44a}$$

$$g_{2H}(\Omega,\omega) = \int\limits_{-\infty}^{+\infty} g_H(t,\omega) e^{i\Omega t} \, dt \quad . \tag{4.44b}$$

We can readily express $g_{2H}(\Omega,\omega)$ in terms of the transfer function $k(i\omega)$, i.e., in terms of the Fourier transform of the pulse response $H(t)$

$$H(t) = \frac{1}{2\pi} \int\limits_{-\infty}^{+\infty} k(i\omega) e^{i\omega t} \, d\omega \quad , \tag{4.45a}$$

hence

$$k(i\omega) = \int\limits_{-\infty}^{+\infty} H(t) e^{-i\omega t} \, dt \quad . \tag{4.45b}$$

From the definition of $B_{2H}(t,\tau)$, making allowance for $H(t)$ being real and using the expansion (4.45a), we find

$$B_{2H}(t,\tau) = \frac{1}{4\pi^2} \iint\limits_{-\infty}^{+\infty} k(i\omega_1) k^*(i\omega_2)$$

$$\exp\left\{ i\left[\omega_1\left(t + \frac{\tau}{2}\right) - \omega_2\left(t - \frac{\tau}{2}\right)\right]\right\} d\omega_1 \, d\omega_2 \quad ,$$

Passing, by (4.31), to the variables Ω and ω, we get

$$B_{2H}(t,\tau) = \frac{1}{4\pi^2} \int\!\!\!\int_{-\infty}^{+\infty} k\left[\mathrm{i}\left(\omega - \frac{\Omega}{2}\right)\right] k^*\left[\mathrm{i}\left(\omega + \frac{\Omega}{2}\right)\right] \mathrm{e}^{-\mathrm{i}(\Omega t - \omega\tau)}\, d\Omega\, d\omega \quad .$$

Comparing this expression with (4.43a) shows that

$$g_{2H}(\Omega,\omega) = k\left[\mathrm{i}\left(\omega - \frac{\Omega}{2}\right)\right] k^*\left[\mathrm{i}\left(\omega + \frac{\Omega}{2}\right)\right] \quad . \tag{4.46}$$

If we forget for the moment that $\xi(t)$ is a *random* function, and $H(t)$ (as we so far believe) a *deterministic* function that is subject, in addition, to the causality condition (4.41), then all the relationships for both functions are formally identical. Expansion (4.25) is similar to the integral (4.45a) for $H(t)$, the moment $B_2(t,\tau)$ is similar to the function $B_{2H}(t,\tau)$, and expression (4.32) for $g_2(\Omega,\omega)$ to expression (4.46) for $g_{2H}(\Omega,\omega)$. With the same reservation, expansions (4.33), (4.37), and (4.40) for the random process are analogous, respectively, to expansions (4.43), (4.42), and (4.44) for the filter. This enables the relationships between the processes at the input and output of the filter to be written in a form that is perfectly symmetrical with respect to the characteristics of the filter and the input process.

These characteristics can be derived from either the Duhamel integral, or the relationship between the spectral amplitudes

$$c_\xi(\omega) = k(\mathrm{i}\omega)c_f(\omega) \quad ,$$

which is valid for the *steady-state* response $\xi(t)$ regardless of whether or not the process $f(t)$ is stationary (see Exercise 3.6.2). It follows from this relationship that

$$\langle c_\xi(\omega_1)c_\xi^*(\omega_2)\rangle = k(\mathrm{i}\omega_1)k^*(\mathrm{i}\omega_2)\langle c_f(\omega_1)c_f^*(\omega_2)\rangle \quad .$$

Passing to the variables Ω and ω, introducing $g_{2f}(\Omega,\omega) = \gamma_f(\omega_1,\omega_2)$ and using (4.46), we obtain for the bivariate spectral density of the process $\xi(t)$ at the filter output the expression

$$g_{2\xi}(\Omega,\omega) = g_{2H}(\Omega,\omega)g_{2f}(\Omega,\omega) \quad . \tag{4.47}$$

If $f(t)$ and $\xi(t)$ are stationary, so that

$$g_{2\xi}(\Omega,\omega) = g_\xi(\omega)\delta(\Omega) \quad , \quad g_{2f}(\Omega,\omega) = g_f(\omega)\delta(\Omega) \quad ,$$

then (4.47) goes over into the known relation (3.10) for the univariate spectral densities

$$g_\xi(\omega) = g_{2H}(0,\omega)g_f(\omega) \quad , \quad g_{2H}(0,\omega) = |k(\mathrm{i}\omega)|^2 \quad . \tag{4.48}$$

According to (4.40b) and (4.44b), the bivariate densities $g_{2\xi}(\Omega,\omega)$, $g_{2f}(\Omega,\omega)$, and $g_{2H}(\Omega,\omega)$ are Fourier transforms (in t) of the instantaneous densities $g_\xi(t,\omega)$, $g_f(t,\omega)$, and $g_H(t,\omega)$. Consequently, from the converse theorem on the convolution spectrum (Sect. 1.5), the instantaneous densities at the filter input and output are

$$g_\xi(t,\omega) = \int\limits_{-\infty}^{+\infty} g_H(t-\theta,\omega)g_f(\theta,\omega)\,d\theta \quad . \tag{4.49}$$

In turn, the instantaneous densities are, by (4.47b) and (4.42b), Fourier transforms (in τ) of the functions $B_{2\xi}(t,\tau)$, $B_{2f}(t,\tau)$, and $B_{2H}(t,\tau)$. Hence, by the same theorem

$$B_{2\xi}(t,\tau) = \iint\limits_{-\infty}^{+\infty} B_{2H}(t-\theta,\tau-\chi)B_{2f}(\theta,\chi)\,d\theta\,d\chi \quad . \tag{4.50}$$

The symmetry of equations (4.47, 49, 50) with respect to the characteristics of the input process and filter allows $H(t)$ to be thought of as a *deterministic process at the input*, and $f(t)$ as a *realization of the random pulse repsonse*, i.e., the response of the random harmonic system characterized by $B_{2f}(t,\tau)$, $g_f(t,\omega)$, $g_{2f}(\Omega,\omega)$. Of course, this situation enables us to dispense with the causality condition for $H(t)$, but forces us to impose this condition on $f(t)$.

By way of illustration, consider now the passage of a nonstationary process $f(t)$ through a band-pass filter whose $|k(i\omega)|$ is nonzero in the intervals $\Delta\omega$, symmetric about the frequencies $\pm\omega_0$ $(\omega_0 > \Delta\omega)$.

The four pass-bands denoted in Fig. 4.4 by A, B, C, and D and having the form of "squares" with a side $\Delta\omega^7$ are separated in the plane (ω_1,ω_2) by the factor $\gamma_H(\omega_1,\omega_2) = k(i\omega_1)k^*(i\omega_2)$. If the filter pass-band is narrow enough $(\Delta\omega \ll \omega_0)$, then at the filter output we obtain a *modulated* nonstationary process. This is seen from a simple comparison of Figs. 4.4 and 4.3.

If the moment $B_{2f}(t,\tau)$ of the input process has the smallest time-scale t_c in t (i.e., the largest time-scale $\sim 1/t_c$ in Ω), then, under the condition $|2\omega_0 - \Delta\omega| > 1/t_c$, there will essentially be no spectral "mass" of the process $f(t)$ in squares C and D, so that the response $\xi(t)$ will be determined by the "content" of squares A and B only. For the *quasi-stationary* input process $f(t)$ the above condition is normally easily satisfied, i.e., the strong inequality $|2\omega_0 - \Delta\omega| \gg 1/t_c$ holds. But the width of the bivariate spectrum $f(t)$ along the

---

[7] Note that the presence of the sharp boundaries of these squares (i.e., the transfer function $k(i\omega)$ of the filter is zero elsewhere) is incompatible with the causality condition. The rectangular band $\Delta\omega$ often used in calculations (see, e.g., Sect. 3.3) corresponds to the response

$$H(t-\theta) = \cos\omega_0(t-\theta) \cdot \frac{\sin[\Delta\omega(t-\theta)/2]}{\pi(t-\theta)/2} \quad ,$$

which is nonzero for $t < \theta$.

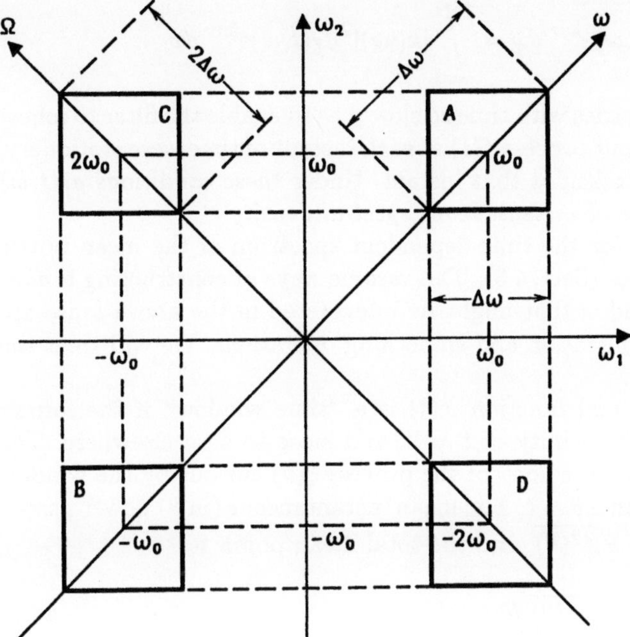

Fig. 4.4. Four pass regions (A, B, C, D) of the band-pass filter in the plane (ω, Ω)

Ω axis may be related to the filter pass-band in a variety of ways. If $1/t_c \ll \Delta\omega$, the spectral "mass" is concentrated near the ω axis, i.e., near the diagonal of squares A and B. The filter here hardly cuts off the spectral "mass" along the Ω axis, therefore the integration over A and B may be extended in Ω to $\pm\infty$, supposing at the same time that the transfer function is equal to its value on the ω axis

$$k\left[i\left(\omega \pm \frac{\Omega}{2}\right)\right] \approx k(i\omega) \quad .$$

By (4.46, 47), we then obtain

$$g_{2\xi}(\Omega,\omega) \approx |k(i\omega)|^2 g_{2f}(\Omega,\omega) \quad ,$$

and therefore, by (4.40a),

$$g_\xi(t,\omega) \approx |k(i\omega)|^2 \int\limits_{-\infty}^{+\infty} g_{2f}(\Omega,\omega) e^{-i\Omega t}\, d\Omega = |k(i\omega)|^2 g_f(t,\omega) \quad . \tag{4.51}$$

Clearly, (4.51) is a *quasi-stationary* form of the conventional (stationary) relation (4.49) between the spectral densities at the input and output of the filter. From (4.51, 37a) we also derive a *quasi-stationary* expression for $B_{2\xi}(t,\tau)$

$$B_{2\xi}(t,\tau) = \int\limits_{-\infty}^{+\infty} g_\xi(t,\omega)e^{i\omega\tau}\,d\omega = \int\limits_{-\infty}^{+\infty} |k(i\omega)|^2 g_f(t,\omega)e^{i\omega\tau}\,d\omega \quad .$$

In other words, $g_f(t,\omega)$ varies with time so slowly as to enable the filter to follow the variation, as if the input process $f(t)$ at each instant of time were stationary, i.e., with characteristics taken at that instant. Under these conditions $g_f(t,\omega)$ actually assumes the role of *instantaneous spectral density*.

Finally, let us consider the time-dependent spectrum of the mean power of a nonstationary process (Sect. 4.3). The various ways of constructing a *nonnegative* function of t and ω that might be interpreted in the above sense are based on the use of the filtration and smoothing operations. We will treat the question following [4.5][8].

We will say that a real function $w(t)$ is a "time window" if the former is positive in a certain T-vicinity of $t = 0$, and close to zero elsewhere. The product $w(t-\theta)\xi(\theta)$ is a "portion" of the process $\xi(\theta)$ cut out by the window in the T-vicinity of the time $\theta = t$. The mean instantaneous (in θ) power passed by the window is $w^2(t-\theta)\overline{\xi^2(\theta)}$, and the total mean power is

$$\overline{E}(t;\omega) = \int\limits_{-\infty}^{+\infty} w^2(t-\theta)\overline{\xi^2(\theta)}\,d\theta \quad . \tag{4.52}$$

For $\overline{E}(t;w)$ to be regarded as the locally-smoothed[9] power of $\xi(t)$ requires that its integral with respect to t be equal to the total mean energy[10], i.e.,

$$\int\limits_{-\infty}^{+\infty} \overline{E}(t;w)\,dt = \int\limits_{-\infty}^{+\infty} \overline{\xi^2(\theta)}\,d\theta \int\limits_{-\infty}^{+\infty} w^2(t-\theta)\,dt = \int\limits_{-\infty}^{+\infty} \overline{\xi^2(t)}\,dt \quad .$$

This requirement is certainly satisfied if the time window is normalized as follows:

$$\int\limits_{-\infty}^{+\infty} w^2(t)\,dt = 1 \quad . \tag{4.53}$$

An arbitrary (and, in the general case, complex) function $F(\theta)$ with the spectral amplitude

$$\tilde{F}(\omega) = \frac{1}{2\pi} \int\limits_{-\infty}^{+\infty} F(\theta)e^{-i\omega\theta}\,d\theta$$

---

[8] The work offers the most consistent and detailed discussion of the problem. Other approaches exist [see e.g., [4.6]], and also treatments of a simlar problem for deterministic signals, but not for random ones (see [4.6–9] and references given there).

[9] Convolution also constitutes a smoothing operation, but only with a weight function. If, in particular, $w^2(t)$ is a rectangular function equal to $1/T$ in the interval $(-T/2, T/2)$ and zero elsewhere, then convolution (4.52) coincides with simple smoothing, i.e., with the time averaging of $\overline{\xi^2(t)}$ over the interval $(t - T/2, t + T/2)$.

[10] Subject to the condition that the total energy is finite (which is not the case with stationary processes).

obeys the following equality (Parseval's theorem):

$$\int\limits_{-\infty}^{+\infty} |F(\theta)|^2 \, d\theta = 2\pi \int\limits_{-\infty}^{+\infty} |\tilde{F}(\omega)|^2 \, d\omega = \int\limits_{-\infty}^{+\infty} d\omega \frac{1}{2\pi} \left| \int\limits_{-\infty}^{+\infty} F(\theta) e^{-i\omega\theta} \, d\theta \right|^2 \quad (4.54)$$

Setting $F(\theta) = w(t-\theta)\xi(\theta)$, we obtain from the ensemble-aveaged equality (4.54)

$$\int\limits_{-\infty}^{+\infty} w^2(t-\theta)\overline{\xi^2(\theta)} \, d\theta = \int\limits_{-\infty}^{+\infty} d\omega \frac{1}{2\pi} \left\langle \left| \int\limits_{-\infty}^{+\infty} w(t-\theta)\xi(\theta) e^{-i\omega\theta} \, d\theta \right|^2 \right\rangle \quad .(4.55)$$

The mean locally-smoothed power of $\xi(t)$ is thus represented in the form of an integral over all the frequencies ω of a *real nonnegative* (and, as is easily seen, even in ω) function

$$g_\xi(t,\omega;w) = \frac{1}{2\pi} \left\langle \left| \int\limits_{-\infty}^{+\infty} w(t-\theta)\xi(\theta) e^{-i\omega\theta} \, d\theta \right|^2 \right\rangle \quad (4.56a)$$

$$= \frac{1}{2\pi} \iint\limits_{-\infty}^{+\infty} w(t-\theta_1)w(t-\theta_2)B(\theta_1,\theta_2) e^{-i\omega(\theta_1-\theta_2)} \, d\theta_1 \, d\theta_2 \quad (4.56b)$$

With the normalization (4.53), the mean total energy of the process is

$$\int\limits_{-\infty}^{+\infty} \overline{E}(t;w) \, dt = \iint\limits_{-\infty}^{+\infty} g_\xi(t,\omega;w) \, dt \, d\omega \quad .$$

In [4.5] the function $g_\xi(t,\omega;w)$ is called the (t,ω)-*density of the physical spectrum* of a process $\xi(t)$.

Expression (4.56) was derived as a *frequency* expansion of the *locally smoothed mean power* $\xi(t)$. We can also obtain another expression for $g_\xi(t,\omega;w)$ based on the "frequency window" $\tilde{w}(\omega)$ which is Fourier-conjugate to $w(t)$

$$w(t) = \frac{1}{2\pi} \int\limits_{-\infty}^{+\infty} \tilde{w}(\omega) e^{i\omega t} \, d\omega \quad . \quad (4.57)$$

Using, in addition, the spectral expansion (4.25) of $\xi(t)$ and the theorem on the spectrum of a convolution [as applied to $w(t)$ and $\xi(t)\exp(-i\omega t)$], we find

$$\int\limits_{-\infty}^{+\infty} w(t-\theta)\xi(\theta) e^{-i\omega\theta} \, d\theta = e^{i\omega t} \int\limits_{-\infty}^{+\infty} \tilde{w}(\omega-\omega_1)c(\omega_1) e^{i\omega_1 t} \, d\omega_1 \quad .$$

Therefore, (4.56) can also be written in terms of the frequency window

$$g_\xi(t,\omega;w) = \frac{1}{2\pi}\left\langle \left| \int\limits_{-\infty}^{+\infty} \tilde{w}(\omega-\omega_1)c(\omega_1)e^{i\omega_1 t}\,d\omega_1 \right|^2 \right\rangle \tag{4.58a}$$

$$= \frac{1}{2\pi} \iint\limits_{-\infty}^{+\infty} \tilde{w}(\omega-\omega_1)\tilde{w}^*(\omega-\omega_2)\gamma(\omega_1,\omega_2)e^{i(\omega_1-\omega_2)t}\,d\omega_1\,d\omega_2 \;, \tag{4.58b}$$

where $\gamma(\omega_1,\omega_2) = \langle c(\omega_1)c^*(\omega_2)\rangle$ is still the bivariate spectral density of $\xi(t)$.

Integrating (4.58b) with respect to time from $-\infty$ to $+\infty$, and taking into consideration that the integral of the exponential function is $2\pi\delta(\omega_1-\omega_2)$, gives

$$\int\limits_{-\infty}^{+\infty} g_\xi(t,\omega;w)\,dt = \int\limits_{-\infty}^{+\infty} |\tilde{w}(\omega-\omega_1)|^2\,\overline{|c(\omega_1)|^2}\,d\omega_1 \;. \tag{4.59}$$

Consequently, the (t,ω)-density of the physical spectrum $g_\xi(t,\omega;w)$ can also be treated as the *temporal* expansion of the *locally smoothed mean spectral density* $\overline{|c(\omega_1)|^2}$. Integrating (4.59) over all ω gives the total mean energy of the process $\xi(t)$

$$\int\limits_{-\infty}^{+\infty} \overline{\xi^2(t)}\,dt = 2\pi \int\limits_{-\infty}^{+\infty} \overline{|c(\omega_1)|^2}\,d\omega_1 \;.$$

It follows that the frequency window should be normalized so that[11]

$$\int\limits_{-\infty}^{+\infty} |\tilde{w}(\omega)|^2\,d\omega = 2\pi \;. \tag{4.60}$$

Smoothing with the help of time and frequency windows is widely used in dealing with nonstationary processes, e.g., in the acoustical analysis of speach. Understandably, these measurements use simple window shapes, i.e., those where the product of the widths [T for $w(t)$ and B for its Fourier transform $\tilde{w}(\omega)$] is not very much larger than its lower boundary: $TB \gtrsim 2\pi$ (Sect. 3.1). Clearly, the same applies to the smoothing intervals: in t (4.56a), and in ω (4.58a).

We shall now see how the density of the physical spectrum $g_\xi(t,\omega;w)$ is related to second moments of $\xi(t)$: the moment $B_{2\xi}(t,\tau)$, the bivariate spectral density $g_{2\xi}(\Omega,\omega)$ and, most interestingly, the instantaneous spectral density $g_\xi(t,\omega)$. Equations (4.56b) and (4.58b) provide the answers to the first two questions. If we pass to new integration variables, namely $\theta_1 = \theta + \tau/2$, $\theta_2 = \theta - \tau/2$ in (4.56b) and $\omega_1 = \omega' - \Omega/2$, $\omega_2 = \omega' + \Omega/2$ in (4.58b), we arrive at

$$g_\xi(t,\omega;w)$$

$$= \frac{1}{2\pi} \iint\limits_{-\infty}^{+\infty} w\left(t-\theta-\frac{\tau}{2}\right)w\left(t-\theta+\frac{\tau}{2}\right)B_{2\xi}(\theta,\tau)e^{-i\omega\tau}\,d\theta\,d\tau \tag{4.61a}$$

---

[11] This also follows immediately from (4.53) and (4.57).

$$= \frac{1}{2\pi} \iint\limits_{-\infty}^{+\infty} \tilde{w}\left(\omega - \omega' + \frac{\Omega}{2}\right) \tilde{w}^*\left(\omega - \omega' - \frac{\Omega}{2}\right)$$

$$\times\, g_{2\xi}(\Omega, \omega') \mathrm{e}^{-\mathrm{i}\Omega t}\, d\Omega\, d\omega' \quad . \tag{4.61b}$$

It follows that for the time window $w(t)$ it would be expedient to introduce the function

$$B_{2w}(t, \tau) = w\left(t + \frac{\tau}{2}\right) w\left(t - \frac{\tau}{2}\right) \quad ,$$

which is similar to the function $B_{2H}(t, \tau)$ for a filter. It will be recalled that the introduction of $B_{2H}(t, \tau)$ and its Fourier transforms (4.42–44) was in no way related to the causality condition for $H(t)$, so that the same equations also hold for $B_{2w}(t, \tau)$. In particular [see (4.42)],

$$B_{2w}(t, \tau) = \frac{1}{2\pi} \int\limits_{-\infty}^{+\infty} g_w(t, \omega) \mathrm{e}^{\mathrm{i}\omega\tau}\, d\omega \quad , \tag{4.62a}$$

$$g_w(t, \omega) = \int\limits_{-\infty}^{+\infty} B_{2w}(t, \tau) \mathrm{e}^{-\mathrm{i}\omega\tau}\, d\tau \quad . \tag{4.62b}$$

In terms of $B_{2w}(t, \tau)$, (4.61a) is written as

$$g_\xi(t, \omega; w) = \frac{1}{2\pi} \iint\limits_{-\infty}^{+\infty} B_{2w}(t - \theta, \tau) B_{2\xi}(\theta, \tau) \mathrm{e}^{-\mathrm{i}\omega\tau}\, d\theta\, d\tau \quad . \tag{4.63}$$

The Fourier expansions (4.62a) for $B_{2w}(t, \tau)$ and (4.37a) for $B_{2\xi}(t, \tau)$ enable us, by the theorem on the spectrum of the product of two functions of τ, to express $g_\xi(t, \omega; w)$ in terms of the instantaneous spectral density $g_\xi(t, \omega)$ of the process $\xi(t)$

$$g_\xi(t, \omega; w) = \frac{1}{2\pi} \iint\limits_{-\infty}^{+\infty} g_w(t - \theta, \omega - \omega') g_\xi(\theta, \omega')\, d\theta\, d\omega' \quad . \tag{4.64}$$

Thus, the (t, ω)-density (as we have seen, always nonnegative) of the physical spectrum of the nonstationary process $\xi(t)$ results from the smoothing of the instantaneous spectral density (that may be locally-negative as well) of $\xi(t)$ *both in time and frequency*. The smoothing intervals T and B are connected by the uncertainty relation. The weight function in such a double smoothing is the instantaneous spectral density $g_w(t, \omega)$ of the time window, which in [4.5] is referred to as the instantaneous spectral smoothing function.

In expression (4.63) for the density of the physical spectrum, the smoothing is in time t and the product of B_{2w} and $B_{2\xi}$ is subjected to the further Fourier transform in the shift τ. As is borne out by calculations analogous to those performed above, we can expect the opposite situation to occur in expression for $g_\xi(t, \omega; w)$ in terms of the bivariate spectral densities $g_{2w}(\Omega, \omega)$ and $g_{2\xi}(\Omega, \omega)$, i.e., smoothing in ω and Fourier transformation in Ω from the product of g_{2w} and $g_{2\xi}$. In fact, this expression has the form

$$g_\xi(t, \omega; w) = \frac{1}{2\pi} \iint\limits_{-\infty}^{+\infty} g_{2w}(\Omega, \omega - \omega') g_{2\xi}(\Omega, \omega') e^{-i\Omega t} \, d\Omega \, d\omega' \quad . \tag{4.65}$$

We will now see that for a *quasi-stationary* process whose moments vary with time t *slowly in the time-scale of the width T of the time window*, the instantaneous density of the physical spectrum $g_\xi(t, \omega; w)$ coincides with the instantaneous density $g_\xi(t, \omega)$. This can be established in a variety of ways, but we shall proceed from (4.64).

As $g_\xi(\theta, \omega')$ varies slowly with θ as compared with $g_w(t - \theta, \omega - \omega')$, we can put $\theta = t$ in g_ξ and then

$$g_\xi(t, \omega; w) \approx \frac{1}{2\pi} \int\limits_{-\infty}^{+\infty} g_\xi(t, \omega') \, d\omega' \int\limits_{-\infty}^{+\infty} g_w(\chi, \omega - \omega') \, d\chi \quad . \tag{4.66}$$

But if we write (4.44a) and (4.46) for the window $w(t)$ rather than for the filter, we have

$$g_w(t, \omega) = \frac{1}{2\pi} \int\limits_{-\infty}^{+\infty} \tilde{w}\left(\omega - \frac{\Omega}{2}\right) \tilde{w}^*\left(\omega + \frac{\Omega}{2}\right) e^{-i\Omega t} \, d\Omega \quad ,$$

and since the time integral of the exponential funtion within the limits $(-\infty, +\infty)$ is $2\pi\delta(\Omega)$, we obtain

$$\int\limits_{-\infty}^{+\infty} g_w(t, \omega) \, dt = |\tilde{w}(\omega)|^2 \quad .$$

Substituting into (4.66) gives

$$g_\xi(t, \omega; w) \approx \frac{1}{2\pi} \int\limits_{-\infty}^{+\infty} g_\xi(t, \omega') |\tilde{w}(\omega - \omega')|^2 \, d\omega' \quad .$$

Most information is contained in the physical spectrum for the extremely narrow frequency window ($B \to 0$ and, respectively, $T \to \infty$). In that case, according to the normalization (4.60), we have $|\tilde{w}(\omega - \omega')|^2 = 2\pi\delta(\omega - \omega')$, which leads to the above result

$$g_\xi(t, \omega; w) \approx g_\xi(t, \omega) \quad . \tag{4.67}$$

It should be noted that (4.67) suggests, in particular, that the instantaneous spectrum of the quasi-stationary process is nonnegative.

The many other interesting issues raised by the spectral analysis of nonstationary processes [including the approximate realization of the (t,ω)-density of the physical spectrum] lie beyond the scope of the present book, and the reader is referred to [4.5]. We would only stress once more the difference between filtration and the action of a time window, confining ourselves to the relations for instantaneous spectral densities.

By (4.49), a filter peforms the *temporal smoothing* of the input instantaneous density $g_f(t,\omega)$ using the weight function $g_H(t,\omega)$, i.e., its own instantaneous spectral function. By contrast, the "output" of a window $g_\xi(t,\omega;w)$ is, by (4.64), obtained as a result of the double smoothing of the instantaneous density $g_\xi(t,\omega)$ at the "input", *both in time and frequency*. Here lies the origin of the differences which occurs in other equations describing the filter and time window.

4.5 Periodically Nonstationary Processes

We have already mentioned processes whose moments depend periodically on time. Consider the process of the form

$$\eta(t) = F(t)\xi(t) \quad ,$$

where $F(t)$ is a periodic deterministic function and $\xi(t)$ is a stationary random process. Its moments are periodic in time

$$\overline{\eta(t)} = F(t)\overline{\xi} \quad B_\eta(t,\tau) = F(t)F(t+\tau)B_\xi(\tau) \quad ,$$
$$\psi_\eta(t,\tau) = F(t)F(t+\tau)\psi_\xi(\tau) \quad .$$

Therefore, separating at the output of a mixer the product of a periodic oscillation and a stationary noise that are applied to the input, we obtain a periodically nonstationary process.

Other examples are magnetic noise when a ferromagnet is alternately magnetized, and shot current in the tube when the mean current varies periodically. Specifically, the latter case is realized in the excited self-oscillatory system (recall that the variance of the shot noise is proportional to the mean current). Clearly, in strongly nonlinear self-oscillatory systems whose limiting cycle does not look like a circle, even small fluctuations will be periodically nonstationary with the period of self-oscillations [4.10].

One more important source of such processes is worth mentioning: linear systems with periodically varying parameters, which also include parametric amplifiers with periodic pumping. If a harmonic oscillation $\exp(i\omega t)$ is fed to the input of such a system, at the output we will obtain the oscillation $k(t,i\omega)\exp(i\omega t)$ where the instantaneous transfer function $k(t,i\omega)$ varies periodically (with the period of parameter variation) with t. The stationary input process

$$f(t) = \int\limits_{-\infty}^{+\infty} e^{i\omega t}\, dC_f(\omega) \quad ,$$

subject to

$$\overline{dC_f(\omega)} = \overline{f}\delta(\omega)\, d\omega \quad ,$$

$$\overline{dC_f(\omega)\, dC_f^*(\omega')} = g_f(\omega)\delta(\omega - \omega')\, d\omega\, d\omega'$$

gives at the output the process

$$\xi(t) = \int\limits_{-\infty}^{+\infty} k(t, i\omega)e^{i\omega t}\, dC_f(\omega) \quad ,$$

whose mean

$$\overline{\xi(t)} = \overline{f}k(t, 0)$$

and mixed moment

$$B_\xi(t, \tau) = \overline{\xi(t + \tau)\xi^*(t)} = \int\limits_{-\infty}^{+\infty} k(t + \tau, i\omega)k^*(t, i\omega)e^{i\omega\tau}g_f(\omega)\, d\omega$$

depend periodically [with the period of $k(t, i\omega)$] on t.

In these examples we have only retained the moments of first and second order, i.e., we stay within the framework of correlation theory. For this theory we may define the periodically nonstationary (for simplicity, real) process as the process whose bivariate distribution function, which varies with t and t', is a periodic function of t for any fixed shift $\tau = t' - t$. Thus

$$w_2(x, t; x', t') = \sum_{-\infty}^{+\infty} v_n(x, x', \tau)e^{in\omega_0 t} \quad . \tag{4.68}$$

where

$$v_n(x, x', \tau) = \frac{1}{2\pi} \int\limits_{-\pi}^{\pi} w_2(x, t; x', t + \tau)e^{-in\omega_0 t}d(\omega_0 t) \quad . \tag{4.69}$$

Since w_2 is real, we have

$$v_n(x, x', \tau) = v_{-n}^*(x, x', \tau) \quad . \tag{4.70}$$

Further, from the symmetry condition for w_2, since

$$w_2(x',t';x,t) = \sum_{-\infty}^{+\infty} v_n(x',x,-\tau)e^{in\omega_0 t'}$$

$$= \sum_{-\infty}^{+\infty} v_n(x',x,-\tau)e^{in\omega_0(t+\tau)} \quad ,$$

we get

$$v_n(x,x',\tau) = v_n(x',x,-\tau)e^{in\omega_0\tau} \quad . \tag{4.71}$$

And from the hierarchy condition

$$w_1(x,t) = \int w_2(x,t;x',t')\,dx' \quad ,$$

subject to (4.70), we have

$$\int v_n(x,x',\tau)\,dx' = e^{in\omega_0\tau} \int v_n(x',x,-\tau)\,dx' \equiv u_n(x) \quad , \tag{4.72}$$

so that

$$w_1(x,t) = \sum_{-\infty}^{+\infty} u_n(x)e^{in\omega_0 t} \quad . \tag{4.73}$$

Here, of course, $u_n(x) = u^*_{-n}(x)$. Lastly, from the normalization condition for w_2 and w_1, it follows that

$$\iint v_n(x,x',\tau)\,dx\,dx' = \int u_n(x)\,dx = \delta_{n0} \quad . \tag{4.74}$$

Furthermore, w_2 and w_1 must be nonnegative at any values of the arguments. It follows immediately that the real Fourier coefficients $v_0(x,x',\tau)$ and $u_0(x)$ feature all the properties of the bivariate and univariate distribution functions of a stationary process. In consequence, the time-averaging of w_2 and w_1, as well as of any moments derived with their help (over the period $T_0 = 2\pi/\omega_0$), gives the distribution functions and the moments of the stationary process.

By (4.73) and (4.68) the moments of the periodically nonstationary process $\xi(t)$ are

$$\overline{\xi(t)} = \sum_{-\infty}^{+\infty} \xi_n e^{in\omega_0 t} \quad , \quad \xi_n = \int x u_n(x)\,dx \quad , \tag{4.75}$$

$$B(t,\tau) = \overline{\xi(t)\xi(t+\tau)} = \sum_{-\infty}^{+\infty} B_n(\tau)e^{in\omega_0 t} \quad , \quad \text{with}$$

$$B_n(\tau) = \iint x x' v_n(x,x',\tau)\,dx\,dx' \quad . \tag{4.76}$$

$$\psi(t,\tau) = B(t,\tau) - \overline{\xi(t)} \cdot \overline{\xi(t+\tau)}$$

$$= \sum_{-\infty}^{+\infty} \left[B_n(\tau) - \sum_m \xi_{m-n}\xi_n e^{im\omega_0\tau} \right] e^{in\omega_0 t} \quad . \tag{4.77}$$

It should be noted that (4.71, 76) give rise to the relationship

$$B_n(\tau)e^{-in\omega_0\tau} = B_n(-\tau) \tag{4.78}$$

which can also be derived from the equality

$$B(t,-\tau) = B(t-\tau,\tau) \quad ,$$

which is the consequence of the very definition of the moment $B(t,\tau)$. From (4.70) and (4.76), or simply from the fact that $B(t,\tau)$ is real, we also have

$$B_n(\tau) = B^*_{-n}(\tau) \quad . \tag{4.79}$$

Let us now discuss the correlation properties of spectral amplitudes in the harmonic expansion

$$\xi(t) = \int_{-\infty}^{+\infty} e^{i\omega t}\, dC(\omega) \quad , \tag{4.80}$$

if $\xi(t)$ is a periodically nonstationary process. If we require that the mean of $\xi(t)$ be expressed by the Fourier series (4.75), we will have

$$\overline{dC(\omega)} = \sum_{-\infty}^{+\infty} \xi_n \delta(\omega - n\omega_0)\, d\omega \quad . \tag{4.81}$$

Next, using (4.80), we write the moment $B(t,\tau) = \overline{\xi(t+\tau)\xi(t)}$ and require that it be expressed by the Fourier series (4.76)

$$\overline{dC(\omega_1)\, dC^*(\omega_2)} = \sum_{-\infty}^{+\infty} g_n(\omega_1)\delta(\omega_2 - \omega_1 + n\omega_0)\, d\omega_1\, d\omega_2 \quad , \tag{4.82}$$

subject to

$$\int_{-\infty}^{+\infty} g_n(\omega)e^{i\omega\tau}\, d\omega = B_n(\tau) \quad . \tag{4.83}$$

Expression (4.82) implies that for a periodically nonstationary process the complex "mass" $\Gamma(\omega_1,\omega_2)$ is distributed in the plane (ω_1,ω_2) along the lines $\omega_2 = \omega_1 - n\omega_0$ (Fig. 4.5), unlike stationary processes whose "mass" is concentrated on the bisector $\omega_2 = \omega_1$ only. Thus, the mean bilinear (specifically, energy-related) characteristics of the periodically nonstationary process are not

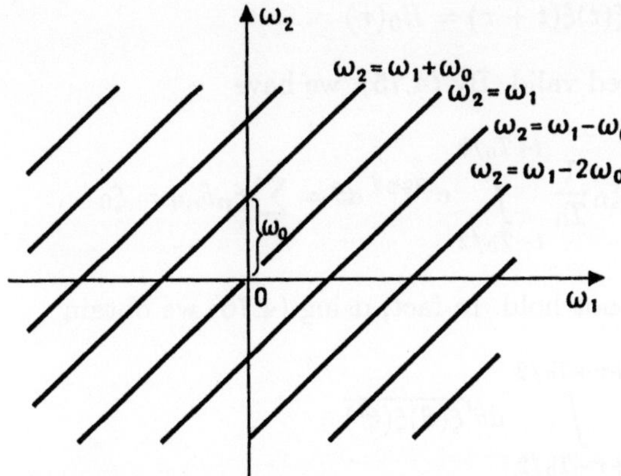

Fig. 4.5. Spectral density of a periodically nonstationary process is localized in the plane (ω_1, ω_2) on the lines $\omega_2 = \omega_1 - n\omega_0$ $(n = 1, \pm 1, \pm 2, \ldots)$

localizable in frequency, but a contribution to the spectral interval $(\omega, \omega + d\omega)$ is made only by those terms in the expansion (4.80) whose frequencies differ by $n\omega_0$ $(n = 0, \pm 1, \pm 2, \ldots)$.

It follows from (4.78, 79, 83) that the spectral densities $g_n(\omega)$ satisfy the relations

$$g_n(\omega) = g_n(n\omega_0 - \omega) = g^*_{-n}(-\omega) \quad , \tag{4.84}$$

whence we obtain, in particular,

$$g_0(\omega) = g_0(-\omega) = g^*_0(-\omega) \quad ,$$

i.e., $g_0(\omega)$ is a real even function. It can also be easily verified that $g_0(\omega)$ is nonnegative[12]. But we have already said that $B_0(\tau)$ and $\psi_0(\tau)$ exhibit all the properties of the mixed moment and covariance of a stationary random process, i.e., these properties are inherent in the quantities $\widetilde{\xi(t)} = \xi_0$, $\widetilde{B(t,\tau)} = B_0(\tau)$, and $\widetilde{\psi(t,\tau)} = \psi_0(\tau)$ smoothed over the period T_0.

What is the implication of these time averages? It might seem that the moments of a periodically nonstationary process $\xi(t)$, averaged over the period T_0, are equal to the moments of the stationary process derived by the moving averaging of $\xi(t)$ over the period, i.e., of the process

$$\eta(t) = \frac{1}{T_0} \int_{t-T_0/2}^{t+T_0/2} \xi(\theta)\, d\theta = \widetilde{\xi(t)} \quad .$$

This assumption consists in that

$$\overline{\eta(t)} = \overline{\widetilde{\xi(t)}} = \widetilde{\overline{\xi(t)}} = \xi_0 \quad ,$$

---

[12] See [4.11], where certain ergodic theorems for periodically nonstationary processes are considered as well.

$$\overline{\eta(t)\eta(t+\tau)} = \overline{\widetilde{\xi(t)}\widetilde{\xi(t+\tau)}} = \overline{\widetilde{\xi(t)\xi(t+\tau)}} = B_0(\tau) \quad .$$

The first of these equalities is indeed valid. By (4.75), we have

$$\overline{\eta(t)} = \frac{1}{T_0} \int\limits_{t-T_0/2}^{t+T_0/2} \overline{\xi(\theta)}\, d\theta = \sum_n \xi_n \frac{1}{T_0} \int\limits_{t-T_0/2}^{t+T_0/2} e^{in\omega_0\theta}\, d\theta = \sum_n \xi_n \delta_{n0} = \xi_0 \quad .$$

However, the second relation does not hold. In fact, using (4.76) we obtain

$$\overline{\eta(t)\eta(t+\tau)} = \frac{1}{T_0^2} \int\limits_{t-T_0/2}^{t+T_0/2} d\theta \int\limits_{t+\tau-T_0/2}^{t+\tau+T_0/2} d\theta'\, \overline{\xi(\theta)\xi(\theta')}$$

$$= \sum_n \frac{1}{T_0^2} \int\limits_{t-T_0/2}^{t+T_0/2} d\theta \int\limits_{t+\tau-T_0/2}^{t+\tau+T_0/2} d\theta'\, B_n(\theta' - \theta)e^{in\omega_0\theta} \quad .$$

Rotating the axes in the plane (θ, θ') by $45°$ allows the double integral to be reduced to a simple one containing the factor δ_{n0}. As a result, after the scale has been changed by a factor $\sqrt{2}$, we find

$$\overline{\eta(t)\eta(t+\tau)} = \frac{1}{T_0} \int\limits_{t-T_0/2}^{t+T_0/2} B_0(u)\, du \neq B_0(\tau) \quad ,$$

i.e., the mixed moment of the process $\eta(t) = \widetilde{\xi(t)}$ averaged over the period is not equal to the smoothed mixed moment $B(t,\tau)$ of the initial periodically nonstationary process $\xi(t)$. To measure $B_0(\tau)$ thus requires a device which with the process $\xi(t)$ at the input, first, forms the product $\xi(t)\xi(t+\tau)$ and then averages this over the period T_0, rather than begins the processing with the averaging of the process $\xi(t)$ itself.

Finally, we shall look at two examples.

1. *Periodically recurring section of stationary process.* Consider a stationary random process $\xi(t)$ with the mean $\overline{\xi}$ and mixed moment $B_\xi(\tau)$. We single out a section with duration T, which then periodically recurs (Fig. 4.6). Clearly, we obtain a periodically nonstationary process $\eta(t)$ with period T, and proceed to find its mean and moment $B(t,\tau)$.

The periodic function $\eta(t)$ may be expanded into a Fourier series. If the time-scale is selected so that the period T is 2π, then

$$\eta(t) = \sum_n c_n e^{int} \quad , \quad c_n = \frac{1}{2\pi} \int\limits_{-\pi}^{\pi} \xi(\theta)e^{in\theta}\, d\theta \quad , \quad \text{or}$$

$$\eta(t) = \frac{1}{2\pi} \int\limits_{-\pi}^{\pi} \xi(\theta)\, d\theta \sum_{-\infty}^{+\infty} e^{in(t-\theta)} = \int\limits_{-\pi}^{\pi} \xi(\theta)\delta_{\text{per}}(t - \theta)\, d\theta \quad , \tag{4.85}$$

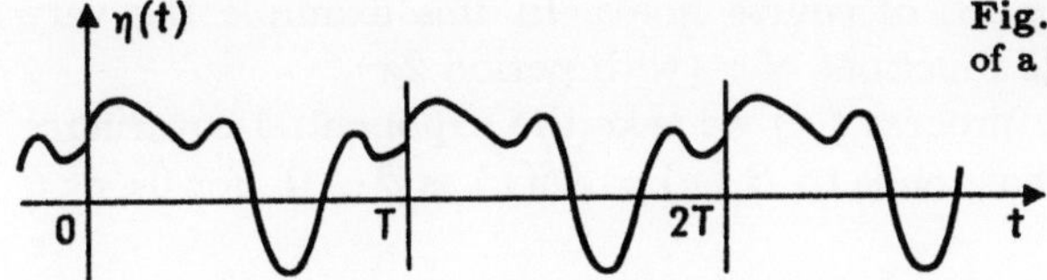

Fig. 4.6. Periodically recurring section of a stationary process

where $\delta_{\mathrm{per}}(x)$ is a periodic (with period 2π) delta-function. Of course, we could immediately write (4.85). Thus, averaging gives the obvious result

$$\overline{\eta(t)} = \overline{\xi} \int\limits_{-\pi}^{\pi} \delta_{\mathrm{per}}(t - \theta)\, d\theta = \overline{\xi} \quad .$$

Further

$$B(t,\tau) = \overline{\eta(t + \tau)\eta^{*}(t)}$$

$$= \iint\limits_{-\pi}^{\pi} B_{\xi}(\theta - \theta')\delta_{\mathrm{per}}(\theta - t - \tau)\delta_{\mathrm{per}}(\theta' - t)\, d\theta\, d\theta' = B_{\xi}(\tau) \quad ,$$

which is valid for $-\pi < t < \pi$ and $-(\pi + t) < \tau < \pi - t$.

This "cell" recurs periodically in the plane (t,τ), i.e., in this plane we obtain a "lattice" as shown in Fig. 4.7.

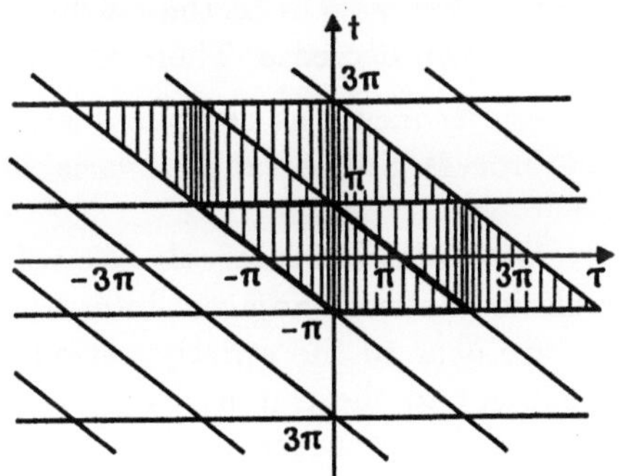

Fig. 4.7. To the derivation of the Fourier coefficients for the second moment of the process depicted in Fig. 4.6

Thus, if we fix τ, say in the interval $0 < \tau < \pi$, then as t varies from $-\pi$ to π the moment $B(t,\tau)$ will take on the value $B_{\xi}(\tau)$ in the interval of t from $-\pi$ to $\pi - \tau$, and $B_{\xi}(2\pi - \tau)$ in the interval from $\pi - \tau$ to π. The coefficients of the Fourier expansion (4.76) of $B(t,\tau)$ will then be

$$B_n(\tau) = \frac{1}{2\pi}\left\{ B_{\xi}(\tau) \int\limits_{-\pi}^{\pi - \tau} e^{-int}\, dt + B_{\xi}(2\pi - \tau) \int\limits_{\pi - \tau}^{\pi} dt \right\}$$

$$= \begin{cases} \left(1 - \dfrac{\tau}{2\pi}\right) B_{\xi}(\tau) + \dfrac{\tau}{2\pi} B_{\xi}(2\pi - \tau) & (n = 0) \quad , \\[2ex] \dfrac{(-1)^n}{2\pi\, in}(1 - e^{in\tau})[B_{\xi}(\tau) - B_{\xi}(2\pi - \tau)] & (n \neq 0) \quad . \end{cases} \tag{4.86}$$

213

For $\tau<0$ the property (4.78) of course holds. In this example the very coefficients $B_n(\tau)$ are periodic functions of τ (with period 2π).

For the initial stationary process $\xi(t)$ we take the exponential covariance $B_\xi(\tau) = A\exp{(-|\tau|/\vartheta)}/2\vartheta$ that tends to $B_\xi(\tau) = A\delta(\tau)$ as $\vartheta \to 0$. For $0<\tau<\pi$ we obtain from (4.86)

$$B_0(\tau) = \frac{A}{2\vartheta}\left[\left(1 - \frac{\tau}{2\pi}\right)e^{-\tau/\vartheta} + \frac{\tau}{2\pi}e^{-(2\pi-\tau)/\vartheta}\right] \quad ,$$

$$B_n(\tau) = \frac{A(-1)^n}{4\pi i n\vartheta}(1 - e^{in\tau})[e^{-\tau/\vartheta} - e^{-(2\pi-\tau)/\vartheta}] \quad .$$

The limiting expressions at $\vartheta = 0$ can, obviously, be used when the correlation time ϑ is fairly small as compared with the recurrence period 2π. In that case, we can ignore the second terms in the brackets, and the first terms in the limit give, in the interval $(-\pi, \pi)$ in question, $B_n = 0$ (for $n\neq0$) and $B_0(\tau) = B_\xi(\tau) = A\delta(\tau)$. For the entire τ axis we can write $B(t,\tau) = B_0(\tau)$ in the form

$$B(t,\tau) = B_0(\tau) = A\delta_{\mathrm{per}}(\tau) \quad .$$

A periodically recurrent section of white noise again gives white noise, but now with a periodic delta-correlation. If we set $A = 2|\xi|^2\vartheta$ [so that $B_\xi(\tau) = \overline{|\xi|^2}\exp{(-|\tau|/\vartheta)}$], where $\overline{|\xi|^2}$ is independent of ϑ, then as $\vartheta \to \infty$ the height of the jumps given by $B(t,\tau) = B_\xi(\tau) - B_\xi(2\pi - \tau)$ will decrease. Therefore, for $\vartheta\gg2\pi$ we obtain $B(t,\tau) = \overline{|\xi|^2}$ up to $1/\vartheta$: very frequent (as compared with ϑ) recurrence of the section of the stationary process gives a random variable independent of t, as was to be expected.

2. *Quasi-periodic pulse process with correlated adjacent intervals.* We will be concerned here with a special case of correlation between adjacent intervals, namely where the pulse generation times t_ν are near to the strictly periodic sequence $\bar{t}_\nu = \nu T_0$. Thus, $t_\nu = \nu T_0 + \varepsilon_\nu$, subject to the conditions $\overline{\varepsilon_\nu} = 0$ $\overline{\varepsilon_\nu\varepsilon_\mu} = \varepsilon^2\delta_{\mu\nu}$. Hence the first notable difference between the present problem and the cited in Sect. 2.4 is the presence of a certain negative correlation between adjacent intervals. Indeed. $\tau_\nu = t_{\nu+1} - t_\nu = T_0 + \varepsilon_{\nu+1} - \varepsilon_\nu$, and hence

$$\overline{\tau_\nu\tau_\mu} - T_0^2 = \begin{cases} 2\varepsilon^2 & (\mu = \nu) \quad , \\ -\varepsilon^2 & (|\mu - \nu| = 1) \quad , \\ 0 & (|\mu - \nu| \geq 2) \quad . \end{cases}$$

It is readily seen that this excludes phase diffusion: the variance of the sum of N intervals $(t_N - t_0)$ is at all times equal to the sum of variances of only two extreme instants of time t_0 and t_N, i.e., it amounts to $2\varepsilon^2$.

A second distinction as compared with the problem in Sect. 2.4 is that here the process is periodically nonstationary. Specifically, if we assume that for identical pulses the variance of the scattering ε_ν tends to zero ($\varepsilon^2 \to 0$), we will arrive at a deterministic periodic process.

Let the shape of the νth pulse be defined, as in Sect. 1.5, by the *random function* $F_\nu(t)$

$$\xi(t) = \sum_\nu F_\nu(t - \nu T_0 - \varepsilon_\nu) \quad , \tag{4.87}$$

where all the $F_\nu(t)$ are independent and distributed in a similar way, so that any means of the variables related to $F_\nu(t)$ are independent of ν. We assume that at $\nu = \mu$ a statistical connection between ε_μ and $F_\nu(t)$ is possible.

Introducing

$$\tilde{F}_\nu(\omega) = \frac{1}{2\pi} \int\limits_{-\infty}^{+\infty} F_\nu(\theta) e^{-i\omega\theta} \, d\theta \quad ,$$

we obtain for a long section of the process (4.87) in the interval $(-T/2, +T/2)$ the amplitude density

$$c(\omega) = \sum_{\nu=-N}^{N} \tilde{F}_\nu(\omega) e^{i\omega(T_0\nu + \varepsilon_\nu)} \quad , \tag{4.88}$$

where $N = T/2T_0$ is the number of mean periods in the interval $T/2$.

First of all, let us find the mean of $c(\omega)$ which controls that of $\xi(t)$

$$\overline{\xi(t)} = \int\limits_{-\infty}^{+\infty} \overline{c(\omega)} e^{i\omega t} \, d\omega \quad . \tag{4.89}$$

Denote

$$\langle \tilde{F}_\nu(\omega) e^{i\omega\varepsilon_\nu} \rangle = I(\omega) \quad , \tag{4.90}$$

then

$$\overline{c(\omega)} = I(\omega) \sum_{\nu=-N}^{N} e^{i\omega T_0\nu} = I(\omega) \frac{\sin\left[\omega T_0(N + 1/2)\right]}{\sin\left(\omega T_0/2\right)} \quad . \tag{4.91}$$

But as $N \to \infty$, the second factor becomes the periodic delta-function

$$\frac{2\pi}{T_0} \sum_n \delta\left(\omega - \frac{2\pi n}{T_0}\right) \quad .$$

Thus

$$\overline{c(\omega)} = I(\omega)\omega_0 \sum_n \delta(\omega - n\omega_0) \quad , \quad \omega_0 = \frac{2\pi}{T_0} \quad .$$

Substituting this into (4.89) gives

$$\overline{\xi(t)} = \omega_0 \sum_n I(n\omega_0)e^{in\omega_0 t} \quad . \tag{4.92}$$

We now find the covariance of $\xi(t)$. Since

$$\tilde{\xi}(t) \equiv \xi(t) - \overline{\xi(t)} = \int\limits_{-\infty}^{+\infty} [c(\omega) - \overline{c(\omega)}]e^{i\omega t}\, d\omega \quad ,$$

we have

$$\begin{aligned}
\psi(t+\tau,t) &= \overline{\tilde{\xi}(t+\tau)\tilde{\xi}^*(t)} \\
&= \int\limits_{-\infty}^{+\infty}\!\!\!\int [\overline{c(\omega_1)c^*(\omega_2)} - \overline{c(\omega_1)}\cdot\overline{c^*(\omega_2)}] \\
&\qquad \times \exp\{i[\omega_1(t+\tau) - \omega_2 t]\}\, d\omega_1\, d\omega_2 \quad .
\end{aligned} \tag{4.93}$$

But, by (4.88),

$$\begin{aligned}
\overline{c(\omega_1)c^*(\omega_2)} &= \sum_{\nu,\mu=-N}^{N} \overline{\tilde{F}_\nu(\omega_1)\tilde{F}_\mu^*(\omega_2)\exp[i(\omega_1\varepsilon_\nu - \omega_2\varepsilon_\mu)]} \\
&\qquad \times \exp[i(\omega_1\nu - \omega_2\mu)T_0]
\end{aligned}$$

Separating the terms with $\mu = \nu$ and denoting

$$\overline{\tilde{F}_\nu(\omega_1)\tilde{F}_\nu^*(\omega_2)e^{i(\omega_1-\omega_2)\varepsilon_\nu}} = K(\omega_1,\omega_2),$$

we get

$$\begin{aligned}
\overline{c(\omega_1)c^*(\omega_2)} &= K(\omega_1,\omega_2) \sum_{\nu=-N}^{N} e^{i(\omega_1-\omega_2)T_0\nu} \\
&\quad + I(\omega_1)I^*(\omega_2) \sum_{\nu=-N}^{N} e^{i\omega_1 T_0\nu} \sum_{\mu=-N}^{N} e^{-i\omega_2 T_0\mu} \quad ,
\end{aligned}$$

where in the second term $\mu\neq\nu$. Adding and substracting similar terms with $\mu = \nu$ and taking (4.91) into account, we get

$$\begin{aligned}
\overline{c(\omega_1)c^*(\omega_2)} &= [K(\omega_1,\omega_2) - I(\omega_1)I^*(\omega_2)] \\
&\quad \times \sum_{\nu=-N}^{N} e^{i(\omega_1-\omega_2)T_0\nu} + \overline{c(\omega_1)}\cdot\overline{c^*(\omega_2)} \quad .
\end{aligned}$$

Hence

$$\overline{c(\omega_1)c^*(\omega_2)} - \overline{c(\omega_1)}\cdot\overline{c^*(\omega_2)}$$
$$= [K(\omega_1,\omega_2) - I(\omega_1)I^*(\omega_2)]\frac{\sin\,(\omega_1-\omega_2)T_0(N+\frac{1}{2})}{\sin\,(\omega_1-\omega_2)\frac{T_0}{2}}\quad.$$

As $N\to\infty$ this gives

$$\overline{c(\omega_1)c^*(\omega_2)} - \overline{c(\omega_1)}\cdot\overline{c^*(\omega_2)}$$
$$= [K(\omega_1,\omega_2) - I(\omega_1)I^*(\omega_2)]\omega_0\sum_n\delta(\omega_1-\omega_2-n\omega_0)\quad.$$

Substitution into (4.93) gives

$$\psi(t+\tau,t) = \omega_0\sum_n e^{in\omega_0 t}$$
$$\int\limits_{-\infty}^{+\infty} [K(\omega_1,\omega_1-n\omega_0) - I(\omega_1)I^*(\omega_1-n\omega_0)]e^{i\omega_1\tau}\,d\omega_1\quad. \tag{4.94}$$

Thus, both the mean (4.92) and covariance (4.94) of the process under consideration are periodic (with period T_0) functions of time t.

The zeroth term of (4.94), equal to the time mean of $\psi(t+\tau,t)$ over period T_0, will be

$$\psi_0(\tau) = \frac{1}{T_0}\int\limits_{-T_0/2}^{T_0/2}\psi(t+\tau,t)\,dt\quad, \tag{4.95}$$

and the spectral density is

$$g_0(\omega) = \omega_0[K_0(\omega) - |I(\omega)|^2]\quad,\qquad\text{where} \tag{4.96}$$

$$K_0(\omega) = K(\omega,\omega) = \overline{|\tilde{F}_\nu(\omega)|^2}\quad.$$

For the mean over period T_0 of the mixed moment

$$B(t+\tau,t) = \overline{\xi(t+\tau)\xi^*(t)} = \psi(t+\tau,t) + \overline{\xi(t+\tau)}\cdot\overline{\xi^*(t)}\quad,$$

we have, by (4.92) and (4.95)

$$B_0(\tau) = \frac{1}{T_0}\int\limits_{-T_0/2}^{T_0/2}B(t+\tau,t)\,dt$$
$$= \psi_0(\tau) + \omega_0^2\sum_{m,n} I(m\omega_0)I^*(n\omega_0)e^{im\omega_0\tau}\delta_{mn}$$
$$= \psi_0(\tau) + \omega_0^2\sum_n |I(n\omega_0)|^2 e^{in\omega_0\tau}\quad.$$

Consequently, the continuous spectral density (4.96) is supplemented by the *discrete* spectrum

$$g_{B_0}(\omega) = \omega_0[K_0(\omega) - |I(\omega)|^2] + \omega_0^2 \sum_n |I(n\omega_0)|^2 \delta(\omega - n\omega_0) \quad .$$

It is this spectrum corresponding to the moment B of the process (3.20), averaged over the period T_0, that was found in *Fortet* [4.12] cited earlier. He also discussed various special cases of pulse definition. If, for instance, $F_\nu(t)$ and ε_ν are independent, then, by (4.90), we obtain

$$I(\omega) = \overline{\tilde{F}_\nu(\omega)}\,\overline{e^{i\omega\varepsilon_\nu}} = I_0(\omega)\varphi(\omega) \quad ,$$

where $\varphi(\omega) = \overline{\exp(i\omega\varepsilon_\nu)}$ is the characteristic function of the random shifts ε_ν. If there is no scattering of pulse generation times, then $\varphi(\omega) = 1$, and the density of the continuous spectrum $g_0(\omega)$ will be

$$g_0(\omega) = \omega_0[K_0(\omega) - |I_0(\omega)|^2] = \omega_0[\overline{|\tilde{F}_\nu(\omega)|^2} - |\overline{\tilde{F}_\nu(\omega)}|^2] \quad .$$

Periodically nonstationary processes of exactly this type, occur, in particular, when dealing with magnetic noise in ferromagnets in a periodically varying magnetic field (alternating magnetization noise) (see [4.13], [Ref. 4.14, Chap. 11] and [4.15]).

4.6 Exercises

4.6.1   Discuss the quasi-stationarity conditions for a normal process with zero mean, if the variance and correlation coefficient vary with time according to the Gaussian law

$$\sigma^2(t) = \sigma_0^2 e^{-\alpha t^2/2t_c^2} \quad , \quad K(t_1, t_2) = \exp\{-[\beta(t_1^2 + t_2^2) - 2\gamma t_1 t_2]\} \quad ,$$

where α, β, and γ are positive and, in accordance with $|K|\leq 1$, subject to the condition $\beta \geq \gamma$.

Solution. From (I.2.28) the bivariate distribution is

$$w_2(t_1, x_1; t_2, x_2)\, dx_1\, dx_2$$

$$= \frac{dx_1 dx_2}{2\pi\sigma_1\sigma_2\sqrt{1-K^2}} \exp\left\{-\frac{1}{2\sqrt{1-K^2}}\left[\left(\frac{x_1}{\sigma_1}\right)^2 + \left(\frac{x_2}{\sigma_2}\right)^2 - 2K\frac{x_1 x_2}{\sigma_1\sigma_2}\right]\right\} \quad ,$$

where $\sigma_{1,2} = \sigma(t_{1,2})$, $K = K(t_1, t_2)$. Hence the univariate distribution is

$$w_1(t, x)\, dx = \frac{dx}{\sqrt{2\pi}\sigma(t)} e^{-x^2/2\sigma^2(t)} \quad .$$

The mixed moment of the second order (i.e., the covariance, since $\bar{x} = 0$) is

$$B(t_1, t_2) = \sigma(t_1)\sigma(t_2)K(t_1, t_2) = \sigma_0^2 \exp\left[-\left(\frac{\alpha}{4t_c^2} + \beta\right)(t_1^2 + t_2^2) + 2\gamma t_1 t_2\right] \quad .$$

Substituting $t_{1,2} = t \pm \tau/2$ gives

$$B_2(t, \tau) = \sigma_0^2 \exp\left[-2t^2\left(\frac{\alpha}{4t_c^2} + \beta - \gamma\right) - \frac{\tau^2}{2}\left(\frac{\alpha}{4t_c^2} + \beta + \gamma\right)\right] \quad .$$

Let us require that the mixed moment $B_2(t, \tau)$ have the form

$$B_2(t, \tau) = \sigma_0^2 \exp\left(-\frac{t^2}{t_c^2} - \frac{\tau^2}{\tau_c^2}\right) \quad . \tag{4.97}$$

It is easily seen that this will be so at

$$\beta = \frac{1}{\tau_c^2} - \frac{1 - \alpha}{4t_c^2} \quad , \qquad \gamma = \frac{1}{\tau_c^2} - \frac{1}{4t_c^2} \quad .$$

For these values of β and γ the correlation coefficient is

$$K(t, \tau) = \exp\left\{-\frac{2 - \alpha}{2t_c^2}t^2 - \left(\frac{1}{\tau_c^2} - \frac{\alpha}{8t_c^2}\right)\tau^2\right\} \quad .$$

The condition $\beta \geq \gamma$ is met for $\alpha \leq 2$ and the condition that β and γ be positive for $t_c \geq \tau_c/2$. By definition, the process will be quasi-stationary for $t_c \gg \tau_c$. This warrants the assumption that $\beta \approx \gamma \approx 1/\tau_c^2$. Hence

$$K(t, \tau) \approx \exp\left(-\frac{2 - \alpha}{2t_c^2}t^2 - \frac{\tau^2}{\tau_c^2}\right) \quad .$$

As $t_c \to \infty$ the process becomes stationary. From (4.97), using (4.33b), we derive the bivariate spectral density

$$g_2(\Omega, \omega) = \frac{\sigma_0^2 t_c \tau_c}{4\pi} \exp\left[-\tfrac{1}{4}(\Omega^2 t_c^2 + \omega^2 \tau_c^2)\right] \quad .$$

Thus, the lines of constant spectral density are ellipses with semiaxes C/t_c in Ω and C/τ_c in ω.

4.6.2 Let a nonstationary process $f(t)$ act on a *nonharmonic* linear system with a pulse response $H(t, \theta)$. The system is supposed to be dissipative (no parametric excitation). With the observation time t sufficiently well separated from the initial moment t_0, the response of the system $\xi(t)$ will be one of a *steady-state* (i.e., independent of the initial conditions), because free oscillations, which might have emerged at the initial moment will already have decayed.

From the Duhamel integral

$$\xi(t) = \int_{t_0}^{t} H(t, \theta) f(\theta) \, d\theta \tag{4.98}$$

express the steady-state response in terms of the spectral amplitude density $c_f(\omega)$ of the force $f(t)$ and the instantaneous transfer function of the system. The latter is to be determined from the steady-state response $k(t, i\omega) \exp(i\omega t)$, to the force $f(t) = \exp(i\omega t)$, by (4:98),

$$k(t, i\omega) = \int_{-\infty}^{+\infty} H(t, \theta) e^{i\omega(\theta - t)} \, d\theta \quad . \tag{4.99}$$

Solution. Substituting the Fourier expansions

$$f(\theta) = \int_{-\infty}^{+\infty} c_f(\omega') e^{i\omega'\theta} \, d\omega' \quad , \quad H(t, \theta) = \frac{1}{2\pi} \int_{-\infty}^{+\infty} k(t, i\omega) e^{i\omega(t-\theta)} \, d\theta \quad ,$$

into the integral (4.98) and taking the integral with respect to θ gives

$$\xi(t) = \iint_{-\infty}^{+\infty} k(t, i\omega) e^{i\omega t} c_f(\omega')$$

$$\times \exp\left[i(\omega' - \omega)(t + t_0)/2\right] \frac{\sin\left[(\omega' - \omega)(t - t_0)/2\right]}{\pi(\omega' - \omega)} \, d\omega \, d\omega' \quad .$$

If t_0 is placed at $-\infty$, the factor with the sine becomes $\delta(\omega' - \omega)$, so that the steady-state response will be

$$\xi(t) = \int_{-\infty}^{+\infty} k(t, i\omega) c_f(\omega) e^{i\omega t} \, d\omega \quad . \tag{4.100}$$

In the special case of the *harmonic* system, (4.99) gives

$$k(t, i\omega) = \int_{-\infty}^{+\infty} H(t - \theta) e^{i\omega(\theta - t)} \, d\theta = \int_{-\infty}^{+\infty} H(\chi) e^{-i\omega\chi} \, d\chi = k(i\omega) \quad , \tag{4.101}$$

and then (4.100) becomes the Fourier integral with the amplitude density

$$c_\xi(\omega) = k(i\omega) c_f(\omega) \quad .$$

This relation was central to Sect. 4.4.

4.6.3 Using the result (4.100), calculate the spectral amplitude density $c_\xi(\omega)$ of the steady-state response and its bivariate spectral density $\gamma(\omega_1,\omega_2) = \langle c_\xi(\omega_1)c_\xi^*(\omega_2)\rangle$.

Solution. To find $c_\xi(\omega)$ it is sufficient to represent the instantaneous transfer function as a Fourier integral with respect to t

$$k(t,\mathrm{i}\omega) = \int\limits_{-\infty}^{+\infty} k_2(\mathrm{i}\omega_1,\mathrm{i}\omega)e^{\mathrm{i}\omega_1 t}\,d\omega_1 \quad , \tag{4.102}$$

hence

$$k_2(\mathrm{i}\omega_1,\mathrm{i}\omega) = \frac{1}{2\pi}\int\limits_{-\infty}^{+\infty} k(t,\mathrm{i}\omega)e^{-\mathrm{i}\omega_1 t}\,dt \quad . \tag{4.103}$$

Note that for the harmonic system, by virtue of (4.103) and (4.101),

$$k_2(\mathrm{i}\omega_1,\mathrm{i}\omega) = k(\mathrm{i}\omega)\delta(\omega_1) \quad . \tag{4.104}$$

Substituting (4.102) into (4.100) gives

$$c_\xi(\omega_1) = \int\limits_{-\infty}^{+\infty} k_2[\mathrm{i}(\omega_1 - \omega),\ \mathrm{i}\omega]c_f(\omega)\,d\omega \quad ,$$

and accordingly,

$$\begin{aligned}
\gamma_\xi(\omega_1,\omega_2) &= \langle c_\xi(\omega_1)c_\xi^*(\omega_2)\rangle \\
&= \iint\limits_{-\infty}^{+\infty} k_2[\mathrm{i}(\omega_1 - \omega),\ \mathrm{i}\omega]k_2^*[\mathrm{i}(\omega_2 - \omega'),\ \mathrm{i}\omega']\gamma_f(\omega,\omega')\,d\omega\,d\omega' \quad .
\end{aligned} \tag{4.105}$$

If the process $f(t)$ at the system input is stationary, so that

$$\gamma_f(\omega,\omega') = g_f(\omega)\delta(\omega - \omega') \quad , \tag{4.106}$$

then it follows from (4.105) that

$$\gamma_\xi(\omega_1,\omega_2) = \int\limits_{-\infty}^{+\infty} k_2[\mathrm{i}(\omega_1 - \omega),\ \mathrm{i}\omega]k_2^*[\mathrm{i}(\omega_2 - \omega),\ \mathrm{i}\omega]g_f(\omega)\,d\omega \quad . \tag{4.107}$$

If the system is harmonic, then using (4.104) we obtain from (4.105)

$$\gamma_\xi(\omega_1,\omega_2) = k(\mathrm{i}\omega_1)k^*(\mathrm{i}\omega_2)\gamma_f(\omega_1,\omega_2) \quad . \tag{4.108}$$

In both cases (4.107, 108) the steady-state response is nonstationary with a bivariate distribution of its spectral "mass". But for the *harmonic* system and *stationary* action the steady-state process at the output is stationary: substitution of (4.104) into (4.107) or (4.106) into (4.108) gives

$$\gamma_\xi(\omega_1, \omega_2) = |k(i\omega_1)|^2 g_f(\omega_1)\delta(\omega_1 - \omega_2) \quad .$$

4.6.4 Show that a stationary process passing through a system with slowly varying parameters yields at the output a steady-state quasi-stationary process.

Solution. According to (4.104), the bivariate transfer function of a harmonic system contains a delta-function. If the system parameters are not constant, but slowly varying, then $k_2(i\omega_1,\ i\omega)$ will be a sharp function that is only nonzero near $\omega_1 = 0$. We write (4.107) in terms of the variables (4.31)

$$g_{2\xi}(\Omega,\omega) = \int\limits_{-\infty}^{+\infty} k_2\left[i\left(\omega - \frac{\Omega}{2} - \omega'\right),\ i\omega'\right]$$
$$k_2^*\left[i\left(\omega + \frac{\Omega}{2} - \omega'\right),\ i\omega'\right] g_f(\omega')\, d\omega' \quad .$$

It is then clear that ω' must be close both to $\omega - \Omega/2$ (otherwise $k_2 = 0$) and to $\omega + \Omega/2$ (otherwise $k_2^* = 0$), which is only possible for sufficiently small Ω. Consequently, $g_{2\xi}(\Omega,\omega) \neq 0$ in a narrow band at $\Omega = 0$.

References

Preface

1. S.A. Akhmanov, A.S. Chirkin: *Statistical Phenomena in Nonlinear Optics* (Moscow University Press, Moscow 1971) (in Russian);
   S.A. Akhmanov, Y.Y. Dyakov, A.S. Chirkin: *Introduction to Statistical Radiophysics and Optics*, Vols. I, II (Springer, Berlin, Heidelberg, New York) in preparation
2. A.M. Yaglom: *An Introduction to the Theory of Stationary Random Functions* (Prentice Hall, Englewood Cliffs, New York 1962)
3. A. Blanc-Lapierre, R. Fortet: *Théorie des fonctions aléatoires* (Masson, Paris 1953)
4. J.L. Doob: *Stochastic Processes* (John Wiley, New York 1953)
5. M.S. Bartlett: *An Introduction to Stochastic Processes* (Cambridge University Press, Cambridge 1956)
6. Yu.A. Rozanov: *Stationary Random Processes* (Holden-Day, San Francisco 1967)
7. A.A. Sveshnikov: *Applied Methods of the Theory of Random Functions* (Pergamon, Oxford 1966)
8. H. Cramér, M.R. Leadbetter: *Stationary and Related Stochastic Processes* (John Wiley, New York 1967)
9. S. Karlin: *A First Course in Stochastic Processes* (Academic, New York 1966)
10. V.L. Lebedev: *Random Processes in Electrical and Mechanical Systems* (Gostekhizdat, Moscow 1958) (in Russian)
11. A. van der Ziel: *Noise* (Prentice-Hall, Englewood Cliffs, New York 1954)
12. B.R. Levin: *Fondaments théorique de la radiotecchnique statistique* (MIR, Moscow) Vols. I, II (1973); Vol. III (1979)
13. W.B. Davenport, Jr., W.L. Root: *An Introduction to the Theory of Random Signals and Noise* (McGraw-Hill, New York 1958)
14. A. van der Ziel: *Fluctuation Phenomena in Semi-Conductors* (Butterworths Scientific Publications, London 1959)
15. R.L. Stratonovich: *Random Noise*, Vols. I and II (Gordon and Breach, New York 1963)
16. D.A. Middleton: *An Introduction to Statistical Communication Theory* (McGraw-Hill, New York 1960)
17. V.I. Tikhonov: *Statistical Radio Engineering* (Sovradio, Moscow 1966) (in Russian)
18. A.N. Malakhov: *Fluctuations in Self-Oscillatory Systems* (Nauka, Moscow 1968) (in Russian)
19. B.R. Levin: *Theoretical Foundations of Statistical Radio Engineering* (Sovradio, Moscow 1974) (in Russian)

Chapter 1

1.1 D. Gabor: J. IEE (London) P.III, **93**, 429 (1946)
1.2 G.S. Agarwal, E. Wolf: J. Math. Phys. N.Y. **13**, 1759 (1972)
1.3 J. Toll: Phys. Rev. **104**, 1760 (1956)
1.4 A.Ya. Khintchine: Math. Ann. **109**, 604 (1934)
1.5 A.N. Kolmogorov: Bul. Mosc. State Univ. **II** (1941) (in Russian)
1.6 S. Bochner: *Vorlesungen über Fouriersche Integrale* (Akad. Verlagsgesellschaft, Leipzig 1932)
1.7 A.G. Mayer, E.A. Leontovich: Dokl. Akad. Nauk SSSR **4**, 353 (1934)
1.8 C. Heiden: Phys. Rev. **188**, 319 (1969)
1.9 R. Fortet: L'onde electrique **34**, 683 (1954)
1.10 I. Korn: Proc. IEEE **58**, 955 (1970)

1.11   I.L. Bershtein: Dokl. Akad. Nauk SSSR **20**, 11 (1938); Zh. Tekh. Fiz. **11**, 305 (1941)
1.12   I.L. Bershtein: Dokl. Akad. Nauk SSSR **68**, 469 (1949); Izv. Akad. Nauk SSSR (ser. fiz.) **14**, 145 (1950)
1.13   W.F. McGee: IEEE Trans. **IT-17**, 149 (1971)
1.14   I.S. Reed: IRE Trans. **IT-8**, 194 (1962)
1.15   A.A. Bobrov, I.N. Verbitskaya: Nauchn. Ezheg. Odessk. Univ. **2**, 79 (1961)
1.16   I.S. Gradshtein, I.M. Ryzhik: *Tables of Integrals, Sums, Series, and Products* (Academic Press, New York 1965)

Chapter 2

2.1   R.C. Bourret: Nuovo Cimento **18**, 347 (1960)
2.2   Y. Kano, E. Wolf: Proc. Phys. Soc. **80**, 1273 (1962)
2.3   L.I. Mandelshtam: *Lectures on Theory of Oscillations* (Nauka, Moscow 1972) (in Russian)
2.4   L.I. Mandelshtam: *Lectures on Optics, Relativity Theory and Quantum Mechanics* (Nauka, Moscow 1972) (in Russian)
2.5   S.M. Rytov: Usp. Fiz. Nauk **29**, 147 (1946)
2.6   J. Dugundji: IRE Trans. **IT-4**, 53 (1958)
2.7   P.R. Karr, R. Wooldrige: IRE Trans. **IT-5**, 33 (1959)
2.8   B.R. Levin: *Fondaments théorique de la radiotechnique statistique* (MIR, Moscow) Vols. I, II (1973); Vol. III (1979)
2.9   Yu.E. Dyakov: Radiotekh. Elektron. **8**, 1812 (1963)
2.10   A. Blanc-Lapierre, M.Savelli, A. Tortrat: Ann. télécomm. **9**, 237 (1954)
2.11   F.V. Bunkin, L.I. Gudzenko: Radiotekh. Electron. **3**, 968 (1958)
2.12   S.M. Rytov: Zh. Eksp. Teor. Fiz. **29**, 702 (1955)
2.13   A.N. Malakhov: Zh. Eksp. Teor. Fiz. **30**, 384 (1956)
2.14   M.I. Rodak: Radiotekh. Elektron. **5**, 1370 (1960)
2.15   V.S. Troitsky, V.V. Khrulev: Radiotekh. Elektron. **1**, 831 (1956)
2.16   N. Wiener: *Non-Linear Problems in Random Theory* (Wiley, New York 1958)
2.17   Yu. E. Aptek, A.M. Gersht: Izv. VUZ, Radiofiz. **6**, 311 (1963)
2.18   A.M. Gersht: Izv. VUZ, Radiofiz. **7**, 701 (1964)
2.19   Ya.I. Khurghin: Nauchn. Dokl. Vyssh. Shk. (Radiotekh. Elektron.) **1**, 96 (1958)
2.20   V.P. Yakovlev: Radiotekh. Elektron. **5**, 1728 (1960)
2.21   B.S. Tsybakov, V.P. Yakovlev: Radiotekh. Elektron. **4**, 543 (1959)
2.22   S.M. Rytov: Izv. VUZ, Radiofiz. **2**, 50 (1959)
2.23   S.M. Rytov: Izv. VUZ, Radiofiz. **2**, 45 (1959)
2.24   A. Blanc-Lapierre, P. Dumontet: Comptes Rendues (Paris) **238**, 1005 (1954)
2.25   E. Wolf: Proc. Roy. Soc. (London) **A230**, 246 (1955)
2.26   P.H. van Cittert: Physica, **24**, 505 (1958)
2.27   G.S. Gorelik: *Oscillations and Waves* (Fizmatgiz, Moscow 1959) (in Russian)
2.28   L. Mandel, E. Wolf: Rev. Mod. Phys. **37**, 231 (1965)
2.29   F. Franson, S. Slansky: *Coherence in Optics* (Nauka, Moscow 1967) (in Russian)
2.30   M. Born, E. Wolf: *Principles of Optics*, 4th ed. (Pergamon, New York 1970)
2.31   S.M. Rytov, Yu.A. Kravtsov, V.I. Tatarskii: *Principles of Statistical Radiophysics 3 — Elements of Random Fields* (Springer, Berlin, Heidelberg, New York) in preparation
2.32   S.M. Rytov, Yu.A. Kravtsov, V.I. Tatarskii: *Principles of Statistical Radiophysics — Wave Propagation Through Random Media* (Springer, Berlin, Heidelberg, New York) in preparation
2.33   B.J. Thompson, E. Wolf: J. Opt. Soc. Am. **47**, 895 (1957)
2.34   H. Nodtvedt: Phil. Mag. **42**, 1022 (1951)
2.35   G.S. Gorelik: Dokl. Akad. Nauk SSSR **58**, 45 (1947); Usp. Fiz. Nauk **34**, 321 (1948); S.I. Borovitsky, G.S. Gorelik: Usp. Fiz. Nauk **59**, 543 (1956)
2.36   R.H. Brown, R.Q. Twiss: Phil. Mag. **45**, 663 (1954)
2.37   R.H. Brown, R.Q. Twiss: Nature **178**, 1046 (1956); Proc. Roy. Soc. **A248**, 199 (1958)
2.38   R.H. Brown: Usp. Fiz. Nauk **108**, 529 (1972)

2.39   L.I. Mandelshtam: *Collected Works*, Vol. II (AN SSSR Press, Moscow 1947), articles
       44, 45, 47 and 48; Vol. III (AN SSSR Press, Moscow 1950), articles 63 and 66
2.40   S.M. Rytov: JETP **29**, 304 and 315 (1955)
2.41   A.N. Malakhov: *Fluctuations in Self-Oscillatory Systems* (Nauka, Moscow 1968) (in
       Russian)
2.42   B.E. Saleh: J. Opt. Soc. Am. **63**, 422 (1972)
2.43   W.H. Carter, E. Wolf: J. Opt. Soc. Am. **63**, 1619 (1973)
2.44   V.V. Ivanov: *Radiation Transport and Spectra of Celestial Bodies* (Nauka, Moscow
       1969) (in Russian)

Chapter 3

3.1   A. Papoulis: IEEE Trans. **IT-18**, 20 (1972)
3.2   L.P. Zachepitskaya: Radiotekh. Elektron. **13**, 1452 (1968)
3.3   N.P. Bobrova, L.P. Zachepitskaya, I.N. Sozinov: Izv. VUZ, Radiofiz. **14**, 103 (1971)
3.4   L.P. Zachepitskaya: Radiotekh. Elektron. **16**, 627 (1971)
3.5   Ya. D. Shirman et al.: *Theoretical Foundations of Radar* (Sovradio, Moscow 1970) (in
       Russian)
3.6   V.I. Tikhonov: *Statistical Radio Engineering* (Sovradio, Moscow 1966) (in Russian)
       S.K. Srinavasan, R. Vasudevan: *Introduction to Random Differential Equations and
       their Applications* (Elsevier, New York 1971)
3.7   R.L. Stratonovich: *Selected Topics in the Theory of Fluctuations in Radio Engineering*
       (Sovradio, Moscow 1961) (in Russian);
       L. Arnold: *Stochastische Differentialgleichungen, Theorie und Anwendungen* (Olden-
       bourg, München, Wien 1973)
3.8   L.A. Vainshtein, V.D. Zubakov: *Selection of Signals from Random Noise* (Sovradio,
       Moscow 1960) (in Russian);
       S.E. Falkovich: *Radar Signal Reception with Fluctuation Noise* (Sovradio, Moscow
       1961) (in Russian)
3.9   P.A. Bakut, I.A. Boshakov, et al.: *Statistical Theory of Radar* (Sovradio, Moscow) Vol.
       I (1963); Vol. II (1964) (in Russian)
3.10   B.R. Levine: *Fondaments théorique de la radiotechnique statistique* (MIR, Moscow)
       Vols. I, II (1973); Vol. III (1979)
3.11   D.A. Middleton: *An Introduction to Statistical Communication Theory* (McGraw-Hill,
       New York 1960)
3.12   S.O. Rice: Bell Syst. Techn. J. **23**, 282 (1944); **24**, 46 (1945)
3.13   S.M. Rytov: Zh. Eksp. Teor. Fiz. **29**, 304 and 315 (1955)
3.14   K. Furutsu: J. Res. NBS $D - 67$, 303 (1963);
       E.A. Novikov: Zh. Eksp. Teor. Fiz. **47**, 1919 (1964)
3.15   L.I. Gudzenko: Radiotekh. Elektron. **1**, 1240 (1956)
3.16   L.I. Gudzenko: Radiotekh. Elektron. **4**, 97 (1959)
3.17   S.Ya. Raevsky, R.V. Khokhlov: Radiotekh. Elektron. **3**, 507 (1958)
3.18   I.Ya. Akopyan, P.L. Stratonovich: Nauchn. Dokl. Vyssh. Shkoly (Ser. Fiz.-Mat.) **1**, 162
       and 187 (1958)
3.19   M.E. Zhabotinsky, P.E. Zilberman: Dokl. Akad. Nauk SSSR **119**, 918 (1958)
3.20   A.N. Malakhov: Izv. VUZ, Radiofiz. **1**, 79 (1958)
3.21   A.N. Malakhov: *Fluctuations in Self-Oscillatory Systems* (Nauka, Moscow 1968) (in
       Russian)
3.22   G.S. Gorelik: Isv. AN SSSR (Ser. Fiz.) **14**, 187 (1950)
3.23   S.M. Rytov, Yu.A. Kravtsov, I.A. Tatarskii: *Principles of Statistical Radiophysics* −
       Elements of Random Fields (Springer, Berlin, Heidelberg) in preparation
3.24   J.B. Johnson: Nature **119**, 50 (1927); Phys. Rev. **29**, 367 (1927); ibid. **32**, 97 (1928)
3.25   H. Nyquist: Phys. Rev. **29**, 614 (1927); ibid. **32**, 110 (1928)
3.26   S.M. Rytov: *Theory of Electrical Fluctuations and Thermal Radiation* (Izd. AN SSSR,
       Moscow 1953) (in Russian)
3.27   R.Q. Twiss: J. Appl. Phys. **26**, 599 (1955)

3.28 H.B. Callen, T.A. Welton: Phys. Rev. **83**, 34 (1951) [see also J.L. Jackson: Phys. Rev. **87**, 471 (1952)]
3.29 H.B. Callen, R.F. Green: Phys. Rev. **86**, 704 (1952); ibid. **88**, 1387 (1952)
3.30 H.B. Callen. M.L. Barasch: Phys. Rev. **88**, 1382 (1952)
3.31 L.D. Landau, E.M. Lifshitz: *Electrodynamics of Continuous Media* (Pergamon, Oxford 1977)
3.32 L.D. Landau, E.M. Lifshitz: *Statistical Physics* (Pergamon, Oxford 1977)
3.33 W. Bernard, H.B. Callen: Phys. Rev. **118**, 1466 (1960)
3.34 F.V. Bunkin: Radiotekh. Elektron. **6**, 3 (1961)
3.35 A.N. Malakhov: Izv. VUZ, Radiofiz. **16**, 1287 (1973)
3.36 R.A. Price: IRE TRans. **IT-4**, 69 (1958)

Chapter 4

4.1 A.N. Kolmogorov: Dokl. Akad. Nauk SSSR **26**, 6 (1940)
4.2 A.N. Kolmogorov: Dokl. Akad. Nauk SSSR **26**, 115 (1940)
4.3 A.M. Yaglom: Matem. Sb. (Nov. Ser.) **37**, 141 (1955)
4.4 R.A. Silverman: IRE Trans. **IT-3**, 182 (1957)
4.5 W.D. Mark: J. Sound Vib. **11**, 19 (1970)
4.6 F. Donati: IEEE Trans. **IT-17**, 7 (1971)
4.7 A.W. Rihaczek: IEEE Trans. **IT-14**, 369 (1968)
4.8 A.W. Rihaczek: IEEE Trans. **IT-18**, 208 (1972)
4.9 M.H. Ackroyd: J. Ac. Soc. Am. **50**, 1229 (1971)
4.10 L.I. Gudzenko: Dokl. Akad. Nauk SSSR **125**, 62 (1959)
4.11 L.I. Gudzenko: Radiotekh. Elektron. **4**, 1062 (1959)
4.12 R. Fortet: L'onde electrique **34**, 683 (1954)
4.13 F.V. Bunkin: Radiotekh. Elektron. **4**, 1913 (1959)
4.14 N.N. Kolachevsky: "Studies of Statistical Phenomena in Processes of Alternating Magnetization of Ferromagnetics"; Ph. D. Thesis, Moscow Physics and Technology Institute (1960)
4.15 N.N. Kolachevsky: *Magnetic Noise* (Nauka, Moscow 1971) (in Russian)

Subject Index

Accumulation (in time) 21, 77, 89
Admittance 173
— matrix 161, 166
Almost periodic function 17
Amplitude density 16, 17, 220
— spectral 20, 40
Amplitude distribution 56
Amplitude modulation 50
Analytical signal(s) 1, 5, 7, 20, 47, 48, 86, 96, 102, 140, 145, 193, 194, 195
— nonstationary 191
— spectral representation of 20
Analytical signal(s), stationary 6, 35, 36, 78, 89
— distribution of 35
— covariance of 21
— "second" covariance of 78
Angular resolution 85
Annihilation operator 5
Anode current 145
Autocorrelation coefficient 80
Averaging, moving 77, 89, 211

Bifurcation points 147
Bivariate density 198
Bershtein experiment 150
Bershtein method 152
Black-body radiation 42
Boltzmann's constant 44, 145, 155
Brownian motion 53, 154, 157

Campbell theorem 70
Cauchy formula 3
Cauchy-Buniakovski inequality 8
Causality condition 197–200, 205
Central limit theorem 118, 119, 120
Characteristic function 26, 35, 36, 56, 57, 70, 71, 82, 123, 181, 218
— multivariate 34
Coefficient
— of mutual correlation 97
— of natural diffusion 149
— of technical diffusion 149
Coherence
— function of the second order, complex 79
— degree of 79
— of the oscillations of two coupled sources 90

— of wave fields 5
— space-time 75
— theory 5, 75
Coherent train 34, 150
— length 150
Complex amplitude(s) 48, 96, 112, 195
— variance for 49
Complex "mass" 14, 210
Conditional probability 53
Conduction current 158
Convergence in probability 78
Convolution 26, 116, 202
Correlation 44, 71, 85, 87, 189, 214
— exponential 53
— of the instantaneous intensities 85
Correlation coefficient(s) 44, 45, 52, 79, 127, 171, 192, 218, 219
Correlation matrix 34, 35, 51, 52, 92, 109, 162
Correlation radius, spatial 46
Correlation theory 1, 5, 11, 12, 42, 75, 85, 96, 110, 139, 140, 142, 147, 208
— application of 42
— of self-oscillations 91
— spectral form of 110
Correlation time 43, 71, 117, 180
— effective 39, 40
Covariance 2, 7, 8, 15, 17, 18–21, 23, 24, 28–30, 32, 38, 42, 49, 78, 79, 82, 85, 86, 102, 112 114–117, 119, 129, 132, , 140, 170, 171, 177, 179, 183–185, 211, 217, 219
— exponential 104
— "first" 140
— mutual, "second" 78
— "second" 140
— of intensities 85
— of thermo-e.m.f. 145
— unconditional 107
Cumulant equations 129, 170
Cumulants 169, 170

Damping time 104
Delay line 23
Delta-correlated
— force 53, 142
— frequency modulation 144
— noise 43, 44
Delta-function, periodic 213, 215
Denormalization phenomena 120

Deterministic function 23, 69, 73
Deterministic processes 31, 46
— periodic 215
Deterministic pulse, covariance of 32
Deterministic signals 202
Derivative, continuity in the mean square 10
Differentiability condition 30
Diffusion coefficient 33, 62, 140
Diffusion law 144, 149
Diffusion time 141
Discrete spectrum 18
Displacement current 158
Dissipative system 111, 113
Distribution 1, 69
— n-variate 1
— of the derivatives 53
— of the luminosity 83
Distribution function(s) 11, 51, 66, 74, 121, 122
— bivariate 209
— for the mean instantaneous power 18
— univariate 209
Doob's theorem 53
Doppler broadening 61, 106
Doppler effect 104, 106
Doppler shift 105
Drift of the gain 135
Duhamel integral 14, 110, 115, 199, 220
Duration of observation 102
Dynamic systems 11, 14, 110

Eigenfunctions 5
Einstein-Fokker-Planck equation 33, 139, 140, 144, 147
Emissivity of the cathode 177
Envelope 27, 48, 52, 53, 79, 94, 153
— deterministic 31
— exponential 80
— Gaussian 31, 32
Ergodicity condition 78, 184
Equilibrium state 154
Equipartition of energy 173
Equipartition theorem 155, 156, 159
Equivalent noise resistance 161
Equivalent noise temperature 160
Exponential distribution 58, 70
External source 81
— consisting of incoherent "point" elements 82
External action 14
— current 171
— e.m.f. 171

Filter 38, 199
— band 116, 117
— Band-pass 42, 43, 200
— Harmonic 197
— matched 118, 119, 120, 167
— passive 197
— pulse response of 197
— settling time 117
Filtration 207
— harmonic 113
— of nonstationary processes 197
— of stationary processes 115
— operation 202
Flicker effect 177, 178, 186
Fluctuation(s)
— amplitude 23, 33, 60, 141, 144, 147
— correlation theory of 139
— equilibrium 165
— harmonic 11
— natural 121, 151
— non-Markovian 147
— of the amplitude and phase 143
— resistance 178
Fluctuation-dissipation theorem (FDT) 110, 153, 163, 164, 165
Fourier expansion 11, 18, 47, 57, 63, 130, 137, 189, 205, 213, 220
Fourier integral 12, 17, 18, 24, 189, 220, 221
Fourier series 17, 128, 210, 212
— generalized 12, 17
Fourier-Stieltjes integral 12, 15, 18, 112, 184, 189
Fourier transform(ation) 23, 32, 53, 63, 69, 80, 84, 115, 124
195, 198, 200, 205, 206
— two-dimensional 84
Fraunhofer zone 75, 81
Frequency dispersion of conductivity 43, 44
Frequency expansion 203
Frequency localization for the second-order moments 182
Frequency-modulated process 40
Frequency spectrum 179
"Frequency-window" 203, 204
Fresnel zone 81
Function with uncorrelated increments 15
Furutsu-Novikov formula 141

Gamma-distribution 71
Gaussian correlation law 24
Gaussian distribution 56, 105, 106
Gaussian noise 129
Gaussian process 53, 54, 128, 153
— narrow-band 128
— stationary 126
Green's function 197
Group velocity 156

Hankel transform 57
Harmonic system(s) 11, 110, 164, 220, 221
— bivariate transfer function of 222
— dynamic equation of 11
— random 200
Harmonizable process 189
Hermitian condition 7

Hermitian matrix 35
Hierarchy condition 209
Hilbert transform(ation) 3, 4, 5, 6, 20, 48

Incoherence, complete 44
Incoherence of sources 90
Illuminance, uniform 78
Impedance 154, 157
Increment, dimensionless 140
Information theory 5
Intensity distribution 76, 78
— observed 77
— instantaneous 88
Interference 27, 87
— fringes 79, 84
— methods for measuring the velocity of
  radiowave 87
— pattern 79–82, 87–89, 93, 94, 107, 108
— pattern intensity of 83
— phenomena 75
Interferometer 34, 81, 82
— intensity, of Brown and Twiss 85, 86
— Michelson 75, 81, 84, 85
— Optical 85
— Young-Rayleigh 75, 81
Intermediate frequency 77
Internal process 14, 111
Inscribed oscillations 31
Inscribed process 32, 38
— correlation time of 32
Instantaneous frequency 39, 40, 48, 119
Instantaneous intensity 76, 77, 78
Instantaneous power 190, 195
Instantaneous pulses, Poisson sequence of 43

Joint distribution 51

Khintchine (Wiener-Khintchine) theorem
  15, 22, 24, 37, 41, 109, 112
Kirchhoff equations (laws) 157, 159, 166,
  172, 174

Lagrange equations 161
Lagrange-Maxwell-equations 161
Langevin equations 144
Langevin functions 45
Life-time of the excited state 104
Limiting cycle 142, 143, 147, 153
Linear
— detector 125
— electric network 166
— Operator 3, 112, 165
— system(s) 11
— system(s) dissipative 90
Linear system(s)
— dynamic 11
— nonharmonic 219
— with constant parameters 14
— with periodically varying parameters 207

Linear vibrator 97
Locally-homogeneous fields 192
Lorentzian line 33, 80, 106

MacDonald Function 58
Magnetic noise 22, 218
Markov process(es) 110, 121
— diffusion 140, 141
— normal 53
Matrices
— generalized susceptibility 164
— of admittance 161
— of impedance 161
— of spectral densities 162
Measurement time 180
Mean
— bilinear quantities 14
— energy 202, 203, 204
— instantaneous power 202
— locally-smoothed power 203
— power spectrum 197
— power, total 202
— square drift 74
Memoryless systems 110
— nonlinear 120, 121, 128
Modulated force, stationary 139
Modulated processes 193, 195
— nonstationary real 194, 195
— periodically nonstationary 195
— stationary 195
Modulating function(s) 48, 77
Modulation 89
Monochromaticity 80
Moving fringes 88
Moment(s) 1, 2, 51, 169, 177, 182
— mixed 1, 2, 7, 9, 15, 18, 19, 112, 170
— of intensity fluctuation 102
— second, of the instantaneous intensities 102

Natural light 98
Nicol's prism 97
Noise measurement 130
Nonequilibrium ensemble 163, 165
Nonlinear operators 165
Nonmonochromaticity 32, 33, 34, 90, 139, 155
Nonstationary actions 113
Nonstationary process(es) 1, 177, 189, 200,
  202, 204, 207, 219
— filtration of 177
— modulated 200
— spectra of 189
Normal distribution 107
— complex 36
Normal process(es) 1, 51, 60, 125, 141
— complex 34
— quasi-stationary conditions for 218
Normal variables 102
Normalization conditions 209

Nyquist formula 110, 155, 156, 157, 160, 161, 162, 171–173
— generalization to branched networks 162

Observation time 77
Optical heterodyning 77
Oscillations
— circularly polarized 99
— frequency-modulated 119
— linearly polarized 98
— modulated 49, 50, 76, 79, 88, 95, 102
— monochromatic 98
— natural 155
— normal 102
— random 51
— settling of 113
— stationary and stationarily coupled 87
— steady-state, forced 112
Oscillator(s)
— coupled 108
— Thomson 139
Oscillatory circuit 47, 53, 188

Parametric amplitudes 207
Parseval's theorem 203
Periodic functions of time 217
Periodic pumping 207
Periodically nonstationary process(es) 53, 207–210, 212, 214, 218
Perturbation method (theory) 121, 165
Phase
— diffusion 33, 60, 94, 140
— diffusion coefficient 147
— distribution 57
— shift, random 59, 60
— velocity 156
Physical spectrum 204, 207
— density of 203, 205
— instantaneous density of 206
— of the nonstationary process 205
Planck formula 44
Point source 81
Poisson process 29, 62, 65, 66, 70, 71, 103, 167, 180
— pulse 24, 28, 29, 40, 119, 181, 187
Polarization 95, 96, 101
— degree of 99, 100, 102
— matrix 96, 98, 101, 102
— statistical properties of 95
Polaroid 97
Positive definite functions 8, 15
Possible values 35
Probabilistic convergence 12
Probabilistic density 25, 26
— bivariate 192
— univariate 192
Processes with independent increments 104
Processes with stationary increments 177

Pulse(s)
— of a complicated shape 31
— random duration of 70
— separations, random 70
— spectral expansion of 27
— square 27
— with a simple shape 118
Pulse generators
— periodic 71
— spectrum of 72
Pulse processes 26, 65, 73, 197
— quasi-periodic, with correlated adjacent intervals 214
— spectra of 65
— with independent increments 65, 66
Pulse response 120, 164, 197, 198, 200, 219

Quality of a circuit 132
Quantum electrodynamics 5
Quantum oscillator, mean energy of 156, 164
Quasi-harmonic representation 47
Quasi-monochromatic process 128, 152
Quasi-periodic process 73
Quasi-stationary (quasi-static) approximation 96
Quasi-stationary electric networks 110, 153
Quasi-stationary processes 177, 189, 192, 193, 200, 206, 207
— instantaneous spectrum of 207
— steady-state 222
— wide-sense 192

Radiation damping 106
Radioastronomy 85, 134
Radioiterferometer 75, 85
Radiometer 130, 134
— modulation 136–138
Radio wave interference 87
Random actions 110
— on dynamic systems 110
— on memoryless nonlinear systems 120
Random amplitudes 70
Random e.m.f. 154
Random function(s) 1, 7, 12, 23, 29, 30, 66, 75, 199, 215
— complex 1, 25
— covariances of 1
— harmonizable 12
— normal 85
— second-order 2, 8, 15
— spectral intensity of 18
— spectral representation of 1, 11
— stationary 9, 10, 14, 15
— with stationary increments 59, 181
Random modulation 78
Random phase 141
Random processe(s) 1, 5, 23, 31, 120, 170, 199
— complex 1, 7
— modulated 46

230

— spectral expansion of 110
— stationary 37, 96, 154, 164
— with delta-correlation 28
— with stationary increments (RPSI) 182, 184–186, 192
Random variables 170, 214
Rayleigh distribution (law) 53, 55, 58, 153
Resonator 34
Response, steady-state 11, 115, 199, 221
Rice distribution 53

Schwarz inequality 37
Self-interference pattern 94
Self-excitation threshold 147
Self-oscillatory system 23, 32, 46, 90, 91, 110, 111, 121, 139, 143, 151–153, 207
— natural nonmonochromaticity of 32
— natural fluctuations of 32
— natural spectral line width 33
— strongly nonlinear 207
Shot current (noise) 43, 71, 113, 114, 139, 140, 145, 165, 177, 178, 207
Signal detection, optimal 119
Signal-to-noise ratio 130
Skin effect 43, 44, 159
Slow perturbation method 139
Smoluchowski equation 53
Smoothing 193, 205, 206
— coefficient 131, 133, 135, 138
— function 205
— operation 202
Space structure of the field 81
Spectral amplitude 31, 112, 167
— apparatus 46, 47
— components 47
Spectral expansion 16, 30, 38, 177, 203
— in the mean square 5
Spectral density 18, 19, 21–23, 26, 27, 29, 33, 40, 42, 43, 44, 50, 59, 74, 80, 83p, 103, 104, 106, 112, 114, 116, 127, 132, 133, 140, 154, 157, 158, 161, 167, 169, 173, 174, 177, 179, 186, 187, 193, 194, 201, 217, 218
— amplitude 26, 67, 112, 174, 199, 220, 221
— bivariate 196, 199, 200, 204, 206, 221
— in positive frequencies 19
— instantaneous 196, 198, 200, 204, 206, 221
— locally smoothed mean 203
— matrix 96, 108, 164, 166
— of natural oscillations 34
— of thermal e.m.f. 43
— total 135
— univariate 199
Spectral distribution function 18
Spectral intensity 23, 46, 189
Spectral line, Gaussian 61, 80, 149
Spectral "mass" 189, 192, 193, 200, 201, 221
— bivariate density of 190
— two-dimensional distribution of 192

Spectral power 88
Spectral representation 14, 16, 164
— of the complex function 20
— of the covariance 500
— of stationary functions 22
— of nonstationary processes 189
Spectrum 30, 65, 83, 116, 117, 189
— bivariate 200
— discrete 23, 218
— continuous 16, 23, 112
— Gaussian 31
— of frequency fluctuations 65
— of frequency deviations 56, 61
— of nonstationary real modulated signal 194
— of oscillations with fluctuating frequency 59
— of pulse self-oscillations 71
— output 127, 128
— rectangular 80
— Shape of 27
Square-law detector 125, 129, 171
Stationarity, wide-sense 3
Stationarity conditions 15, 37
— wide-sense 193
Stationary action(s) 11, 113, 222
Stationary function(s) 3, 15, 28, 180
Stationary process(es) 1, 35, 42, 51, 78, 169, 177, 179, 180, 182, 184, 186, 189, 202, 207, 210, 222
Stationary process(es)
— ergodic 184
— non-Gaussian 56
— non-Poisson 56
— periodically recurring section of 212
— wide-sense 6, 15, 19, 22, 192
Statistical independence 21
Steady-state process 222
Stochastic differential equations 16, 139, 147
Stokes parameter(s) 1011, 102
Structure function 177, 181, 183–188
Superposition principle 11
Symmetry condition 208
System with slowly-varying parameters 222

Technical drift 142, 147, 148, 151
Technical frequency drift 33, 34, 139
Telegraph signal, generalized 23
Threshold sensitivity 135
Time-averaging 89, 130, 209
Time-window 197, 202, 205–207
Theory of random functions 16, 79, 130
Theory of turbulence 177
Thermal e.m.f. 159
— generalized 172
— local 173
— random 157
Thermal equilibrium 110, 173, 175, 176

Thermal fluctuations 157
— theory of 154
Thermal noise 43, 71, 139, 140, 142, 145, 153,
    158, 161
Thermal radiation 44, 153
— equilibrium (black-body) 44
Thévenin theorem 162
Transfer function(s) 14, 38, 112, 114, 117,
    167, 198, 201
— Two-terminal network 157, 173
— Two-wire line 154

Uncertainty relation 23, 24, 31, 32, 80, 205

Van der Pol equations 139
Van der Pol plane 49, 91
Van der Pol variables 51

Variables, mean bilinear 11
Variance 2, 42, 43, 71, 130, 207, 215, 218
Vector oscillation 95
Visibility 79, 81, 82, 84–86, 107, 108
— of fringes 89, 108
Voigt function, normalized 106

Wave fields 75
White light 44
White noise 42, 44, 53, 108, 115, 121, 140,
    193, 214
— stationary 192
Wien law 46

Zero drift 135, 138
Zero-modulation technique 138

Principles of Statistical Radiophysics 1

Elements of Random Process Theory

Misprints and corrections (b: from bottom, t: from top)

pp.	line	printed	must be
3	14 b	at the point of observation	at any point of observation
17	3 t	$\overline{J} = \overline{(2n - N)}^2 a^2 = \ldots$	$\overline{J} = \overline{(2n - N)^2} a^2 = \ldots$
17	11 t	$p(\partial/\partial\varrho)$	$p(\partial/\partial p)$
20	5 t	$\ldots = e^2/T^2 D[n] = \ldots$	$\ldots = e^2 D[n]/T^2 = \ldots$
20	7 t	$\ldots = q/NP.$	$\ldots = q/Np.$
20	16 b	$\overline{(\Delta I_T)^2}\,\overline{I}^2 = \ldots$ (2.10)	$(\Delta I_T)^2/\overline{I}^2 = \ldots$ (2.10)
21	8 t	$\ldots = \dfrac{n^{-n}}{n!} e^{-n}$. (2.11)	$\ldots \dfrac{\overline{n}^n}{n!} e^{-n}$. (2.11)
22	8 t	$W(t) = \ldots$ (2.12)	$w(t) = \ldots$ (2.12)
31	11 b	of sin $2\omega t$ in (2.29)	of sin $2\omega t$ and cos $2\omega t$ in (2.29)
46	4 t	$\displaystyle\int^{+\infty}$ (3.23)	$\displaystyle\int_{-\infty}^{+\infty}$ (3.23)
47	4 b	$\overline{I}_{\text{s.e.}} = n_1 \overline{m}_e,$	$\overline{I}_{\text{s.e.}} = n_1 \overline{m} e,$
86	14 b	$p\{x_i < \ldots\}$	$P\{x_i < \ldots\}$
92	13 t	N_n-variate	Nn-variate
94	8 b	of correlations	of a correlation
95	11 t	compounds $\xi^k(t)$	components $\xi^k(t)$
96	4 b	$\dfrac{\xi(t)\xi(t+\tau)}{\overline{\xi}^2}$	$\dfrac{\overline{\xi(t)\xi(t+\tau)}}{\overline{\xi}^2}$
120	1 b	$\overline{\overset{\cdots}{\xi}\xi}$ is nonzero.	$\overline{\overset{\cdot\cdot}{\xi}\xi}$ is nonzero.
137	5 b	Therefore, in terms of R and φ (5.34) gives	Therefore, the transition to variables R and φ in (5.33) gives
139	9 b	$\ldots = I_0[\ldots]^2 \exp\ldots$ (5.42)	$\ldots = I_0[\ldots] \exp\ldots$ (5.42)
143	1 t	For $\phi \geq 0$ the $\ldots$	For $\Phi \geq 0$ the $\ldots$
145	12 b	notable	noticeable
156	1 b	$\ldots - \cos\psi\delta\alpha.$	$\ldots - \cos\psi\Delta\alpha.$
158	13 t	(softself-excitation)	(soft self-exitation)

pp.	line	printed	must be
158	12 t	with a Thomson oscillator	with a Thomson circuit
176	12 t	$\overline{\theta^2} = 2BT$, $\qquad$ (5.135)	$\overline{\theta^2} = 2Bt$, $\qquad$ (5.135)
178	t	$\xi(r)\xi(r')$ is ...	$\overline{\xi(r)\xi(r')}$ is ...
179	12 t	... the section Δz ...	... the segment Δz ...
188	Caption to Fig. 5.16	The notation for the derivation of the integral equation (5.12.6) is specified	The notation for (5.161) is specified
192	23 t	Denoting by $y_i = 1$ the ...	Denoting by $y_i = i$ the ...
194	5 b	$v(x_2\|x_1) = \dfrac{w_2(x_1, x_2)}{w_1 w_2} = \ldots$ (5.174)	$v(x_2\|x_1) = \dfrac{w_2(x_1, x_2)}{w_1(x_1)} = \ldots$ (5.174)
202	16 b	"Corrections for discreteness" of $v^{(1)}$, $v^{(2)}$, ... are ...	"Corrections for discreteness", i.e., $v^{(1)}$, $v^{(2)}$, ..., are ...
207	6 t	$\xi(t)$	$\boldsymbol{\xi}(t)$
207	7 t	$f(t)$	$\boldsymbol{f}(t)$
207	11 t	$\xi(t)\ldots f(t)$	$\boldsymbol{\xi}(t)\ldots\boldsymbol{f}(t)$
207	14 t	$\xi(t)$	$\boldsymbol{\xi}(t)$
207	15 t	$f(t)$	$\boldsymbol{f}(t)$
207	19 t	$r = (x, y, z)$	$\boldsymbol{r} = (x, y, z)$
209	3 b	$\Delta t_i = t_{i+1}$ $(t_0 = 0, t_n = t)$,	$\Delta t_i = t_{i+1} - t_i$ $(t_0 = 0, t_n = t)$,
213	6 b	... of aRC-circuit ... currentI	... of a RC-circuit ... current I
213	5 b	IR-circuit	LR-circuit
216	7 b	(Equation number is omitted)	(6.22)
238	13 t	$4h(w_0^2 + 3h^2) + w_0^2 D > 0.$	$4h(w_0^2 + 3h^2) + w_0^4 D > 0.$
229	5 b	... of (6.54).	... of (6.53).
242	3 t	$\dot{\overline{x}} + p(t)\dot{\overline{x}} + q(t)\overline{x} = 0.$	$\ddot{\overline{x}} + p(t)\dot{\overline{x}} + q(t)\overline{x} = 0.$

Subject Index

pp.	printed	must be
249	Additivity axion 8, 25	Additivity axiom 8, 25
249	Concept frequeny 9	Concept of frequency 9
249	Convergence of the mean square of probability	Convergence in the mean square in probability
249	Corrections 225	Corrections for discreteness 202
249	for discreteness 202	to diffusion approximation 225
251	Makov sequence(s)	Markov sequence(s)
252	Random functions 1, 2, ..., 231	Random functions 1, 2, ..., 231
	deterministic 207	
252	Random refraction 178	Random refraction of a ray 121, 175, 178
	of a ray 121, 175	